ELEMENTARY SOIL AND WATER ENGINEERING

3rd EDITION

ELEMENTARY SOIL AND WATER ENGINEERING

GLENN O. SCHWAB

Professor of Agricultural Engineering
The Ohio State University
Columbus, Ohio

RICHARD K. FREVERT

Director Emeritus, Agricultural Experiment Station
Professor Emeritus of Agricultural Engineering
The University of Arizona
Tucson, Arizona

JOHN WILEY & SONS

New York • Chichester • Brisbane • Toronto • Singapore

Library of Congress Cataloging in Publication Data:

Schwab, Glenn Orville, 1919-
Elementary soil and water engineering.

Includes index.
1. Soil conservation. 2. Hydraulic engineering.
3. Water conservation. I. Frevert, Richard K.
II. Title.

S623.S39 1985 631.4 85-7280
ISBN 0-471-82587-5

Printed in the United States of America

10 9 8 7 6 5 4 3 2 1

PREFACE

As in the first and second editions, this elementary textbook on soil and water engineering has been written primarily for classroom and laboratory instruction at the college level for students in agriculture and in related nonengineering fields. The purpose of this book is to present up-to-date information that has been gathered through experience and research in a form that will be useful for the beginning student of this subject. The engineering phases of soil and water conservation are emphasized, but we realize that all aspects of these problems must be considered, including agronomic, economic, and others. Vocational agriculture teachers in high schools, instructors for adult training classes, engineers, county extension directors, contractors, farm managers, farmers, and others who face engineering problems on farm lands may find the information of value.

The book includes subject matter on simple surveying and its application to field problems, in addition to information on the design and layout of conservation practices. The first chapter relates soil-and-water-conservation engineering to broader world problems of food and fiber production; Chapters 2, 3, and 4 cover surveying and the use of equipment; Chapter 5 deals with elementary hydrology; Chapters 6 through 9 include various aspects of soil erosion and its control; Chapters 10, 11, and 12 cover water supply, structures, and storage on the farm; Chapters 13 and 14 include subsurface and surface drainage; Chapters 15, 16, 17, and 18 discuss irrigation with an emphasis on sprinkler and trickle systems; and Chapter 19 covers watershed and farm planning. Sample instrument survey field notes are given in applicable chapters to illustrate and to standardize the recording of data. Many example problems have been included to explain design procedures. A list of references is given at the end of each chapter.

We were most gratified by the general acceptance of the first and second editions of *Elementary Soil and Water Engineering.* This third edition includes the latest recommendations, which have been developed from experience and from an expanded research program in recent years. A new chapter has been added on trickle irrigation.

We are deeply indebted to many individuals and organizations for the preparation of this material. Credit for illustrations is given after the figure title. Among the many individuals who have made constructive criticisms or contributed valuable suggestions are D. L. Bassett, P. Buriak, W. Clyma, D. K. Frevert, J. Gerkin, T. J. Logan, L. Lyles, B. H. Nolte, M. L. Palmer, J. Richey, J. van Schilfgaarde, R. K. White, and D. L. Widrig. Mrs. Donna Calhoun typed several chapters and copied the manuscript.

We also express appreciation to our wives for their sympathetic understanding during the preparation of this edition.

We are dedicating this book to the memory of our colleagues Kenneth K. Barnes and Talcott W. Edminster, who contributed much to the earlier editions.

Glenn O. Schwab
Richard K. Frevert

CONTENTS

CHAPTER 1

INTRODUCTION

Soil and water conservation practices are here considered to be those measures that provide for the management of water and soil in such a way as to insure the most effective use of each. These conservation practices involve the soil, the plant, and the climate, each of which is of utmost importance. The soil phase of conservation requires an inventory of the soil, quantitative measurements of its physical characteristics, and information on soil response to various treatments. The plant phase brings to light questions of adequate, but not excessive, water for plant growth and optimum utilization of the plant as a means of preventing erosion and of increasing the rate of movement of water into and through the soil. The plant is also greatly influenced by its biotic environment, which includes all living things and the interrelated actions and reactions that individual organisms directly or indirectly impose on each other. The climate phase is an overwhelming aspect, involving water which can be partly controlled by *man with appropriate drainage and irrigation practices. Temperature, wind, humidity, chemical constitutents in the air, and solar radiation are factors over which man has little or no control.

The engineering approach to soil and water conservation problems involves the physical integration of soil, water, and plants in the design of a coordinated water management system based upon the best physical information available. It is important that specialists in the various aspects of conservation have an appreciation of one another's techniques, as there are few problems that can be solved within the limits of any one field.

1.1 Population and Food Requirements

The ever-increasing problem of food production must be considered on a world-wide scale. No longer can any country afford to isolate itself or ignore the problem. The President's Science Advisory Committee (1967) concluded that

*The word man appears occasionally in this book for reasons of style and accepted English usage and is intended to refer to human beings collectively. Similarly, the impersonal he, him, or his is intended to refer to both males and females.

the United States must be concerned for the hungry countries of the world because of humanitarian reasons, national security, and long-range economic expansion.

According to the United Nations, the 1984 world population of 4.7 billion is expected to be more than 6 billion in the year 2000. Presently, there is no serious worldwide shortage of food, but critical nutritional problems arise from the uneven distribution of food among countries, within countries, and among families with different levels of income. The world population growth rate is caused by an increasing birthrate and a declining deathrate. The most critical areas are in the developing countries. A low per capita income and a low per capita gross national product in these countries are their worst economic problems.

For the immediate future, the world's food supply is critical in many areas of the world. The estimated need for food in the developing countries continues to grow as their populations expand. For the period 1980 to 1985, the most rapidly growing countries were in Africa and the Middle East, regions that have an estimated population growth rate of 3.4 percent per year. This rate would result in a doubling of the current population of these regions in 20 years. Only a few countries now have food production rates sufficient to meet future needs. The supply and demand for foodstuffs is strongly interdependent with a country's gross national product.

In the United States, the population has grown to 92 million in 1910, 151 million in 1950, and 227 million in 1980. A population of 260 million is expected by the year 2000. Up to 1984, food produciton has exceeded demand, leaving a considerable amount for export. The potential for production was demonstrated in the early 1980s, when high farm prices were partially responsible for large surpluses in the following years. Because of a highly competitive market, the desired goal for agriculture should be for greater production efficiency.

1.2 Land Use and Crop Production Trends in the United States

Major land uses in the 48 states from 1900 to 1982 are shown in Table 1.1. Since 1920, cropland, pasture, and forests have decreased as land is diverted to nonagricultural and other uses. Urban land, roads, and the like have increased about 56 percent from 1920 to 1982. Over time, land in agriculture may change from one use to another, that is, cropland may change to pasture, and other pasture may change to cropland. For urban land and developed land to change back to agriculture is not likely.

The distribution of land in the United States in 1977 is shown in Fig. 1.1. Nonfederal land represents about two thirds of the total. Prime farmland, which is about 25 percent of all agricultural land, is considered our best, producing high crop yields with the least damage to the soil. The loss per day of prime farmland by regions is shown in Fig. 1.2 for an eight-year period. The average loss to urban and water areas per year for the United States was one million acres or 2740 acres per day. The Corn Belt had 81 million acres of prime farmland or about 23 percent of the total, the highest in the country. For all

Table 1.1. Major Land Uses in 48 Contiguous States

Land Use	*1900*	*1920*	*1940*	*1959*	*1969*	*1978*	*1982*
			(In million acres)				
Cropland[a]	389	480	467	457	472	470	474
Pasture and nonforest grazing land	761	652	650	630	601	585	591
Forest and woodland	600	602	608	614	603	583	567
Urban, roads, and other land[b]	153	169	180	201	221	259	264
Total	1903	1903	1905	1902	1897	1897	1896

[a]Cropland harvested, crop failure, cropland idle or fallow, and cropland used only for pasture.
[b]Special land uses, such as urban areas, highways and roads, farmsteads, parks, wildlife refuges, and military reservations.
Source: USDA, Economic Research Service.

rural land, three million acres (U.S. Soil Conservation Service, 1981) were lost to nonfarm purposes each year.

One of the leading causes of farmland loss is poorly planned suburban development. Efforts to reduce this type of use have not been successful. Land for agriculture cannot compete with land for urban and industrial development. Land once taken for housing and factories will never revert back to agriculture.

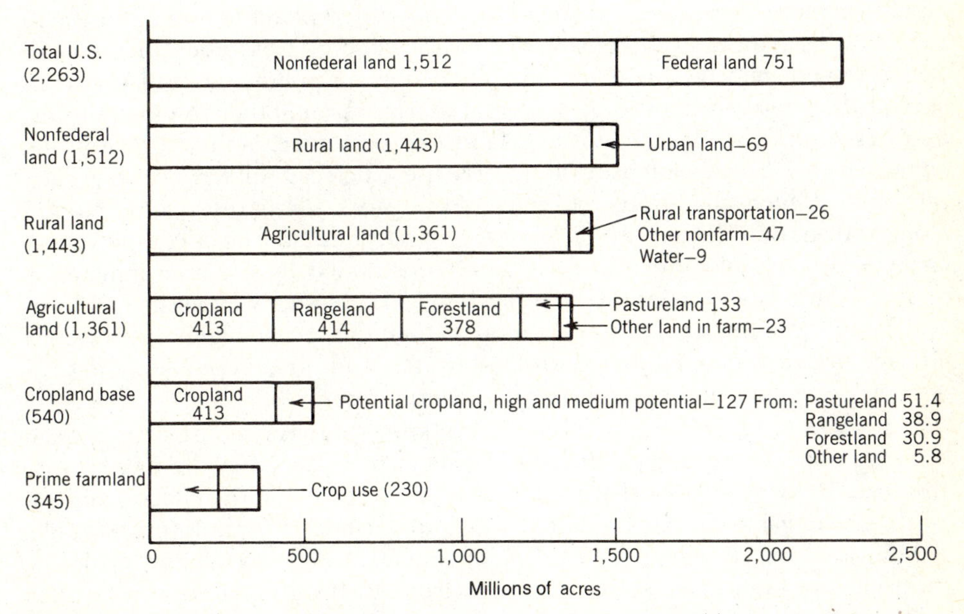

Figure 1.1. Distribution of agricultural land in the 50 states in 1977. *Source*: U.S. Department of Agriculture and the President's Council on Environmental Quality (1981).

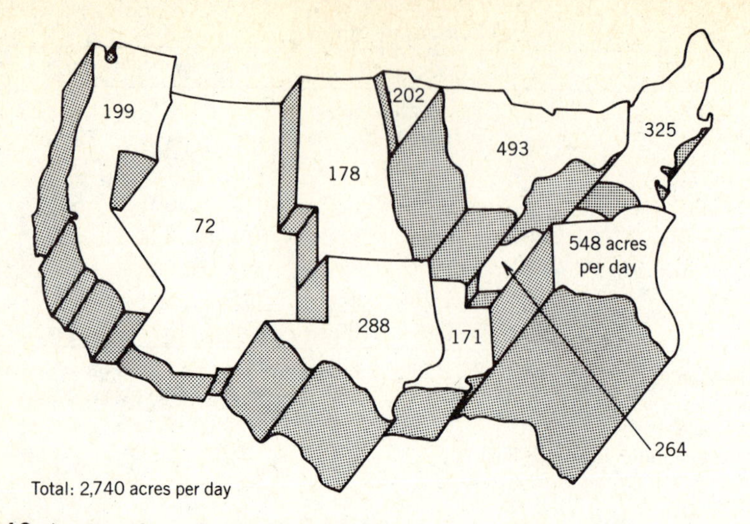

Figure 1.2. Average loss of prime farmland in acres per day to urban and water uses by regions for the period 1967–1975. *Source*: Sampson (1978).

Many private and government agencies are concerned with agricultural land preservation, which greatly affects our long-range potential to produce food.

Crop and livestock production has been gradually increasing in the United States for more than three decades, even though cropland acreage planted to crops has been nearly constant or decreasing slightly. Continued increases in per acre yields may be threatened by soil erosion, air pollution, regulatory constraints, the increasing cost of fertilizers, water, fuel, and other resource inputs, and less productive newly cultivated land. In the future, land will have to be cultivated with greater intensity, and water use efficiency will have to be greatly improved. Genetically improved crops and livestock will have to be developed along with more efficient use of fertilizers, pesticides, and farm equipment. A challenging technological transformation, constrained by resource limitations and environmental concerns, will be necessary. Increasingly expensive and uncertain energy supplies will make it more difficult to secure higher crop yields. Greater pressure may be placed on land, water, and other inputs for the production of biomass and other renewable energy resources.

The existence of resources and technical knowledge is not sufficient in itself to bring about increased production. Farmers must be given the incentive to produce by being provided with a satisfactory return for their labor and investment. Reliable markets, good prices and facilities for transport, storage, handling, and processing must be developed. Channels of supply for fertilizers, pesticides, and improved seeds must be expanded. Institutional reforms, which influence factors such as credit and tenure, are often necessary. Research programs must be supported and new technology passed on to the farmer. Capital must be obtained and invested in the development of land resources, water management systems, agricultural equipment, industrial plants, schools, hospitals, and other essential facilities for social and economic progress. In addition to all of these possibilities, research must continue in order to develop new synthetic foods and to utilize resources in the oceans.

Increasing the production of available cropland will be one of the most important ways of meeting food needs in the future. Proper management of soil and water resources can have an important role in solving this problem. The solution of some of the engineering aspects of soil and water management practices is the field of interest to which this textbook is directed.

1.3 Soil and Water Conservation Problems and Needs

From an engineering point of view, the major problems in agriculture that involve soil and water are (1) soil erosion by wind or water, (2) drainage for removal of excess water or for leaching soluble salts, (3) irrigation for applying water when rainfall is not sufficient, (4) conserving soil moisture by reducing evapotranspiration or increasing infiltration, (5) water resource development by increasing the efficiency of storage, conveyance, and use, and (6) flood reduction by proper watershed management, increasing stream channel capacity, and building flood storage reservoirs. Except for flood reduction (6), ways to reduce or solve these problems will be discussed in this book. Many other aspects of crop production are involved, such as unfavorable soil and adverse climate, over which man has little or no control.

A national inventory of the soil and water conservation needs in 50 states, made in 1967, showed that nearly two thirds of the nonfederal rural land of the United States was in need of some kind of conservation treatment. The results of this inventory are shown in Fig. 1.3 by regions. The erosion hazard was by far the most serious problem, affecting about 49 percent of the land. Corrective measures include contouring, strip cropping, terracing, tillage practices, shelterbelts, and grassed waterways. Erosion is especially serious in the Mountain, Northern Plains, Southern Plains, and Corn Belt regions. About 24 percent of the land has unfavorable soil. Included in this classification are soils with shallow topsoils, stoniness near the surface, low moisture-holding capacity, high salinity, and high sodium. Surface drainage, deep tillage, and erosion-control measures may be required. Ditch or subsurface drainage may be installed to leach out excess salts.

Land having an excess water problem represents about 19 percent of the total. Poor internal soil drainage, wetness, high water table, and overflow from streams are the major causes of this problem. Corrective measures include open ditches, field surface drains, tile drains, and flood reduction measures, such as levees, reservoirs, and land-treatment practices. Major concentrations of land with this problem occur in the Lake States, Corn Belt, Delta States, and southeast regions.

The engineering design for many of the above-mentioned practices involves physical quantities, including prediction of rates and volumes of water to be handled. Determination must be made not only of the frequency of occurrence

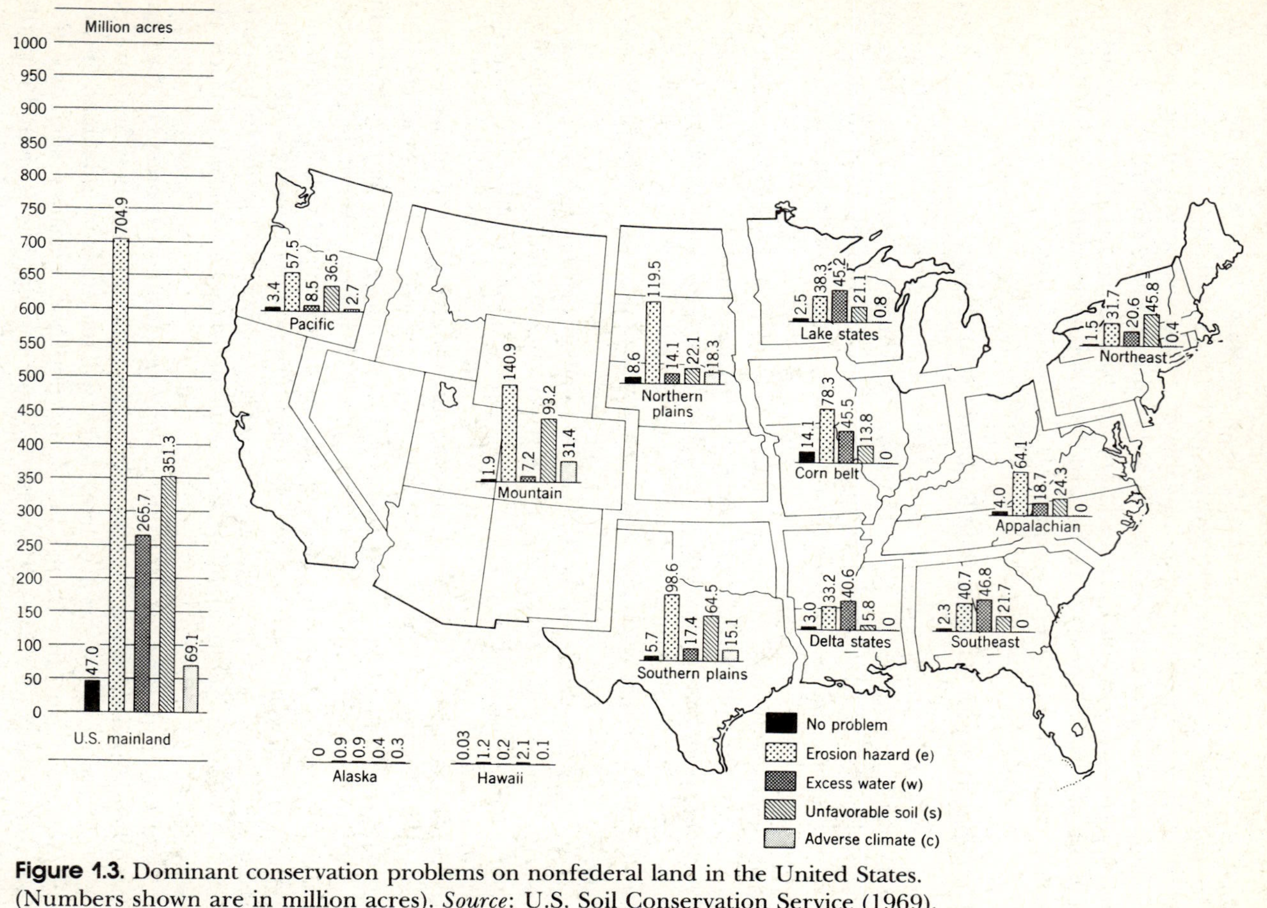

Figure 1.3. Dominant conservation problems on nonfederal land in the United States. (Numbers shown are in million acres). *Source*: U.S. Soil Conservation Service (1969).

of these rates and volumes, but also of the disposition that is to be made of the runoff. The topography of the land is probably of first importance as it will to a large extent determine which of the various methods of water handling may be utilized. The rate at which the water moves into and through the soil is also critical, because this will determine the necessity for drainage and the types of water control facilities that can be used; for example, level or graded terraces, gravity or sprinkler irrigation. Also to be included are the characteristics of plants for stabilizing channels, as well as their requirements for water and their ability to tolerate high water tables. By utilizing all of this information along with the principles of hydraulics, it is possible to complete the design of the facility needed to solve the problem.

In the design of a system varying degrees of engineering skill are involved, from those required in the simple process of planting the crop on the contour to the more complicated procedure of designing involved drainage systems, water-conserving and erosion-preventing structures, and irrigation systems. Some of the simpler jobs can be carried out by individuals having only general training in field techniques, but a qualified engineer should be retained for the more complicated problems.

1.4 Investment in Soil and Water Conservation Practices

The net investment of public and private funds in soil and water conservation practices from 1910 to 1975 is shown in Fig. 1.4. Although considerable government support has been provided since 1940, private funds are far greater than public funds. The greatest increase has been in irrigation, but this trend may not continue in the future because of limited water supply, especially in the West.

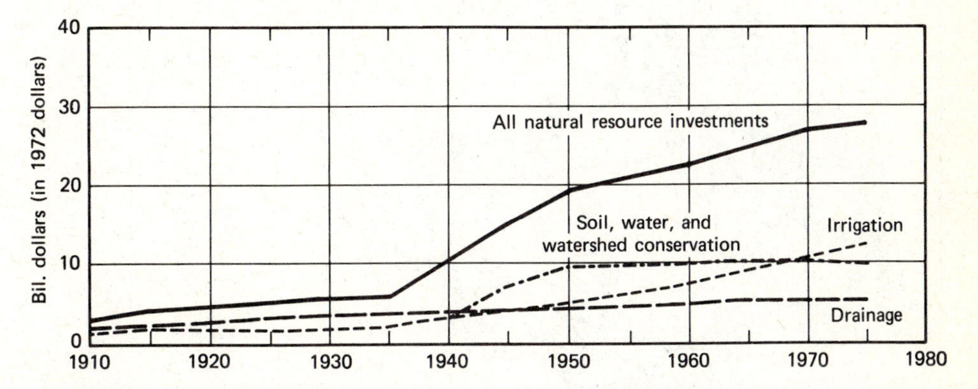

Figure 1.4. Net investments, both public and private, in soil and water conservation practices (5-year averages). (Investments are on a net or depreciating basis as of the year shown.) *Source*: U.S. Department of Agriculture (1977).

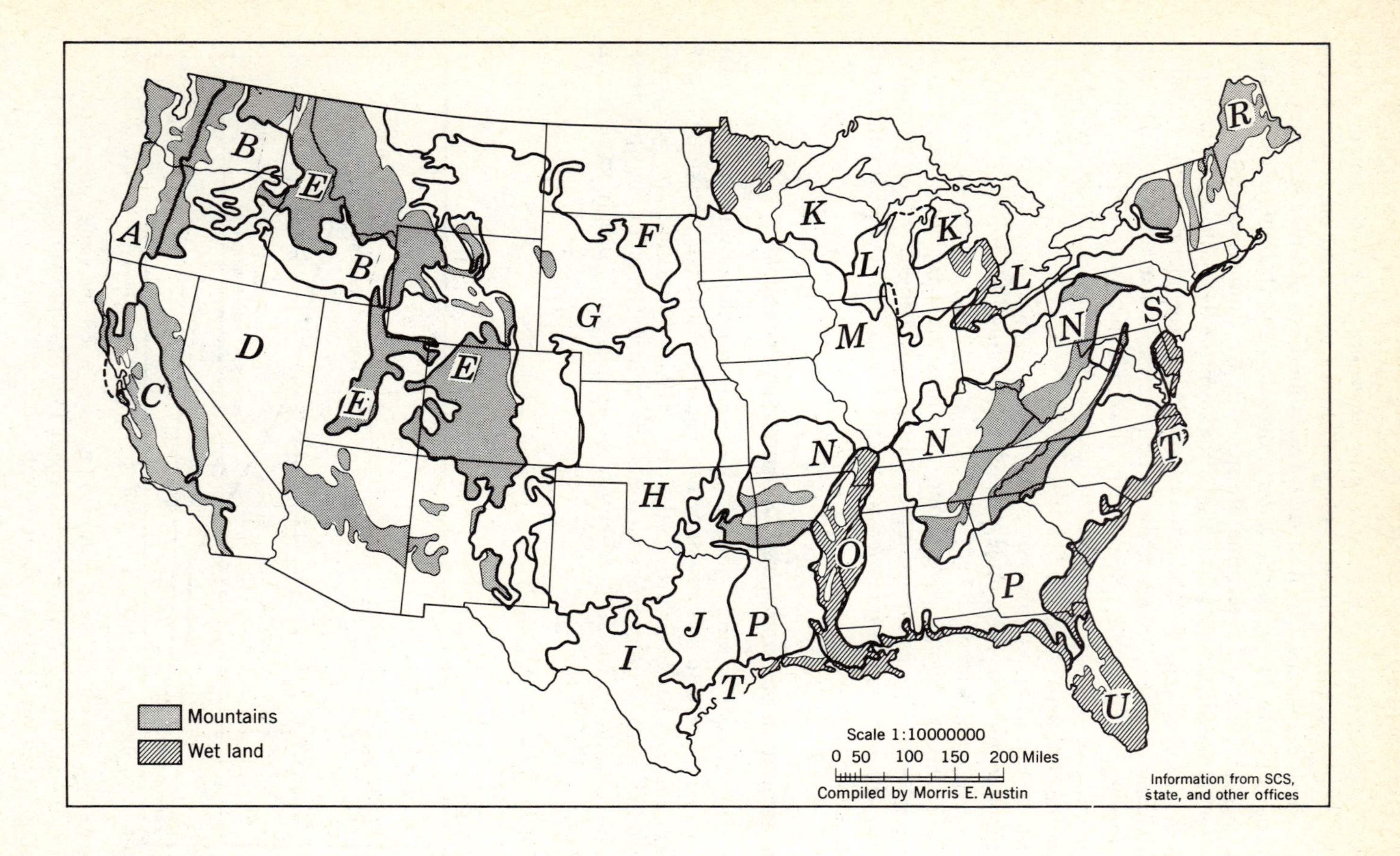
A
B
E
B
D
C
E
E
F
G
H
I
J
P
T
K
L
K
L
M
N
N
N
O
S
R
T
P
U
Mountains
Wet land
Scale 1:10000000
0 50 100 150 200 Miles
Compiled by Morris E. Austin
Information from SCS,
state, and other offices

1.5 Land Resource Regions

For purposes of making recommendations for various conservation practices, it is helpful to subdivide the United States into a series of major land resource regions, as shown in Fig. 1.5. These regions have been delineated on the basis of broad physiographic characteristics and are the basis for much of the current study and planning of conservation needs in both research and action programs. These regions will be referred to in later chapters.

REFERENCES

Council for Agriculture Science and Technology (CAST) (1981) *Preserving Agricultural Land: Issues and Policy Alternatives*, Ames, Ia.

Dunford, R. W. (1982) "The Evolution of Federal Farmland Protection Policy, *J. Soil and Water Conservation* **37**(3):133–136.

President's Science Advisory Committee (1967) *The World Food Problem*, U.S. Government Printing Office, Washington, D.C.

Sampson, N. (1978) *Preservation of Prime Agricultural Land, Environmental Comment*, Urban Land Institute, Washington, D.C.

U.S. Department Agriculture (1977) *Handbook of Agriculture Charts*, Agr. Handb. 524, U.S. Government Printing Office, Washington, D.C.

U.S. Department Agriculture and the President's Council on Environmental Quality (1981) *Final Report, National Agricultural Lands Study*, U.S. Government Printing Office, Washington, D.C.

U.S. Soil Conservation Service (1981) *America's Soil and Water: Condition and Trends*, U.S. Government Printing Office, Washington, D.C.

U.S. Soil Conservation Service (1969) "Soil and Water Conservations Needs Inventory," *Soil Conservation* **35**(5):99–109.

Figure 1.5. Major land resource regions of the United States. (A) Northwestern forest, forage, and specialty crop region. (B) Northwestern wheat and range region. (C) California subtropical fruit, truck, and specialty crop region. (D) Western range and irrigated region. (E) Rocky Mountain range and forest region. (F) Northern Great Plains spring wheat region. (G) Western Great Plains range and irrigated region. (H) Central Great Plains winter wheat and range region. (I) Southwestern plateau and plain, range and cotton region. (J) Southwestern prairie cotton and forage region. (K) Northern lake states forest and forage region. (L) Lake states fruit, truck, and dairy region. (M) Central feed grains and livestock region. (N) East and Central general garming and forest region. (O) Mississippi Delta cotton and feed grain regions. (P) South Atlantic and Gulf Slope cash crops, forest, and livestock region. (R) Northeastern forest and forage region. (S) Northern Atlantic Slope truck, fruit, and poultry region. (T) Atlantic and Gulf Coast lowlands forest and truck crop region. (U) Florida subtropical fruits, truck crops and range region. (*Courtesy Soil Conservation Service.*)

CHAPTER 2

DISTANCE AND AREA MEASUREMENT

On the farm it is often necessary to measure field distances and areas, and to lay out buildings. These operations may be simplified if a few of the principles are thoroughly understood.

MEASUREMENT OF DISTANCE

2.1 Pacing

Distances may be measured by pacing when the desired accuracy is not greater than 2 ft in 100 ft (1 in 50). Although the length of a pace is approximately 3 ft, each individual should check his pace with an accurately measured distance. A common way of remembering the length of pace is to determine the number of paces per 100 ft. It must be kept in mind that the length of a pace may vary when walking through short or tall vegetation, in going up or downhill, on wet or dry ground, and on plowed or firm soil.

2.2 Measuring Equipment

The 100-ft steel tape is commonly used for measuring distances in the field. Metallic cloth tapes are also suitable for this purpose, but they are not as durable as the steel tape. Most tapes 50 to 100 ft in length can be rolled up in a case that will fit into a pocket. For ease of reading, some tapes are graduated in tenths and hundredths of a foot (or meter) for their entire length. For most agricultural purposes a 100-ft steel tape $\frac{5}{16}$ in. in width, with markings for each foot and with the last foot at the end divided into tenths, is quite satisfactory. Other equipment needed for measuring distances consists of a set of taping pins, range poles, and a plumb bob. A set of taping pins consists of 11 pins with a suitable carrying ring (Fig. 2.1). Each pin is about 12 in. in length with a loop on one end. These

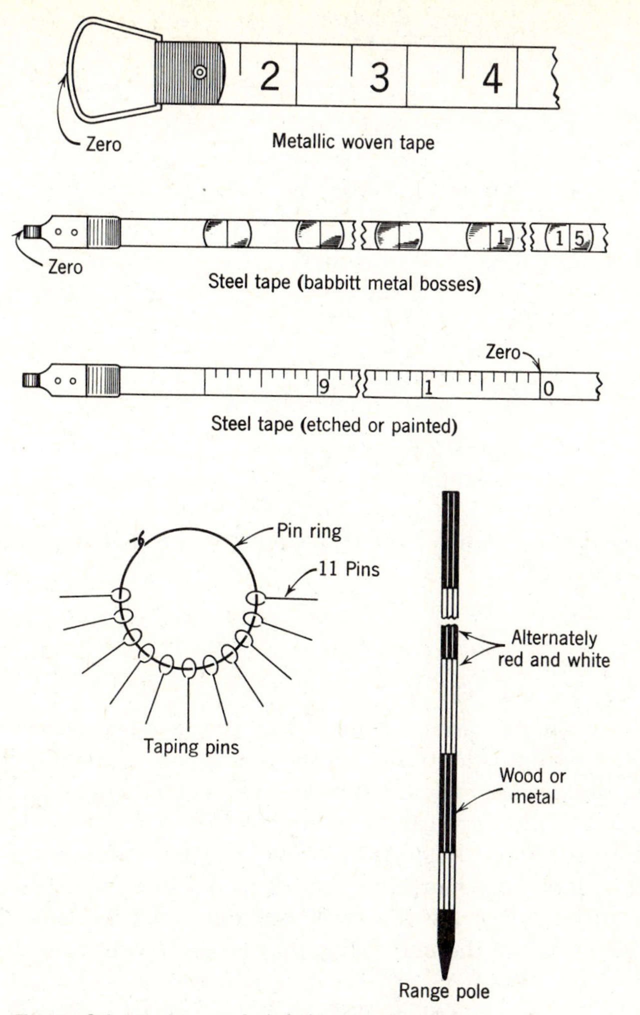

Figure 2.1. Taping and sighting equipment.

may be purchased or made from heavy gage wire. Range poles are about 8 ft in length with a steel point on one end and painted alternately red and white. They may be seen from a considerable distance. Four-foot lath stakes may serve as a good substitute when the vegetation is small and the distances are short.

An odometer is a simple device which measures distance by registering the number of revolutions of a wheel. It is convenient to make the wheel circumference 10 ft and to divide the circumference into 10 equal parts. Such a device, as shown in Fig. 2.2, may be pushed by hand or attached to a vehicle. On smooth ground the precision may be within 1 percent, but on rough ground or in rank vegetation measurements are less accurate. The distance indicated by the odom-

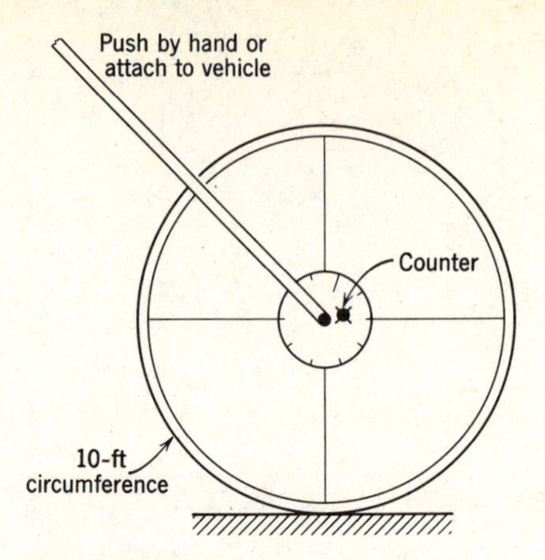

Figure 2.2. Odometer for one-man measurement.

eter or by taping on sloping land is always greater than the true horizontal distance (see Table 2.1).

2.3 Taping

Taping refers to the operation of measuring the distance between two points. Land surveys, building layout, and most other applications require horizontal distances. On sloping land the horizontal measurements may be obtained with a plumb bob so that the tape can be held level (Fig. 2.3), or by taping the slope distance, taking the slope angle, and converting to a horizontal distance. On land that has a slope less than 2 percent (2 ft in 100 ft) slope distance may be sufficiently accurate (error less than 0.02 percent, see Table 2.1). If the slope is not more than 5 percent, the plumb bob may be used with the full 100-ft length

Table 2.1 Conversion of Slope Distance to True Horizontal Distance*

Slope (percent or ft per 100 ft)	*Error in Slope Distance (ft per 100 ft)*	*True Horizontal Distance (ft)*
1	0.005	99.995
2	0.020	99.980
3	0.045	99.955
4	0.080	99.920
5	0.125	99.875
10	0.50	99.50
15	1.13	98.87
20	2.02	97.98
30	4.61	95.39

*Horizontal distance = $[100^2 - (\text{percent slope}^2)]^{1/2}$.

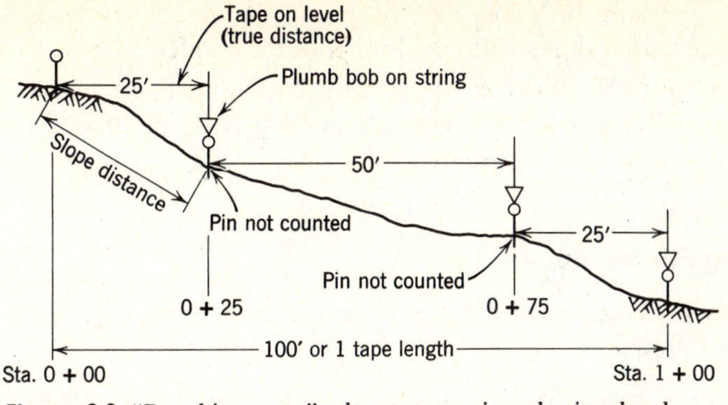

Figure 2.3. "Breaking tape" when measuring sloping land.

of the steel tape. On slopes greater than 5 percent, it is necessary to "break tape" (Fig. 2.3). The extra pins required when breaking tape for distances less than the length of the tape should be returned to the head tapeman so as not to cause the pin count to be in error.

Example 2.1. On a uniform slope of 5 percent (5 ft per 100 ft), the slope distance was measured to be 420.20 ft (128.08 m). Determine the true horizontal distance between the two points.

Solution. From Table 2.1 read an error of 0.125 ft per 100 ft (0.125 m per 100 m). The total error to be subtracted from the slope distance = (420.20 per 100) × 0.125 = 0.53 ft (0.16 m) The true horizontal distance = 420.20 − 0.53 = 419.67 ft (127.92 m), or an alternate procedure from Table 2.1:

$$\begin{aligned}\text{True horizontal distance per 100 ft} &= (100^2 - 5^2)^{1/2}\\ &= 99.875 \text{ ft per 100 ft}\\ \text{Total distance} &= 99.875/100 \times 420.20\\ &= 419.67 \text{ ft } (127.92 \text{ m})\end{aligned}$$

When a series of marks are set along a prescribed line, a standard system of numbering helps to distinguish these distances from other measurements either horizontal or vertical. The zero point is designated 0 + 00 and each 100-ft point is called a station, which is the number before the plus sign (see Fig. 2.3). For example, 4 + 00 is station 4. A point 550 ft from the starting point is written 5 + 50, 646.2 ft is 6 + 46.2, and so forth. The +50 and +46.2 are called pluses.

2.4 Care in Handling Measuring Equipment

The following precautions should be observed when handling measuring equipment.

1. When unrolling the steel tape, be careful not to leave kinks in it as stretching the tape may easily break it.

2. If a steel tape becomes wet, it should be wiped dry with a cloth before storing. Some tapes will rust and should be oiled.
3. Taping pins should not be left in the ground or lying around loose, but should be placed on the ring. A small strip of cloth tied to the ring end of the pin will aid in seeing them.

2.5 Taping Procedure

The proper procedure for taping (chaining) over nearly level ground between two points follows:

1. The head tapeman with the zero end of the tape unrolls it as he walks toward the distant point. The rear tapeman holds the 100-ft end of the tape at the starting point.
2. For the first 100-ft measurement, the head tapeman should have 10 pins and the rear tapeman 1 pin, regardless of whether the 1 pin is needed as the starting point.
3. The rear tapeman should align the head tapeman by sighting on the distant point. The tape should be straight and stretched with a pull of about 15 lb.
4. When the rear tapeman has the 100-ft mark exactly on the starting point (or on a pin in the ground), he will call "*stick.*" The head tapeman will place a pin exactly at the "0" mark on the tape and call "*stuck.*"
5. The rear tapeman picks up the pin and walks forward, while the head tapeman pulls the chain to the next point. When the end of the tape is about 5 ft from the pin in the ground, the rear tapeman calls "*chain*" to signal the head tapeman to stop.
6. The procedure in (3), (4), and (5) is repeated until the head tapeman has no more pins (or the distant point is reached). The rear tapeman hands the head tapeman the 10 pins. Both tapemen should count the pins. The distance to the 11th pin, which is in the ground, is 10 tape lengths or 1000 ft. The number of pins held by the rear man is always the distance from the starting point in hundreds of feet.
7. Taping is continued until the distant point is reached. Care should be taken in measuring the distance from the last 100-ft pin to the final point, especially with tapes that have subdivisions of a foot only at the end foot.

2.6 Units of Measurement

In surveying, tenths and hundredths of a foot are used in place of inches. Since there are 12 in. or 10 tenths in a foot, it is quite confusing to convert from one system to the other. The two scales are shown in Fig. 2.4 and may be used to convert tenths to the nearest $\frac{1}{8}$ in. and vice versa. To convert values above 6 in., add 0.5 ft or 6 in., for example 0.75 (0.5 plus 0.25) converts to 9 in. (6 plus 3).

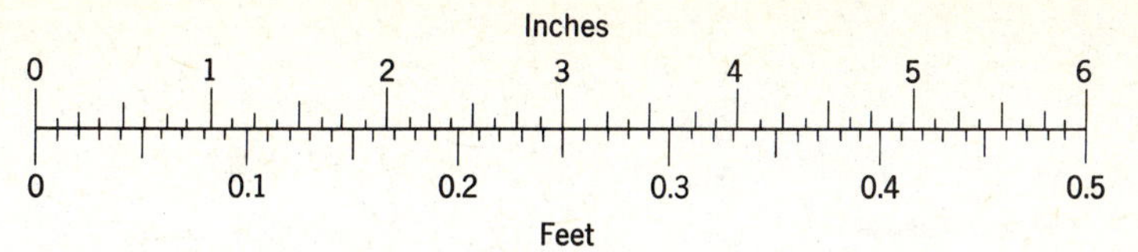

Figure 2.4. Conversion scale for inches to tenths of a foot and vice versa.

2.7 Common Errors in Measurement

Some of the more common errors in measuring distance are listed below:

1. Tape not pulled tight enough.
2. Tape not in proper alignment.
3. Pins not carefully placed at proper marks on tape.
4. Mistake in counting pins or lost pins.
5. Mistake in determining the number of feet less than 100. This happens when measuring the fractional tape length at the end of the line.
6. Plumb bob not used when measuring on a slope.
7. Wrong points on the tape used for the zero or the 100-ft mark.
8. Reading or recording the wrong numbers.

2.8 Stadia

To measure distance by stadia, a surveying instrument must be equipped with stadia hairs. In addition to the leveling cross hair, such an instrument (Fig. 2.5) has two additional cross hairs called *stadia hairs*. The stadia cross hairs are so placed that when the level rod is 100 ft from the instrument the interval between the stadia hairs as read on the level rod is 1 ft. (This conversion factor is correct only when the correction constant in the instrument is zero and when the telescope is level.) The accuracy of stadia measurement depends upon the accuracy of the instrument, the distance from the instrument to the level rod, and the ability of the instrument man. Normally, an accuracy of $\frac{1}{2}$ ft per 100 ft can be obtained with a good instrument. Levels do not ordinarily come equipped with stadia hairs; however, most manufacturers will install them, if desired.

2.9 Electronic Distance Measurement (EDM)

These instruments for electronic distance measurement (EDM) send out a beam of light or high-frequency microwaves to the far end of a line to be measured, and a reflector or transmitter-receiver reflects the light or microwaves back to the instrument where they are analyzed electronically to give a digital readout of the distance between the instrument and the reflector. The second generation of EDM instruments was developed and perfected in the 1960s as solid-state

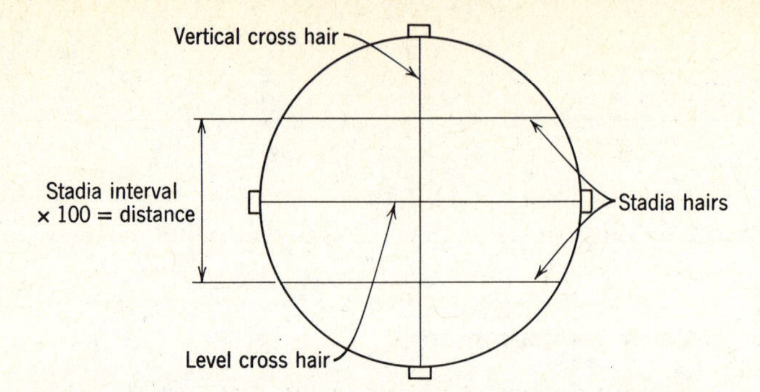

Figure 2.5. Leveling and stadia cross hairs in a telescope.

electronics miniaturization took place. They are easily portable, highly accurate for long distances, and operate at night or through fog or rain. Both EDM and stadia methods are especially suitable for measuring across water, rough terrain, or other land features where taping would be difficult or impossible. EDM instruments are much more accurate than stadia and are available for short to long distances. They measure slope distance, but some will give a direct readout in horizontal length. For topographic mapping the EDM instruments are set up next to a transit, or some can be mounted directly on a standard transit. Although EDM devices are generally too expensive for the average user, they may begin to decrease in price much like computers.

2.10 Laying Out Right Angles

For laying out right angles without the use of a compass or a transit, the steel tape is quite convenient.

Three-Four-Five Method. To lay out a right angle with this method it is necessary to lay out a triangle whose sides are in the proportion of 3, 4, and 5. Convenient lengths are 30, 40, and 50 or 60, 80, and 100 ft. From some base line *AB* (Fig. 2.6*a*) and a starting point *D*, measure the distance *DC* and swing an arc on the ground at *C*. Next measure the distance *DE* and then from *E* swing another arc near *C* which will intersect the previous arc at that point and locate the point *C*. It is always desirable to check the distances to be sure that no mistake has been made. A simple, approximate method is shown in Fig. 2.6b.

Chord Method. If a perpendicular is to be erected through some point *A* (Fig. 2.7*a*) within 100 ft of the base line *EF*, swing an arc from *A* toward *C* until it crosses the base line *EF*, likewise swing the same length arc so that it will intersect the base line again at *D*. Measure the distance *CD* along the base line and locate *B* so that it is midway between *C* and *D*. This will establish the point *B*. The arc length *AC* must be greater than the distance *AB*.

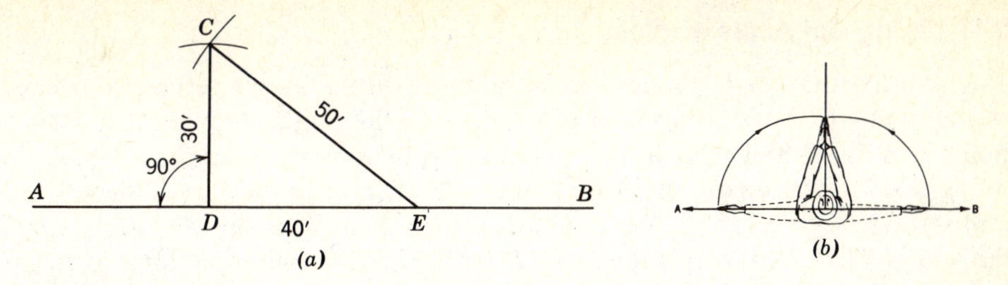

Figure 2.6. (*a*) Three-four-five method of laying out a right angle, and (*b*) an approximate method.

When a base line *EF* is given (Fig. 2.7*b*) and when the perpendicular line must be constructed through *D*, measure off two equal distances from *D* to *A* and from *D* to *B*. As shown in Fig. 2.7*b*, swing an arc from *B* to *C* and then the same length arc from *A* to *C*. *AC* and *BC* should be greater than the distance *AD* or *DB*.

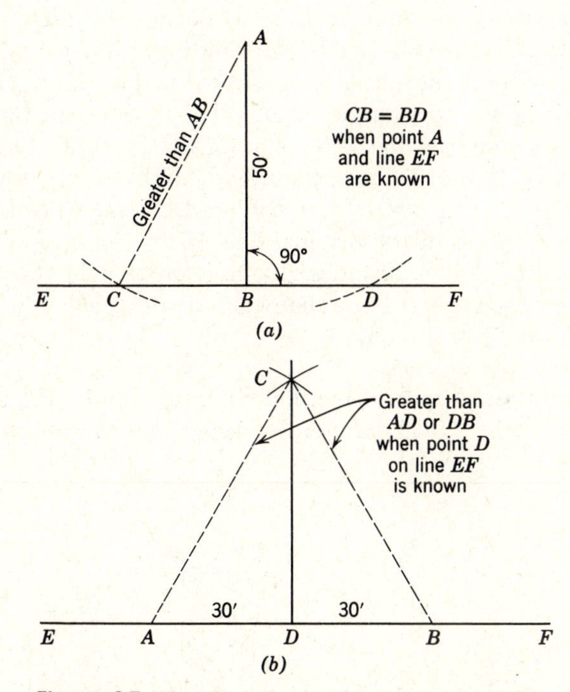

Figure 2.7. Chord methods of laying out a right angle (*a*) when a line and point *A* are known, and (*b*) when a line and point *D* on a line are known.

2.11 Laying Out Acute Angles

Angles other than right angles can be laid out with a tape if reference is made to trigonometric tables. For example, suppose a 30° angle is to be established at point *A* in Fig. 2.8. Tables of trigonometric functions show that the tangent of 30° is 0.5774. Measuring off *AB* equal to 100 ft, erecting a perpendicular at *B* and setting *BC* as 57.74 ft will establish the angle at *A* as the 30° angle whose tangent is 57.74/100. Any values of *AB* and *BC* may be used so long as *BC*/*AB* is 0.5774. Larger values will give greater accuracy.

Trigonometric functions are given on most hand calculators. The tangent of an angle is defined as the opposite side/adjacent side and written "tan." Two other useful functions are the sine and cosine, written "sin" and "cos," respectively. The "sin" is the opposite side/hypotenuse. In Fig. 2.8, $\sin A = 57.74/AC = 0.5$, thus $AC = 57.74/0.5 = 115.48$. The "cos" is the adjacent side/hypotenuse. In Fig. 2.8, $\cos A = 100/115.48 = 0.866$.

2.12 Extending Straight Lines Through Obstacles

Perpendicular Offsets. When an obstacle such as a building lies in a tape line, it is possible to establish an auxiliary line to permit the measurement of the portion of the taped distance blocked by the building and to permit the extension of the taped line beyond the obstacle, as shown in Fig. 2.9*a*. This may be accomplished by laying out right angles at *A* and *B* and measuring $AA' = BB'$ of sufficient length to establish line $A'B'$ clear of the obstacle. Line $A'B'$ is then extended and points C' and D' are established. Right angles are laid out at C' and D'. Measurement of $CC' = DD' = BB' = AA'$ establishes points *C* and *D*, which lie on the extension of the original line *AB*.

Since distance $B'C'$ equals distance *BC*, that portion of the taped distance obstructed by the obstacle has been determined. Accuracy is improved if distances *AB* and *CD* are 100 ft or more.

Equilateral Triangles. A second method for extending a line through an obstacle is illustrated in Fig. 2.9*b*. Equilateral triangle *AEB* is established by swinging

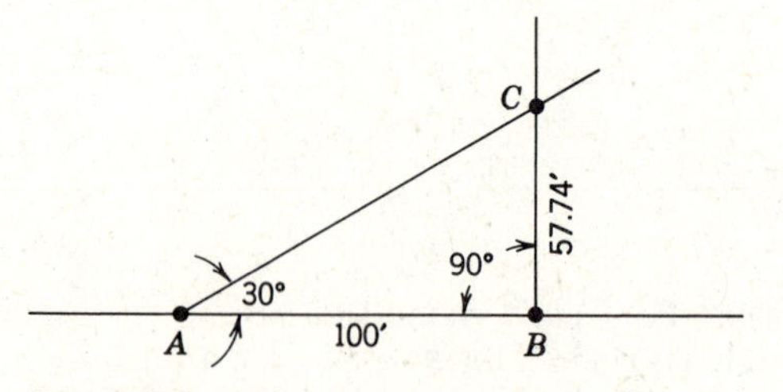

Figure 2.8. Laying out an acute angle from a line.

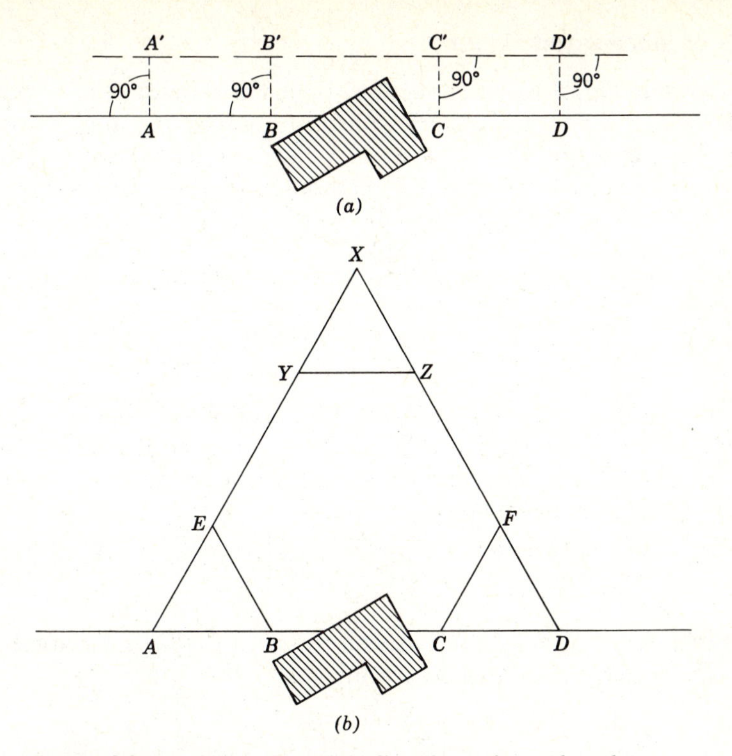

Figure 2.9. Extending a straight line through an obstacle (*a*) by perpendicular offsets, and (*b*) by equilateral triangles.

arcs with a tape. Line *AE* is extended to *X* and equilateral triangle *XYZ* is established to locate line *XZ*. Distance *XD* is laid off equal to *XA*. Equilateral triangle *CDF* is then established to define *CD* as the extension of *AB*.

Sight Method. A straight line may be laid out between two points that are separated by a ridge and cannot be seen from either, by setting range poles or tall stakes between the points and by aligning them by sight. Starting from one point, set two poles in approximate alignment. Proceed to the first pole and set a third pole in line with the first two. Continue in this manner until the second point can be seen. Reset the poles until they are all in alignment.

MEASUREMENT OF AREAS

On a farm the area of fields is frequently desired. In general, most irregular-shaped areas may be measured with the steel tape by dividing the field into smaller divisions so as to form triangles, trapezoids, or rectangles.

2.13 Four- or More-Sided Figure

The field shown in Fig. 2.10 is an irregular-shaped area that has been subdivided into smaller areas. It is not necessary to measure any of the angles in order to determine the shape and size of the area. Where the length of the three sides of a triangle is known, the area may be computed by the formula:

$$A = \sqrt{s(s - a)(s - b)(s - c)} \tag{2.1}$$

where a, b, c = length of the three sides and

$$s = \frac{a + b + c}{2}$$

The area of triangles ABF and BCF may be found in this manner. Since the above formula requires considerable calculations, it may be desirable to measure the perpendicular distances AG and HC from the line BF and thus compute the area of the triangles by the following formula:

$$A = \frac{(\text{base} \times \text{height})}{2} \tag{2.2}$$

Since the lines CD and EF are parallel, the area $CDEF$ is a trapezoid. The area may be computed by using the formula:

$$A = h\frac{(a + b)}{2} \tag{2.3}$$

where h = perpendicular distance between the two parallel sides and a, b = length of the two parallel sides.

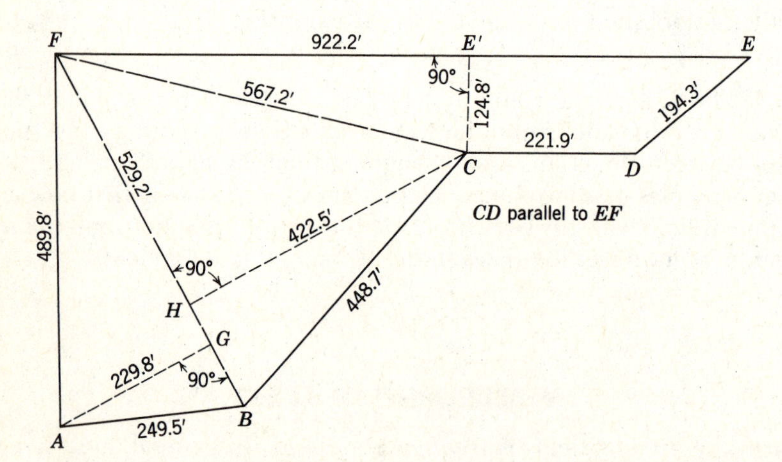

Figure 2.10. Measurements to obtain the area of an irregular-shaped field with straight boundaries.

Example 2.2. Determine the area of the field ABCDEF in Fig. 2.10 in acres.

Solution. Subdivide the field into two triangles and a trapezoid and compute the areas as follows:

Area			*Square Feet*
ABF	$\frac{FB \times AG}{2}$	$\frac{529.2 \times 229.8}{2}$	60,805.1
BCF	$\frac{FB \times HC}{2}$	$\frac{529.2 \times 422.5}{2}$	111,793.5
CDEF	$\frac{(CD + EF)}{2} \times CE'$	$\frac{(221.9 + 922.2)}{2} \times 124.8$	71,391.8
		Total area	243,990.4

Since one acre is 43,560 sq ft, the total area in acres is 5.60 (2.27 ha).

2.14 Curved Boundary Areas

If a field is bounded on one side by a straight line and on the other by a curved boundary, as shown in Fig. 2.11, the area may be computed by the trapezoidal rule:

$$A = d\left(\frac{h_0}{2} + \sum h + \frac{h_n}{2}\right) \tag{2.4}$$

where d = distance between offsets, h_0, h_1, etc.

h_0, h_n = end offsets

Σh = sum of the offsets, except end offsets

In the above formula, the curved boundary is assumed to be a straight line between offsets h_0 and h_1; h_1 and h_2, etc.

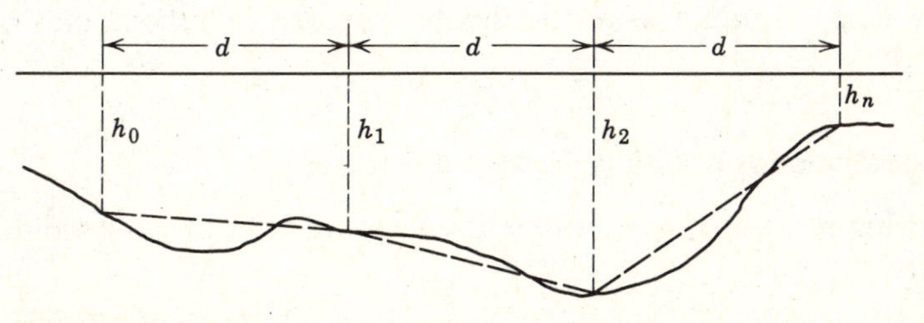

Figure 2.11. Measurements to obtain an irregular area with a curved boundary.

If the end offsets h_0 and h_n were zero, which would occur when the curved boundary crosses the base line at the end points, the above formula would become

$$A = d \times \Sigma h \tag{2.5}$$

Example 2.3. Determine the cross-sectional area of a stream 10 ft wide in which water depths at each 2-ft intervals were 0, 0.8, 1.1, 1.5, 0.3 and 0 ft.

Solution. Substitute in Eq. 2.5 with $d = 2$,

$$A = 2(0.8 + 1.1 + 1.5 + 0.3) = 7.4 \text{ ft}^2 \ (0.69 \text{ m}^2)$$

RECORDING FIELD NOTES

An important part of surveying with levels or transits is the proper recording of field data in a notebook. Standard forms for keeping notes have been devised to provide a systematic procedure for recording field data.

Field notes are those recorded in the field at the time the work is done. Notes made later, from memory or copied from other field notes, may be useful, but they are not field notes. It is *not easy* to take good notes. The recorder should realize that field notes are regarded as a permanent record and are likely to be used by other persons not familiar with the locality who must rely entirely on what has been recorded. For this reason the field book should contain all the necessary information with a good sketch showing the location of the various points.

2.15 Field Book

The standard field book has horizontal lines extending across the double page as shown in Fig. 2.12. The left-hand side of the double page has six columns for tabulating numerical data, and the right-hand side has vertical columns about $\frac{1}{8}$ in. wide for recording explanatory notes and sketches. These books are made with flexible or hard back covers and may have looseleaf, spiral, or permanent bindings. They are of a size that will conveniently slip into a coat pocket. The hard-covered field book with the permanent binding is generally desired for field records. For student work the flexible-covered field book may be satisfactory.

2.16 Suggestions for Keeping Good Field Notes

The following is a list of suggestions for keeping a set of good field notes (see Fig. 2.12):

1. The first few pages of the field book should be reserved for an index, which should be kept up-to-date as the work is completed.

TYPE OF SURVEY
Legal Description or Location
Numerical data
such as distances, rod
readings, elevations, and
point designations.
(Specific examples to be
given in later chapters.)
Weather
Names of Party 3
Date
Explanatory notes on same line
as point designation in first column
on left-hand side of page.
Sketch of layout
for data shown
on left-hand side
of page.
(Show north arrow,
point designations,
fences, roads and
other topographic
features.)

Figure 2.12. Proper location for field information to be recorded in a surveying field book.

2. The double sheet is considered as one page. Pages should be numbered in the extreme upper right-hand corner of the right-hand side of the page.
3. A descriptive title should be printed at the top of the first page for each job. It should show the type of survey, the legal description, and the name of the job.
4. At the top of the right-hand side of the page the names of the survey party should be recorded along with the job assigned to each. For example, the instrument person is designated by $\overline{\wedge}$, the rodman by ϕ, the note recorder by REC, the head tapeman by H.T., and the rear tapeman by R.T. The date and weather conditions at the time the data were taken should be recorded at the top right-hand side of the page as shown in Fig. 2.13.
5. A 3H or 4H hard lead pencil, well pointed, should be used in recording data. Ink should *never* be used as it will smear if the field book gets wet.
6. Titles, descriptions, and words should be lettered in the best form possible, usually capital letters for headings and titles, and lowercase letters for other information.
7. Numbers recorded should be neat and plain and one figure should never be written over another. In general, numerical data should not be erased; if a number is in error, a line should be drawn through it, and the

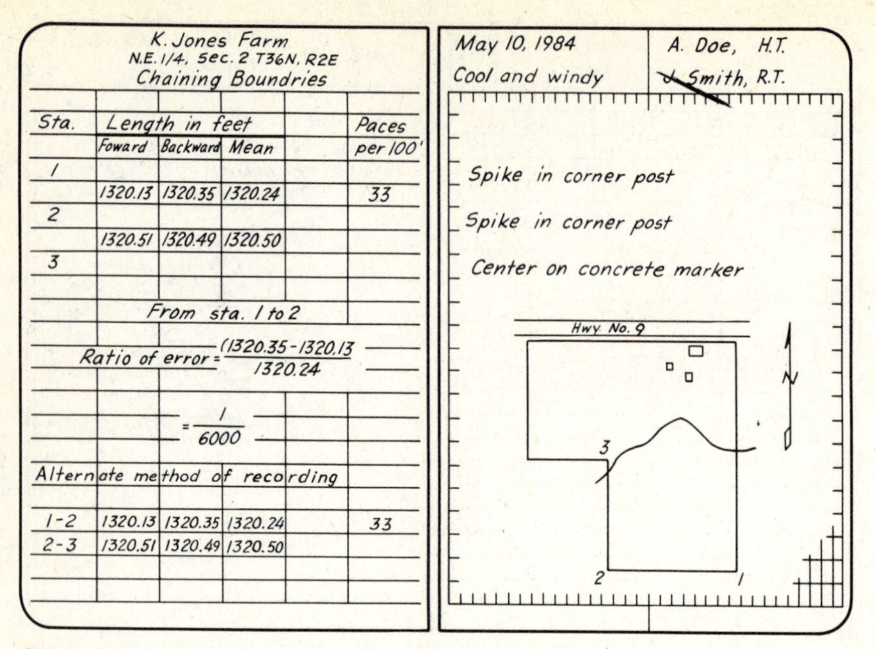
K. Jones Farm
N.E. 1/4, Sec. 2 T36N, R2E
Chaining Boundries

Sta.	Length in feet			Paces
	Foward	Backward	Mean	per 100'
1				
	1320.13	1320.35	1320.24	33
2				
	1320.51	1320.49	1320.50	
3				

From sta. 1 to 2

Ratio of error = (1320.35 - 1320.13) / 1320.24

= 1 / 6000

Alternate method of recording

1-2	1320.13	1320.35	1320.24	33
2-3	1320.51	1320.49	1320.50	

May 10, 1984 | A. Doe, H.T.
Cool and windy | J. Smith, R.T.

Spike in corner post
Spike in corner post
Center on concrete marker

Figure 2.13. Sample field-book page for recording distances.

corrected value written above. Portions of sketches and explanatory notes may be erased if there is a good reason for doing so.

8. In tabulating numbers, all figures in the tens column, for example, should be in the same vertical line. Where decimals are required, the decimal point should never be omitted. The number should always show to what degree of accuracy the measurement was taken; thus a rod reading to the nearest 0.01 ft would show 7.40 rather than 7.4 without the zero.
9. Sketches should be neat and large enough to show details without crowding the figures together. Rod readings, distances, and other numerical data are generally not shown on the sketch, provided they are recorded on the left-hand side of the page; however, reference points, a north arrow, names of streams, roads, landowners, and property lines are shown. The sketch is drawn on the right-hand side and should correspond to the data recorded on the left-hand side or on succeeding pages.
10. Explanatory notes, such as the approximate location of reference points, should be shown on the same horizontal line, but on the right-hand side. They are needed to make clear what the numerical data and sketches fail to bring out.
11. If a page of notes becomes illegible or erroneous, the data should be retained and usable notes reentered in the book before writing the word "void" in large letters diagonally across the page. The page number of

the continuation of the notes should be indicated. Voiding portions of a page may be done in the same manner as voiding a full page.

12. Scribbling should not be done in the field book. A piece of scratch paper held in the field book with a rubber band is convenient for making calculations. If this is not available, the back pages of the field book will suffice. Particular attention should be paid to neatness and arrangement of the data.

2.17 Field Notes for Recording Distances

Example field notes are given in Fig. 2.13 for a farm boundary survey. In the left-hand column the abbreviation "Sta." refers to the station or the point from which the measurement is taken. The number of paces per 100 ft is recorded for later use and it is not a part of the survey. The distances are recorded on a line between the stations from which the measurements were taken. This method is often preferred so that each station can be easily described directly across the page on the right-hand side. An alternate method of recording is shown at the bottom of Fig. 2.13.

2.18 Accuracy in Surveys

The degree of accuracy in any survey will depend upon the nature of the survey and the use that will be made of the data. There is little reason for reading the rod to the nearest 0.01 if the work can be done only to the nearest 0.1 ft. It is necessary for the surveyor to use good judgment and common sense. The degree of accuracy will depend upon the nature of the work, but it should be consistent. No measurement should be considered correct until it is verified. This should be done preferably by some other method than that used in the original measurement. In no case should figures be juggled so as to make the data check.

When justified, distances are measured twice and the mean distance computed as shown in Fig. 2.13. The actual or desired accuracy of taping is often expressed as a ratio of error. It is the difference between the two measurements divided by the mean or average distance. For convenience it is converted to a fraction whose numerator is always 1. As shown in Fig. 2.13, the ratio of error is 0.22/1320.24 or 1/6000, meaning that the error is 1 unit in 6000 units. Each unit may be inches, feet, or meters.

REFERENCES

Edwards, D. M., G. D. Bubenzer, and J. K. Mitchell (1977) *Surveying Fundamentals—A Tutorial Approach*, American Printing & Publishing, Inc., Madison, WI.

Kissam, P. (1978) *Surveying Practice*, 3rd ed., McGraw-Hill Book Co., New York.

Moffitt, F. H., and H. Bouchard (1982) *Surveying*, 7th ed, Harper & Row, New York.

PROBLEMS

2.1 In taping between two points the slope distance was 1840.3 ft (560.9 m) and the average slope was 1 percent. Compute the horizontal distance. If the average slope was 10 percent, what would be the horizontal distance?

2.2 Convert 3, 5, and 10 in. to a decimal part of a foot.

2.3 Convert 0.25, 0.1, and 0.85 ft to inches.

2.4 In laying out a right angle by the 3-4-5 method, what should be the distance *DC* and *CE* in Fig. 2.6*a* if *DE* is 96 ft (29.26 m)?

2.5 Compute the distance *AB* in Fig. 2.8 if *BC* is 100 ft (30.48 m) and the angle *CAB* is to be 40°. Compute *AC*.

2.6 Determine the area in acres of *CDEFH* in Fig. 2.10 if *CD* is 200 ft (61.0m), *CE'* is 112.5 ft (34.3 m), *FE* is 840 ft (256.0 m), *FH* is 400 ft (121.9 m), and *HC* is 380 ft (115.8 m).

2.7 Determine the area in acres (hectares) for Fig. 2.11 if d is 100 ft (30.5 m), h_0 is 300 ft (91.4 m), h_1 is 500 ft (152.4 m), h_2 is 800 ft (243.8 m), and h_n is 50 ft (15.2 m).

2.8 A distance was measured twice and found to be 820.11 ft (249.97 m) and 819.95 ft (249.92 m). What is the ratio of error?

2.9 If the desired accuracy for a survey is such that the ratio of error should not exceed 1/4000, what is the maximum error in feet (m) for a distance of one-half mile? One mile is 5280 ft (1609.34 m).

2.10 In chaining with a 100-ft steel tape, pins were exchanged twice and the rear tapeman had 6 pins in his hand. How far was the distance measured if the last pin in the ground was 38.7 ft from the point to be measured?

2.11 Determine the area in sq ft (m^2) of a triangular-shaped field if the sides are 500 ft (152.4 m), 600 ft (182.9 m), and 900 ft (274.3 m).

2.12 Compute the acreage (hectares) of a contour strip of wheat if the perpendicular distance (d in Fig. 2.11) between consecutive parallel offsets is 100 ft (30.5 m). The offset distances, which are the widths of the strip, are 100, 180, 210, 190, 141.2, 100, and 0 ft (30.5, 54.9, 64.0, 57.9, 43.0, 30.5, and 0 m).

2.13 Determine the area of a field from a sketch and measurements supplied by your instructor.

CHAPTER 3

LEVELS AND LEVELING

Much of the work in soil and water conservation is devoted to control of water movement. Water must be moved in desired directions at controlled velocities, and to accomplish this it is necessary to measure accurately differences in elevation. The level is an instrument used to determine such differences. Its use is essential to satisfactory work in drainage, erosion control, and irrigation.

TYPES OF LEVELS

There are many types of levels, ranging in cost from a few dollars to many hundreds of dollars. The type selected for a given job depends upon the accuracy required.

3.1 Hand Levels

Instruments of this type are suitable for rough leveling, such as the running of guide lines for contour farming or for rough determinations of land slope. The simplest type, represented by Fig. 3.1, consists of a metal tube 6 in. long, upon which is mounted a level vial. A prism in the tube reflects the image of the level vial into the eyepiece end of the tube, so that the position of the bubble may be observed as a sight is taken through the tube. Since most hand levels do not have magnification, sight distance is limited to about 100 ft. When held in the hand without a stabilizing stick, readings at 50 ft cannot be taken more accurately than about 0.2 ft. For this reason hand levels are not suitable for accurate surveys, such as tile drainage and terrace layout.

The Abney level shown in Fig. 3.2 is a modified hand level with a level vial attached to a vertical arc, which gives slope in degrees or percent directly. When sighting on the desired point, the moveable arm is rotated until the reflection of the bubble in the eyepiece is centered. The line of sight must be parallel to the land slope. This slope can be obtained by sighting on a second person at a point equal to the eye height of the observer.

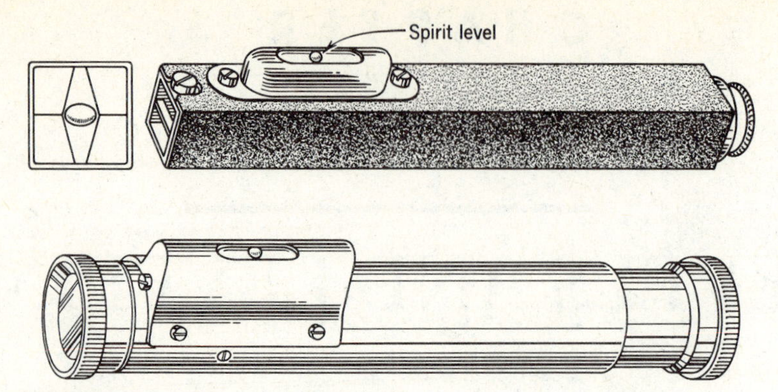

Figure 3.1. Two types of hand levels. (*Courtesy Keuffel and Esser.*)

3.2 Tripod Levels

Levels designed for use with a tripod are available in a wide range of accuracies. They are given the names farm level, builder's level, architect's level, or engineer's level, according to the use for which they are primarily designed. Instruments of this type are well suited for the layout of conservation and water management systems. Based on their construction, these instruments may be referred to as dumpy, tilting, or self-leveling levels. The dumpy level is shown in Fig. 3.3. Its telescope is attached rigidly to the frame of the instrument. This type was originally shorter than other types; hence, the name, "dumpy." The name has little present significance. Builder's and architect's levels have a circular horizontal scale for reading angles. Engineer's levels are usually 18 in. or more in length and are highly accurate. Most are of the dumpy type.

The tilting level and the self-leveling level (Fig. 3.4) often use only three screws for rough leveling of the instrument upon the base plate. In the tilting level, a bubble tube affixed to the telescope tube is viewed through a prism system with

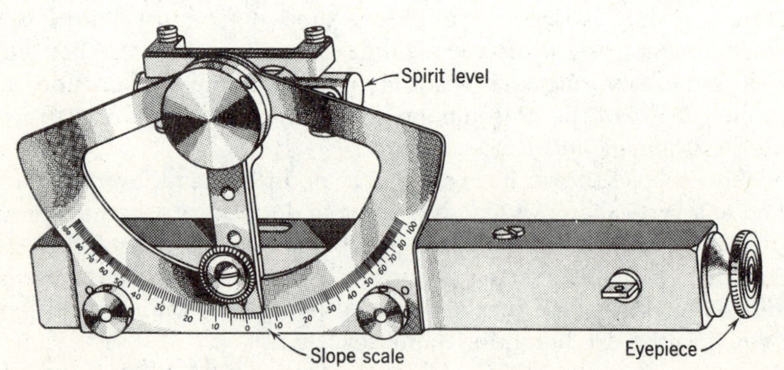

Figure 3.2. Abney level or clinometer. (*Courtesy Keuffel and Esser.*)

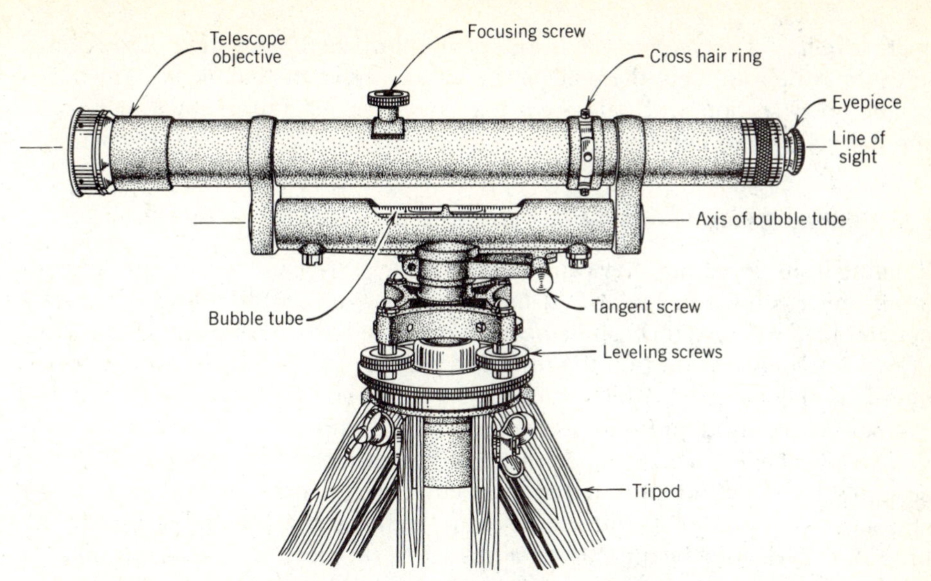

Figure 3.3. Engineer dumpy level. (*Courtesy Keuffel and Esser.*)

an eyepiece adjacent to the telescope eyepiece. The prisms permit both ends of the bubble to be viewed simultaneously in a split image. The instrument is leveled for each reading. It is level when the images of the two bubble ends are brought into coincidence by adjusting a micrometer screw that tilts the telescope. The self-leveling level automatically sets its own level line of sight through a prism system that includes a prism suspended by wires from the upper portion of the telescope barrel and free to swing as a damped pendulum acting under the force of gravity. In the three-screw levels, a spherical bubble vial is used for approx-

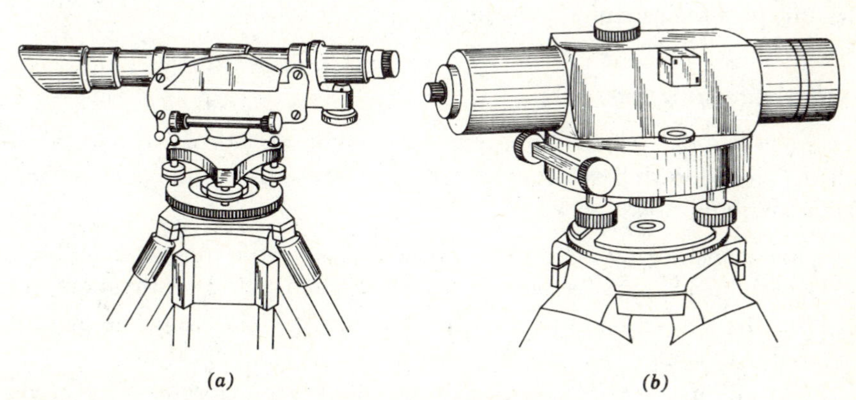

Figure 3.4. (*a*) Tilting level, and (*b*) self-leveling level.

imate leveling. This bubble vial is independent of the line of sight. The leveling screws may be rotated individually or in pairs to move the bubble into the target.

The cross section and parts of a telescope and level are shown in Figs. 3.3 and 3.6. The cross hairs are focused by rotating the eyepiece or by moving it.

3.3 Rotating Beam Levels

Rotating beam levels may have an electronic or a laser beam. A rotating electronic beam unit is shown in Fig. 3.7. It has a range of about 400 ft and the beam is not visible to the naked eye. The beam is projected as a level plane by a rotating prism. The detector unit on the level rod is moved up or down until the beam intersects the detector, which makes an audible signal. Only one person (the rodman) is required, and the unit will shut off automatically if it is out of level.

The laser beam level developed for earthwork with heavy equipment is more accurate, heavier, and more expensive than the electronic beam level. Readings at distances up to 1500 ft with the laser are accurate to within a few hundredths of a foot. The laser beam may be adjusted to the desired height and tilted to produce a sloping plane. Such a slope is desirable when it is used with trenching or other earth-moving equipment. A special detector unit on the machine maintains automatically the same slope as the laser plane. Any number of machines can be operated from one laser unit. Surveys can be made at night with either the electronic or laser instruments.

3.4 Transits

A transit is a more versatile and generally a more expensive surveying instrument than a tripod level. A transit can measure accurately both vertical and horizontal angles as well as stadia distance. As shown in Fig. 3.5, the transit has a telescopic sight and a bubble tube similar to those on tripod levels. The sight and bubble tube are fastened rigidly together and rotate in the vertical plane. However, if the bubble tube is centered, the transit serves as a tripod level. Many engineers prefer this type of instrument for mapping.

3.5 Care of Instruments

To secure continued reliable service, levels must be used properly and carefully handled. The following suggestions for the care and handling of surveying instruments should be observed.

1. When transporting instruments, protect them from impact and vibration. Place them on a firm base in vehicles rather than on top of other equipment.
2. Place the lens cap and tripod cap in the instrument box while the instrument is in use.
3. Leave the instrument box closed and in a place where it will not be disturbed while the instrument is in use.

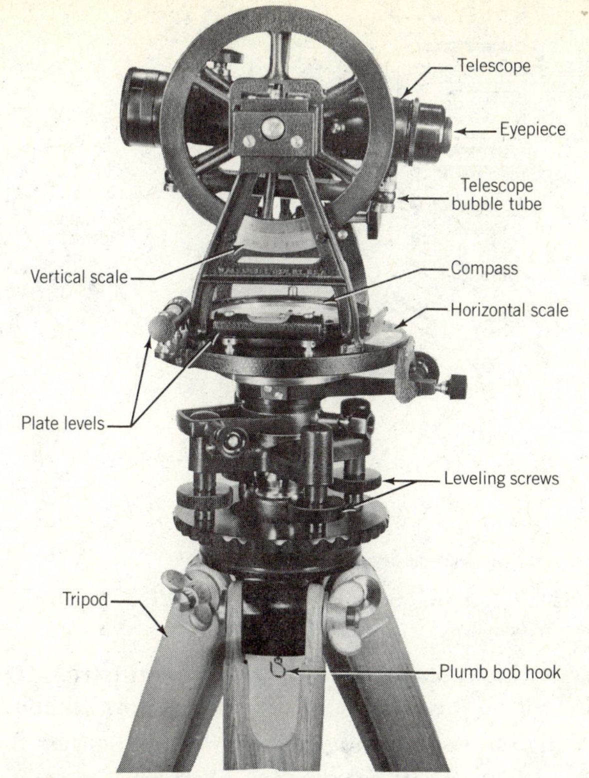

Figure 3.5. Transit or theodolite.

4. Avoid running or otherwise taking a chance on falling while carrying an instrument.
5. Never force screws or other moving parts of an instrument.
6. Protect the lenses from the direct rays of the sun while in use. Use the sunshade regardless of the weather.

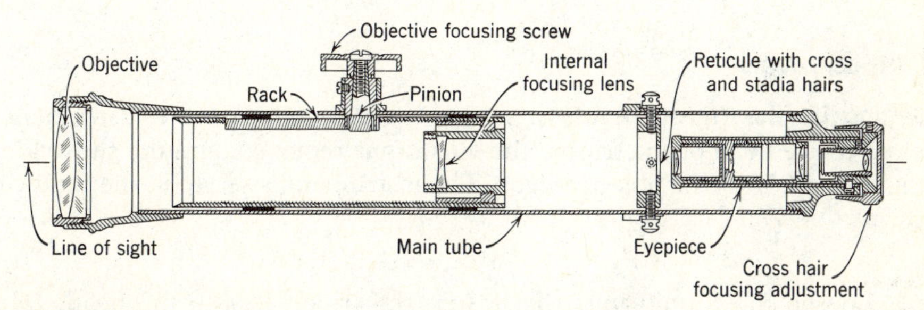

Figure 3.6. Cross section of an internal focusing telescope.

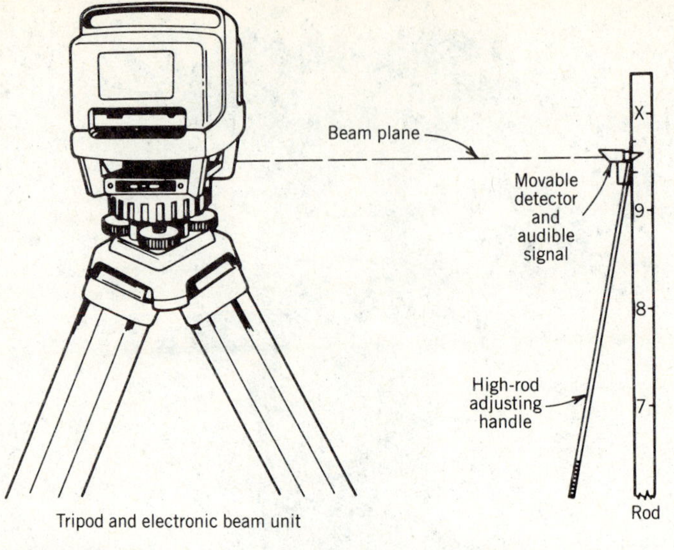

Figure 3.7. Rotating beam level and survey rod. (*Courtesy Spectra-Physics.*)

7. Cross fences with an instrument by spreading the tripod legs and placing the instrument on the far side of the fence before climbing.
8. Bring leveling screws to a snug bearing, but do not jam them.
9. Rub lenses only with soft tissue, not with fingers or rough cloth. Do not remove lenses.

LEVELING

3.6 The Hand Level

The hand level is simply held to the eye with the hand. Resting the hand against the cheek aids in holding the instrument steady. Some users rest the instrument on a staff of convenient length. Its operation is basically the same as that of the tripod level, though measurements are not as accurate.

3.7 Tripod Levels

Setting Up the Tripod Level. In setting up a tripod level the instrument is screwed to the head of the tripod, the lens cap is removed, and the sunshade is put in its place over the objective lens. The instrument is set up in the following steps:

1. Loosen the thumb nuts that fasten the tripod legs to the head. (Most modern levels have friction joints.)

2. Spread the legs 3 or 4 ft, push them firmly into the ground, and adjust the legs so that the tripod head is level in both directions.
3. Tighten the thumb nuts holding the legs to the tripod head.
4. Make sure that the telescope clamp screw is loose and swing the telescope to a position directly over an opposite pair of leveling screws. With a three-screw level, place the bubble vial over one of the screws.
5. Move the leveling screws as shown in Fig. 3.8 and turn them simultaneously to level the bubble tube. These screws should be turned so that the thumbs move toward one another or away from one another. The bubble will follow the left thumb. Keep the leveling screws working firmly against one another, but not tight enough to jam.
6. Turn the telescope to a position across the other opposite pair of leveling screws and again level the bubble.
7. Repeat over each opposite pair of screws until the bubble stays level or nearly level throughout a 360° circuit. With a three-screw level and circular bubble vial (Fig. 3.9), leveling screws may be turned one at a time or in pairs to bring the bubble into the target circle on the vial.
8. Turn the telescope to bring the rod into the field of vision.
9. Focus on the cross hairs by adjusting the eyepiece.
10. Focus on the rod by turning the objective focusing knob.
11. When using a level, check the bubble immediately before and after each reading and make any necessary adjustment of the leveling screws. Do this while you are standing in position to sight through the telescope, as movement that shifts your weight may affect the level. With a tilting level, adjustment of the telescope-tilt micrometer screw should be made for each observation. No adjustment need be made between observations with a self-leveling level so long as the bubble remains within the circle. Some

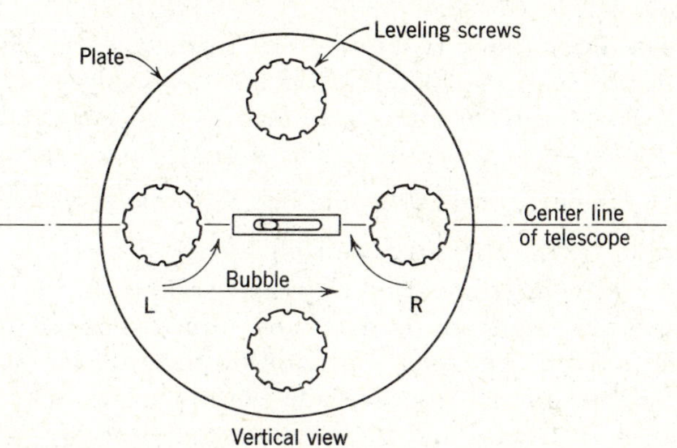

Figure 3.8. Leveling an instrument with four leveling screws. Note the bubble follows the left thumb.

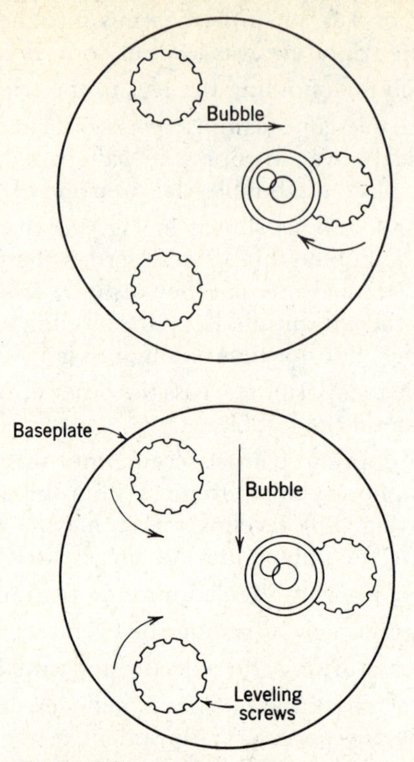

Figure 3.9. Leveling an instrument with three leveling screws.

levels will display a red light when the level is not within the automatic range.

Types of Level Rods Many types of direct reading level rods with different scales are available. The Philadelphia rod has two sections about 7 ft long as shown in Fig. 3.10*a*. It can be extended to read 13 ft and has a target. The target is convenient for laying out a contour line, a grade line as for a terrace, or for precise leveling. The Frisco rod shown in Fig. 3.10*b* has three sections $4\frac{1}{2}$ ft long. Either English or metric scales are available.

The Lenker rod with an inverted moveable scale is shown in Fig. 3.10*c*. Elevations are read directly from the rod. Low numbers are at the top and high ones are at the bottom. An index elevation on the rod, for example, 10 ft, is moved to the line of sight having the same elevation. It is especially suitable for mapping relatively flat fields and for rotating beam levels.

Reading the Rod. Most rods in English units are graduated in feet and tenths of feet as shown in Fig. 3.11*a*. The width of one black line is 1/100 (0.01) ft, and

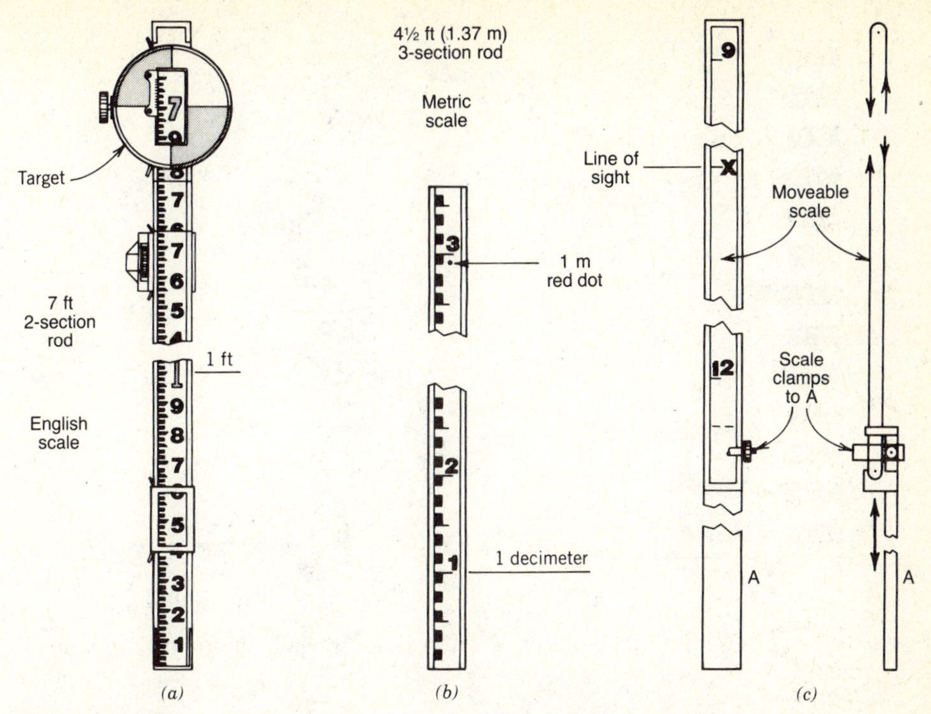

Figure 3.10. Three types of leveling rods: (*a*) Philadelphia, (*b*) Frisco, and (*c*) Lenker (scale inverted to read elevations directly).

the width of a white space between black lines is 1/100 ft. Several rod readings are indicated on the figure. The reading gives the distance in feet from the lower end of the rod.

Figure 3.11*b* illustrates a common source of error in rod readings. Notice that if the rod is slanted in any direction, the reading will be too large. The rodman should hold the rod lightly letting it balance in the vertical position. The fingers should not obscure the face of the rod, for in this position they will frequently obscure the instrument man's view of the rod. Many surveyors find it helpful if the rodman waves the rod back and forth through the true vertical position. If this is done, the instrument man can read the minimum reading, which is the true reading.

To save time in leveling, the instrument man and the rodman should communicate with each other by means of signals. Some standard signals are given in Fig. 3.12.

3.8 Differential or Bench Mark Leveling

The operation of leveling to determine the relative elevations of points some distance apart is known as differential leveling. It consists of making a series of

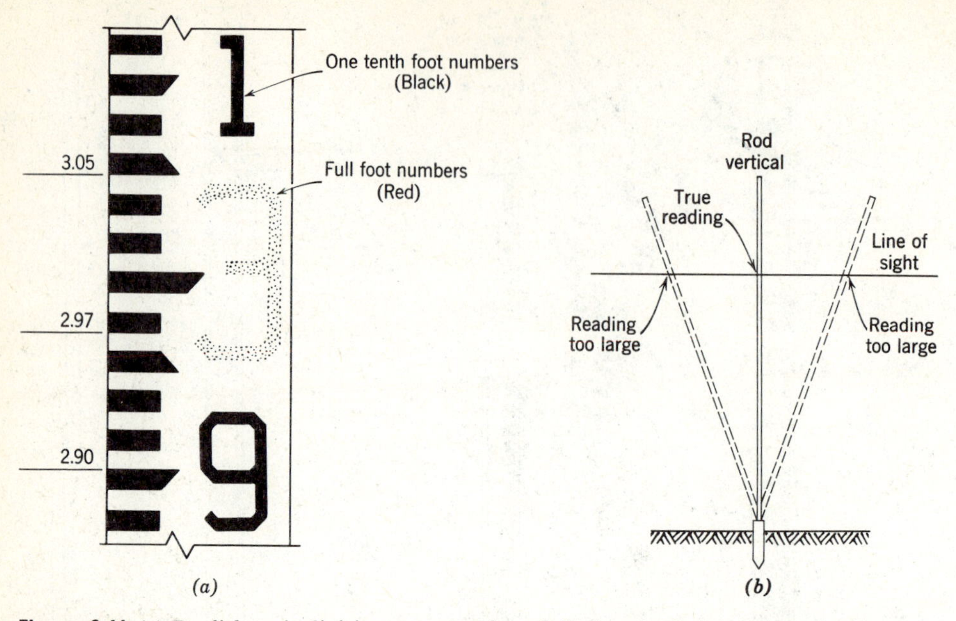

Figure 3.11. (*a*) English unit divisions on a rod, and (*b*) "waving" the rod to obtain a true reading.

instrument setups along the general route between the points and from each setup taking a rod reading back to a point of previously determined elevation and a reading forward to a point of unknown elevation. These points at which elevations are known or determined are called *bench marks* or *turning points*. A bench mark is a permanently established reference point, the elevation of which is assumed or is accurately measured. A turning point is a temporarily established reference point having its elevation determined as an intermediate step in a differential leveling traverse.

For example, referring to Figs. 3.13 and 3.14, if the elevation of bench mark number 1 (B.M. 1) is known to be 100.00, the elevation of B.M. 2 can be found by differential leveling. The first setup of the instrument is made at some point a convenient distance away from B.M. 1 and along the general route to B.M. 2. This distance will depend on the magnifying power and accuracy of the instrument, but in most instances it will be from 75 to 200 ft. The rod is held on B.M. 1, and the rod reading (in this case 5.62) is noted in the field book (Fig. 3.14). This reading is a *backsight* (B.S.); it is a reading taken on a point of *known* elevation.

Addition of the backsight on B.M. 1 to the elevation of B.M. 1 gives the elevation of the line of sight. (In this case 100.00 + 5.62 = 105.62.) This is the *height of instrument* (H.I.) and is entered in the H.I. column of the field notes. Notice that in the field notes the H.I. is entered on the line between the B.M. 1 line and the *turning point* line (T.P. 1), indicating that the instrument is set up between B.M. 1 and T.P. 1. This form of notes is very convenient, although it is not necessary to leave a line for the H.I. between the B.M. and T.P. lines.

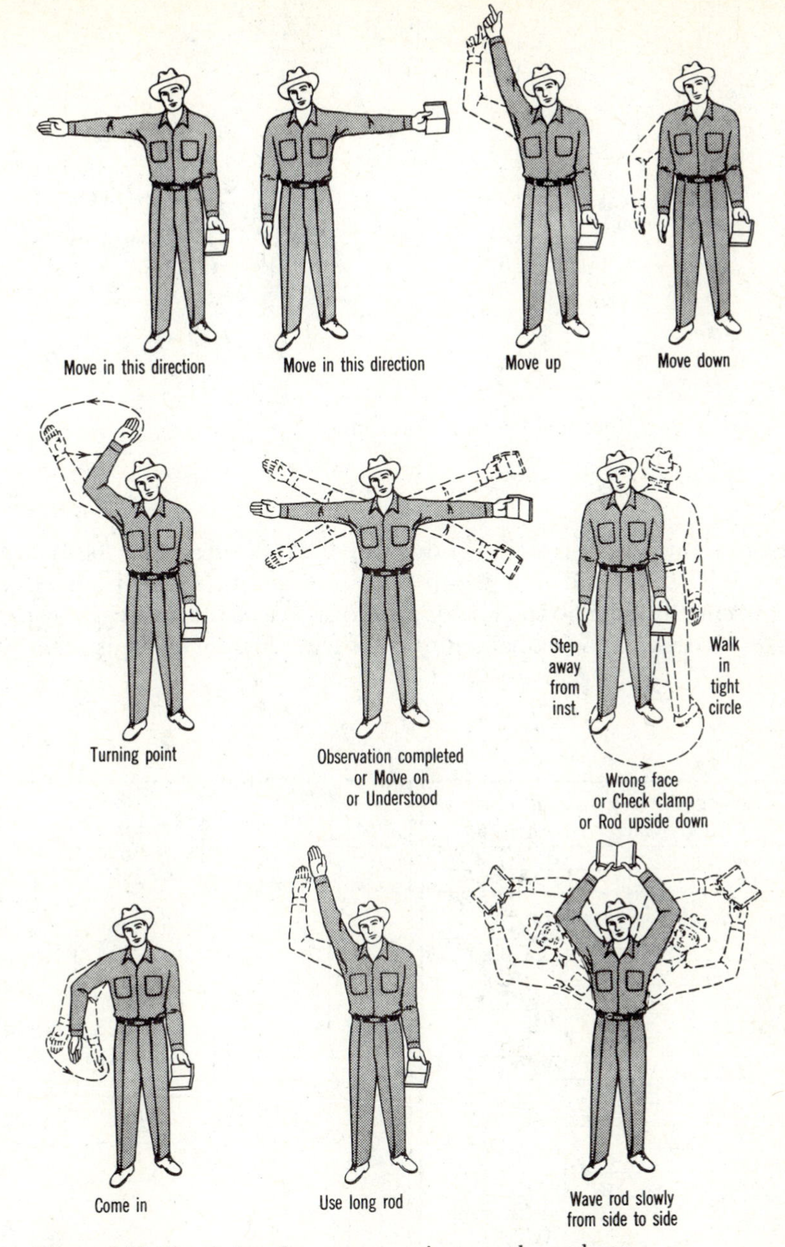

Figure 3.12. Hand signals to communicate to the rodman.

After the backsight on B.M. 1 has been recorded, the rodman moves to turning point number 1 (T.P. 1) and sets a hub or finds some other firm object on which to rest the rod. With the instrument still at setup 1, a rod reading is taken on T.P. 1. This reading is a *foresight* (F.S); it is a reading taken on a point of

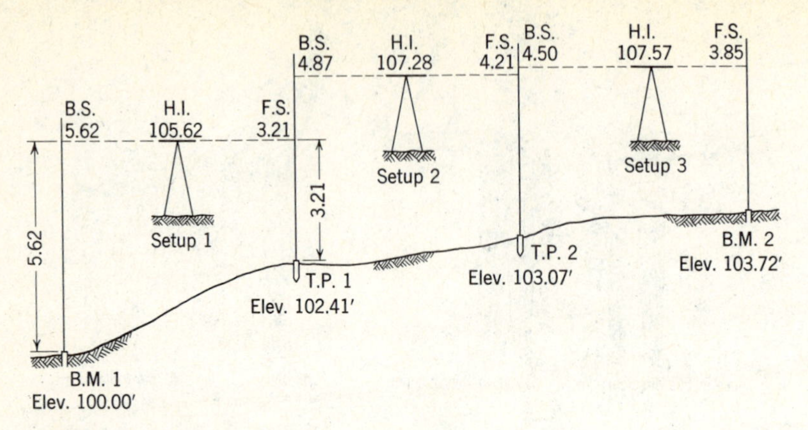

Figure 3.13. Differential leveling procedure.

unknown elevation. The foresight (in this case 3.21) is entered in the field book opposite T.P. 1. It is good practice to make the line of sight distance for the foresight essentially equal to the line of sight distance of the backsight by pacing. The reason for this will be explained under the discussion of adjustment of the instrument.

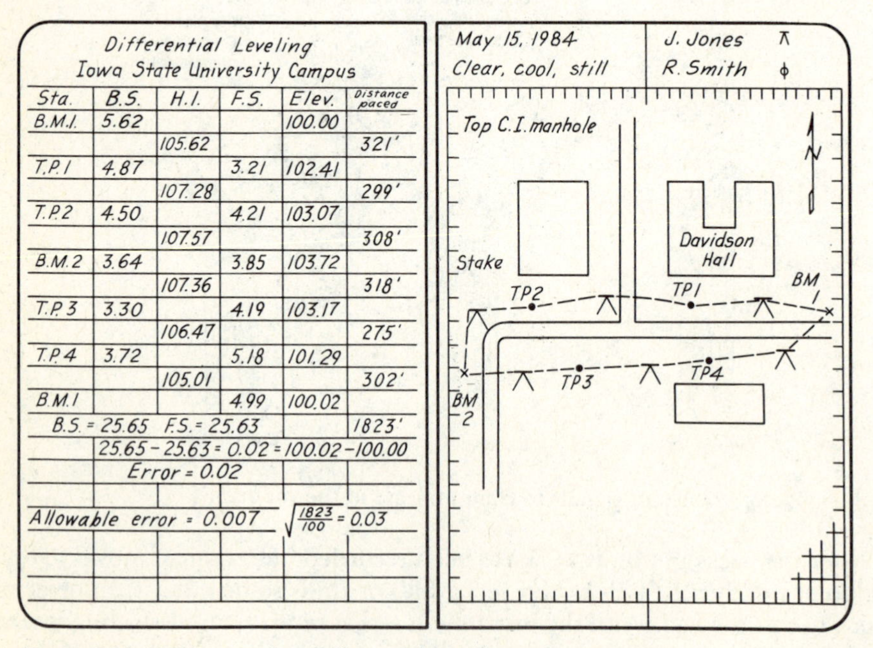

Differential Leveling
Iowa State University Campus

Sta.	B.S.	H.I.	F.S.	Elev.	Distance paced
B.M.1	5.62			100.00	
		105.62			321′
T.P.1	4.87		3.21	102.41	
		107.28			299′
T.P.2	4.50		4.21	103.07	
		107.57			308′
B.M.2	3.64		3.85	103.72	
		107.36			318′
T.P.3	3.30		4.19	103.17	
		106.47			275′
T.P.4	3.72		5.18	101.29	
		105.01			302′
B.M.1			4.99	100.02	
B.S. = 25.65		F.S. = 25.63			1823′

25.65 − 25.63 = 0.02 = 100.02 − 100.00
Error = 0.02
Allowable error = 0.007 $\sqrt{\frac{1823}{100}}$ = 0.03

Figure 3.14. Field notes for differential leveling.

The elevation of T.P. 1 can now be computed by subtracting the foresight on T.P. 1 from the instrument height (in this case 105.62 − 3.21 = 102.41), and this elevation is entered in the notes opposite T.P. 1. Note that T.P. 1 now becomes a point of known elevation.

The rodman remains at T.P. 1 while the instrument man moves to setup 2. From this position a backsight is taken on T.P. 1, the new H.I. is determined, and T.P. 2 is established and its elevation is determined as was previously done for T.P. 1. Careful study of Figs. 3.13 and 3.14 should make this procedure clear. The same sequence of events is carried through (i.e., determine H.I. from a backsight on a T.P. of known elevation and determine the elevation of a new T.P. by foresight) until B.M. 2 is reached.

The standard form for keeping level notes is shown in Fig. 3.14, but for the beginner they may be written out as below and kept on a separate sheet:

Elevation B.M. 1	100.00	
B.S. on B.M. 1	+5.62	
H.I.	105.62	
F.S. on T.P. 1	−3.21	
Elevation T.P. 1	102.41	etc.

The standard format takes less time to record the data and shows the information more clearly.

As a check on the accuracy of the work, a line of differential levels is then run from B.M. 2 back to B.M. 1, and the difference between the original elevation of B.M. 1 and its calculated elevation is the error. In the example given, this is 100.02 − 100.00 = 0.02. A reasonable allowance for the error is given by

$$\text{allowable error} = 0.007\sqrt{\frac{\text{length of traverse in feet}}{100}}$$

In the sample given, the length of the traverse is 1823 ft and the allowable error is 0.03. Thus, 0.02 is within reasonable accuracy.

The length of the traverse shown in Fig. 3.14 should be obtained by the rodman. The distance of 321 ft from B.M. 1 to T.P. 1 should be the sum of the foresight and the backsight distances rather than the straight-line distance between the two points. Normally, the traverse length and ratio of error are not required, unless the accuracy of the survey is specified.

A check on the arithmetical accuracy of differential leveling notes may be made by subtracting the sum of all the foresights from the sum of all the backsights and comparing this difference with the difference between the final and initial elevations. If the two differences are equal, the notes are arithmetically correct. However, this does not check errors in rod readings or in instrument adjustment. A thorough understanding of differential leveling is basic to the use of the level for other work.

3.9 Profile Leveling

Profile leveling is the process of determining the elevations of a series of points at measured intervals along a line. This is particularly important in drainage and terrace layout. Essentially, it is a process of differential leveling with a number of intermediate foresights between turning points. As illustrated in Figs. 3.15 and 3.16, the foresights taken from setup 1 are each subtracted from the height of instrument at this setup to determine the elevations of the intermediate points between B.M. 1 and T.P. 1. The elevations of points between T.P. 1 and T.P. 2 are determined in a similar manner. The location of the instrument at setups 1 and 2 may be at any convenient point so long as the rod can be read accurately. In Fig. 3.15 the elevation of each of the stations is plotted and a line is drawn between the points, thus giving a profile along the proposed ditch. From such a profile the depth of the ditch can be determined, as well as the amount of soil to be moved. Further applications of profile leveling will be given in later sections of this book.

In surveying, stations along a line are generally designated as 0 + 00, 3 + 50, etc. (see Fig. 3.15), where the numeral preceding the + sign is the number of hundred feet from the starting point and the two numerals following the + sign are the additional number of feet, but they must be less than a hundred. Thus, 3 + 50 station means that it is 350 ft from the initial point designated as 0 + 00. The advantages of this system of marking are that station numbers cannot be confused with rod readings and that designation of stations may be simplified. For example, a station 300 ft from an initial point is referred to as station 3 rather than station 300.

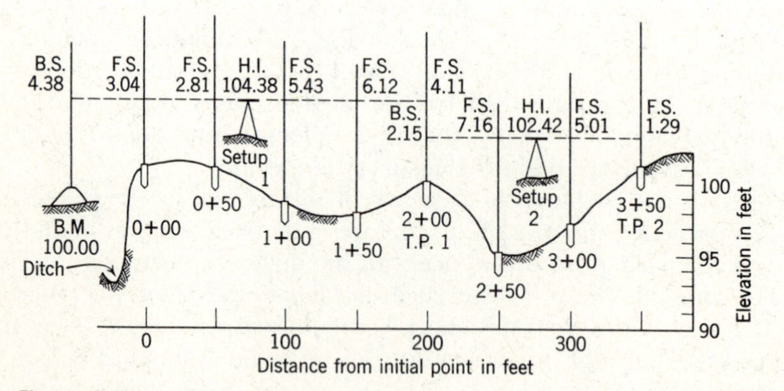

Figure 3.15. Profile leveling procedure.

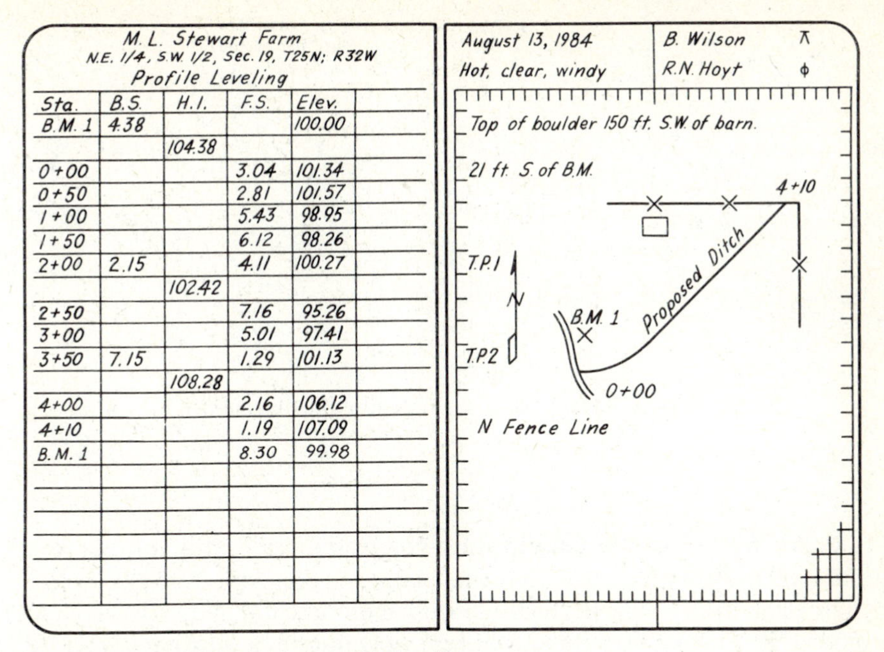

M. L. Stewart Farm
N.E. 1/4, S.W. 1/2, Sec. 19, T25N; R32W
Profile Leveling

Sta.	B.S.	H.I.	F.S.	Elev.	
B.M. 1	4.38			100.00	
		104.38			
0+00			3.04	101.34	
0+50			2.81	101.57	
1+00			5.43	98.95	
1+50			6.12	98.26	
2+00	2.15		4.11	100.27	
		102.42			
2+50			7.16	95.26	
3+00			5.01	97.41	
3+50	7.15		1.29	101.13	
		108.28			
4+00			2.16	106.12	
4+10			1.19	107.09	
B.M. 1			8.30	99.98	

August 13, 1984
Hot, clear, windy
B. Wilson
R.N. Hoyt
Top of boulder 150 ft. S.W. of barn.
21 ft. S. of B.M.

Figure 3.16. Field notes for profile leveling.

ADJUSTMENT OF LEVELS

Measurements made with a level can be applied with confidence only if the instrument used is adjusted properly for accurate work. It is thus essential that the surveyor be thoroughly familiar with the methods of field checking and adjusting his instrument.

3.10 Adjustment of the Hand Level

The accuracy of a hand level may be checked by applying the principles which underlie the following discussions of adjustment of the dumpy level. However, the extent of adjustment possible varies with the individual model of hand level.

3.11 Adjustment of the Dumpy Level

To Make the Horizontal Cross Hairs Horizontal When the Instrument Is Level. Set up and level the instrument and sight on some well defined point, such as a tack head on a tree or post. Turn the telescope about its vertical axis so that the point appears to traverse the field of view. If the point does not remain on the horizontal cross hair, loosen two adjacent screws holding the cross hair ring (Fig. 3.17) and tap lightly on one of the screws to rotate the ring enough to bring the horizontal cross hair into position. Tighten the screws carefully.

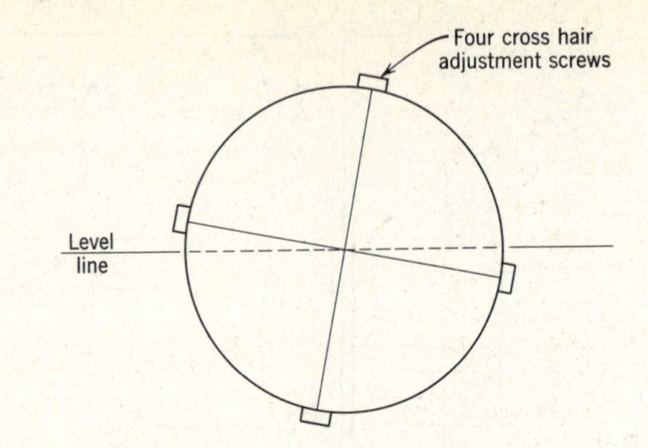

Figure 3.17. Cross-hair ring and adjusting screws.

To Make the Axis of the Bubble Tube Perpendicular to the Vertical Axis. Set up and level the instrument. With the telescope over one opposite pair of leveling screws center the bubble perfectly. Swing the telescope about its vertical axis through 180°. If the bubble does not remain centered, adjustment is necessary. Bring the bubble halfway to the center by adjusting the screws at the end of the bubble tube. After the adjustment is made, relevel the instrument and check again.

This adjustment is not essential to accurate work, but it enables the instrument man to rotate the instrument without having to level between shots.

To Make the Line of Sight Parallel with the Axis of the Bubble Tube. Set up the instrument exactly midway between two hubs that are 150 ft apart, as shown in the field notes of Fig. 3.18. From the midway position take a rod reading on hub A and on hub B, and enter these readings in the notes as a and b. The difference between these readings is the true difference in elevation whether or not the instrument is in adjustment. It is because of this that the suggestion was earlier made that foresight and backsight distances be of equal length.

Move the instrument to A (or B) as in Fig. 3.18 and set up so that the eyepiece just clears the rod held on the hub. Sighting backwards through the telescope, taking reading a', and then sighting in the normal manner take reading b'. If $b' - a' = b - a$, this adjustment is correct. If this equality is not satisfied, raise or lower the cross hair ring by carefully loosening and tightening the top and bottom screws until reading b' is given by $b' = b - a + a'$, or in Fig. 3.18 is equal to 5.24. After the adjustment is made, the checking procedure should be repeated. It is good practice to run through the entire checking procedure twice and only adjust the cross hairs after the same error is observed on two consecutive checks. This will prevent errors in operation of the instrument from being construed as errors in adjustment.

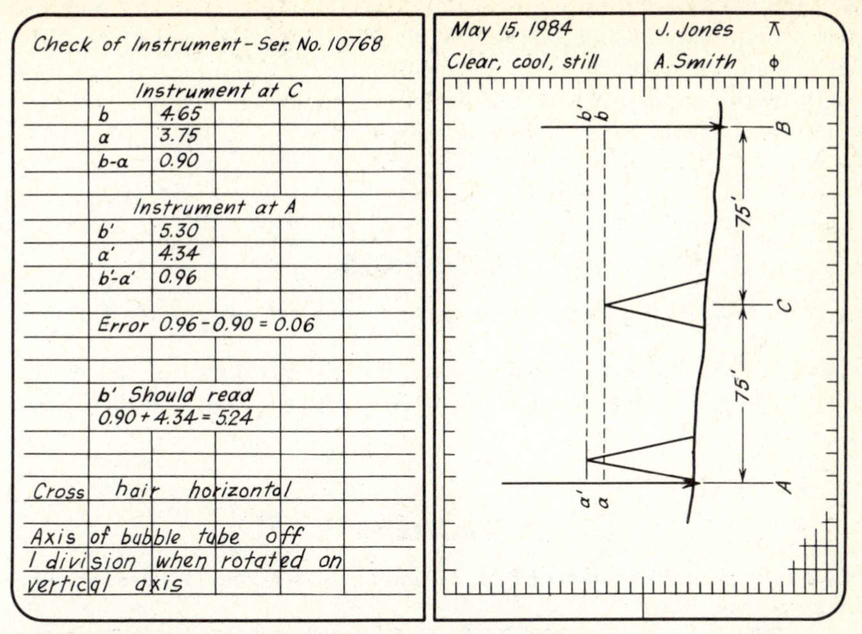

Figure 3.18. Field notes for checking and adjusting the level.

3.12 Adjustment of the Tilting Level

The object of adjustment is to make the ends of the bubble coincident when the line of sight is horizontal. This may be done by using the alternate peg test of Section 3.11 to establish a level line of sight. The telescope tilt is adjusted by the micrometer screw to bring the cross hairs onto the correct rod reading for the level line of sight. After this is accomplished, adjust the screws holding the bubble housing to the telescope until the bubble ends are coincident when viewed through the prism.

3.13 Adjustment of the Self-Leveling Level

To Make the Circular Bubble Centered When the Axis is Vertical. Center the bubble with the leveling screws. Then rotate the telescope 180°. If the bubble does not remain centered, correct one half of the deviation of the bubble by manipulating the three bubble-adjusting screws. These screws are usually found under a circular bubble vial cover or observation prism. Relevel and repeat the check and adjustment until the bubble remains centered through 360° of rotation of the telescope.

To Make the Line of Sight Level Parallel to the Axis of the Bubble Tube. This test and adjustment are the same as making the line of sight parallel with the axis of the bubble tube for a dumpy level (Section 3.11). The alternate peg test is performed to establish a level line. With the circular bubble centered, the line of sight is automatically leveled. The cross hairs are brought in to this level line of sight by the adjustment of reticule screws, which are behind a screw cover, around and in front of the telescope eyepiece.

REFERENCES

Davis, R. E. et al. (1981) *Surveying Theory and Practice*, 6th ed., McGraw-Hill Book Co., New York.
Kissam, P. (1978) *Surveying Practice*, 3rd ed., McGraw-Hill Book Co., New York.
Moffitt, F. H., and H. Bouchard (1982) *Surveying*, 7th ed., Harper & Row, New York.

PROBLEMS

3.1 Determine the elevation of B.M. 2 from the following notes. Check arithmetic by adding F.S.'s and B.S.'s

Sta.	*B.S.*	*H.I.*	*F.S.*	*Elevation*
B.M. 1	1.21			50.00
T.P. 1	6.20		4.65	
T.P. 2	4.82		3.11	
T.P. 3	3.03		5.22	
B.M. 2			3.16	

3.2 An error of 0.08 ft (0.02 m) was made in leveling a distance of 4900 ft (1494 m). Is this within the allowable error?

3.3 Rod readings when adjusting a level were as follows, using the notation in Fig. 3.18: a is 4.13, b is 6.14, a' is 3.85, and b' is 5.90. What is the error and what should the reading b' have been?

3.4 If B.S.'s of 3.71, 4.36, and 6.13 and F.S.'s of 5.68, 6.50, 5.23, 5.09, 5.02, 4.03, 3.42, 3.04, 5.34, and 5.21 were recorded in Fig. 3.16 instead of those shown, what would be the new elevation of each station? Tabulate the data in the standard form for field notes.

3.5 From the following rod readings compute the correct elevation of B.M. 2 if all sight distances were 100 ft and the level had an error of 0.02 ft per 100 ft (rod reads too high). Assume that the level bubble was always centered and that no errors were made in reading the rod. *Hint*: Correct each rod reading before computing.

Sta.	*B.S.*	*H.I.*	*F.S.*	*Elevation*
B.M. 1	3.20			50.00
T.P. 1	0.92		4.60	
B.M. 2			8.14	

3.6 If both B.S. sight distances in Problem 3.5 were 200 ft and both F.S. sight distances were 100 ft, compute the correct elevation of B.M. 2.

3.7 From the survey in Fig. 3.16 check the arithmetic by summing the B.S. and the F.S. *Hint:* Select only F.S. applicable to the differential survey. What is the error?

3.8 Compute the allowable error from the survey in Fig. 3.16 if the total length of the traverse back to B.M. 1 was 802 ft. Is the actual error within the allowable?

3.9 In a differential level survey from B.M. 1 to B.M. 2 the sum of all B.S. was 34.62 ft and the sum of all F.S. was 39.66 ft. If the elevation of B.M. 1 was 50.00 ft, what is the elevation of B.M. 2? What procedure should be followed to verify that the elevation of B.M. 2 is correct?

3.10 Determine the elevation of all T.P.'s and other points of unknown elevation from data supplied by your instructor. Record the data according to the standard form in a field book. Prove that you have made no mistake in your arithmetic. What is the error in the survey?

CHAPTER 4

LAND SURVEYS AND MAPPING

Location and identification of specific tracts of land are based on public land surveys. Details of topography and works of man are recorded on various types of maps. Aerial photographs contain details not often obtained by ground survey. Surveying and locating farmland, map preparation, and use of aerial photographs are common, useful, and often essential activities in the planning and recording of conservation practices.

PUBLIC LAND SURVEYS

4.1 Metes and Bounds

A tract of land identified by giving the direction and length of its several sides is said to be described by *metes and bounds*. The location of property corners may be described by map coordinates and other terrain features. Tracts of land so described may be relocated provided that at least one of the original corners can be identified, and the true direction of one of the sides can be determined. This method of land survey, which originated prior to 1785, is found in the eastern states and in a few isolated areas in other parts of the United States. This system of survey was developed by Thomas Jefferson.

4.2 Rectangular System of Public Land Survey

In 1785 the Continental Congress passed a law that provided for the subdivision of public lands into townships, sections, and quarter sections. As shown in Fig. 4.1, these townships are located with respect to some *initial point* through which passes a true north–south line, called the *principal meridian*, and an east–west line, a true parallel of latitude, called the *base line*. Standard parallels of latitude are located at intervals of 24 miles north and south from the base line, and they intersect the principal meridian at right angles. Guide meridians are located at

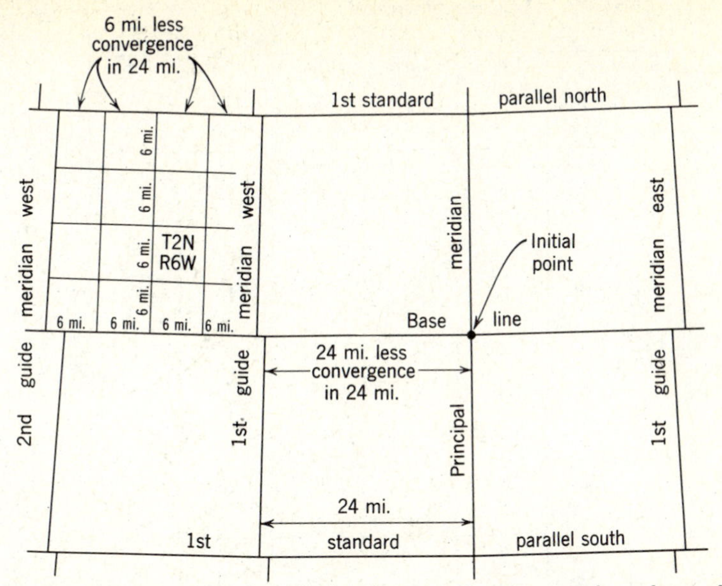

Figure 4.1. Rectangular system of public land survey showing quadrangles and townships.

intervals of 24 miles east and west of the principal meridian measured along the base line or along one of the standard parallels. Because the guide meridians converge, the north side of each 24-mile quadrangle is less than 24 miles in length. Likewise, the north side of each township (6 miles square) is less than 6 miles. In Fig. 4.1 only the upper left 24-mile quadrangle is divided into townships. The true meridian lines which subdivide the 24-mile quadrangles into townships are called *range lines*, and the latitudinal lines at 6-mile intervals are called *township lines*. Townships are designated by numerals corresponding to the number of townships (also called tiers) from the initial point. For example, township T2N, R6W is in the second row of townships north of the base line and in the sixth column of townships west of the principal meridian.

The subdivision of a township into sections one mile square (640 acres) is shown in Fig. 4.2. Sections are numbered by starting in the northeast corner and continuing east and west across the township as shown. If the survey were made without error, all sections would be 1 mile square except those along the west boundary of the township. These fractional sections are less than 1 mile in width because of convergence of the range lines. If errors were made in the survey, sections 1 through 6 may be less or more than 1 square mile.

Each section may be further subdivided into as small tracts as necessary, usually into fourths or halves of sections and of quarter sections (40, 80, 160, or 320 acres). The legal description of these subdivisions, as shown in Fig. 4.3*a*, always begins with the smallest unit. For example, the NE $\frac{1}{4}$, Sec. 23, T2N, R6W, —— county contains 160 acres in the 2nd tier of townships north of the base

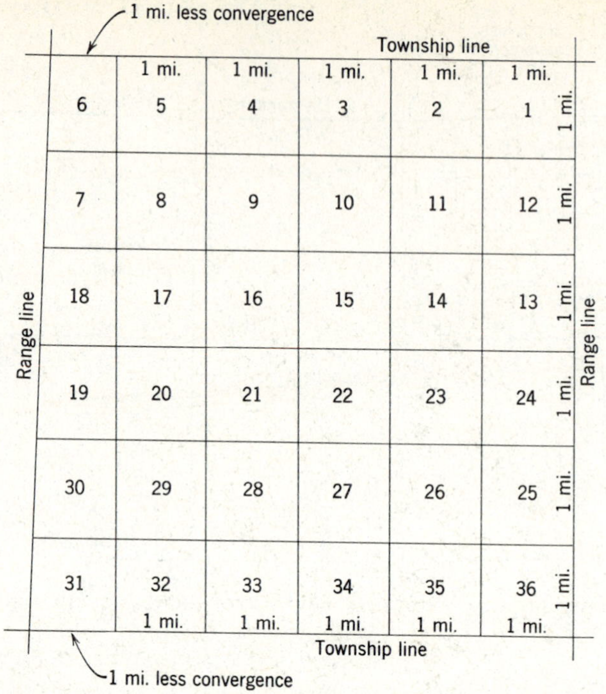

Figure 4.2. Subdivision of a township into sections.

line and in the 6th range of townships west of the principal meridian. Section 23 would be the 9th mile (6 + 3) north and the 32nd mile (24 + 6 + 2) west from the initial point (see Figs. 4.1 and 4.2). As shown in Fig. 4.3*a*, the SE$\frac{1}{4}$, SW$\frac{1}{4}$, SW$\frac{1}{4}$, SW$\frac{1}{4}$, Sec. 23 contains 10 acres or $\frac{1}{64}$ of a section.

Distances between two tracts of land within the same survey can be determined from the legal description as in the following example.

Example 4.1. Find the distance between the southeast corner of Section 33, T2N, R6W and the southeast corner of Section 36, T2N, R4E.

Solution. Since the two points are in T2N, they are directly east and west.

SE corner Section 33 to SE corner Section 36, R6W =	3 miles
SE corner Section 36, R6W to principal meridian =	30 miles
principal meridian to SE corner Section 36, R4E =	24 miles
Total distance	57 miles

Corrections for errors made in the survey or caused by convergence of the range lines occur in the west and north rows of sections in the township. In these sections corrections fall in the west or north rows of quarter sections, as

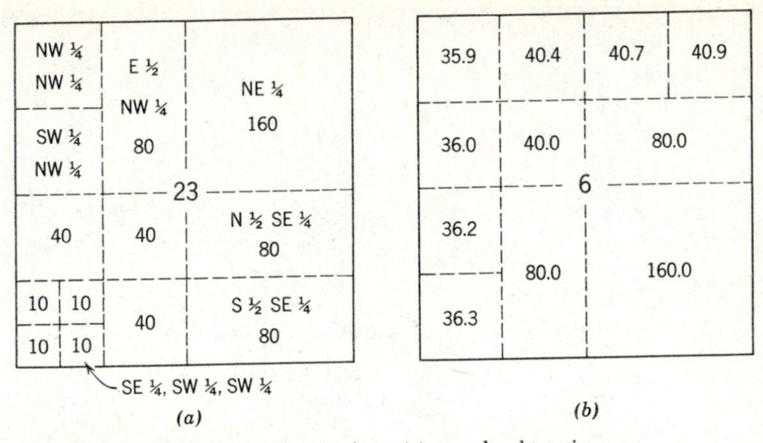

Figure 4.3. Subdivision of a section: (*a*) standard section; (*b*) fractional section.

shown in Fig. 4.3*b* for section 6. Since Section 6 is the northwest section of the township, the corrections occur both in the west and north quarter sections. Sections 1 to 5 have corrections on the north row of quarter sections, which may result in areas greater or less than the standard 40 acres. Sections 7, 18, 19, 30, and 31 have corrections on the west row of quarter sections, and these areas are usually less than 160 acres because of convergence.

TYPES OF MAPS

A map is a useful and frequently an essential means of recording information for selecting soil and water conservation practices and for recording the practices themselves. Several types of available maps are described in the discussion that follows.

4.3 Aerial Photographs

Effective use of aerial photographs requires understanding of their characteristics. In taking the pictures 40 percent to 60 percent overlap between adjacent pictures is allowed. The center portion of each photograph generally has a usable area of 4 square miles. The tick marks shown in Fig. 4.4 aid in locating the center of the photograph. Outside the 4-square mile area the scale is generally too distorted for accurate measurements.

The scale on aerial photographs is only approximate because of slight distortion in taking the picture (except where the camera is directly above) and because of variation in shrinkage of photographic paper. The 10 by 10 contact prints of USDA aerial photos have a scale of about 1667 ft per in. Enlargements

Figure 4.4. Government aerial photograph flown June 30, 1949.

can be obtained with scales of 1320, 1000, 660 and 400 ft per in. If the scale of an aerial photograph is not known, it can be obtained by measuring both the ground and map distance between two points that can be easily defined and identified. Preferably, the two points should be at least 500 ft apart. For example, if the ground distance is 600 ft and the map distance is 1.5 in., the scale is 400 ft per in. Where accurate distances are to be obtained from aerial photographs, the scale should be verified with a field measurement as just described. In reading an aerial photograph it should be held so that shadows of objects fall toward the reader, otherwise valleys appear as ridges and vice versa. Identifying objects on the photograph can best be achieved by noting (1) the tone or shade of gray, (2) shape, (3) shadow, and (4) relative size. Tone depends on the amount of light reflected by the object, and it is affected by the color and texture of the surface and by the angle the sun's rays strike the surface. In fields the color and texture change with seasons, which means that the date the picture was taken is important. Fields, roads, bridges, buildings, and contoured fields can be identified principally by shape. Shadows of trees and buildings not only help in recognition, but they also indicate the height of such objects. Relative size of unknown objects when compared to others of known size may help to identify them.

Much progress is being made by applying remote sensing techniques to aerial mapping. With infrared and special color films with various combinations of color filters, plant and soil features can be detected and differentiated on the ground. Such techniques are being developed for mapping and inventorying crop cover; wet, dry, and other soil conditions; tile drains; insect damage; incidence of plant disease; and certain types of polluted water. Although still in the process of development, such techniques promise to provide powerful new tools for large-scale and timely specialized surveys. On a much larger scale, mapping can be done with orbiting earth satellites.

Up-to-date aerial photographs, if properly made, provide details that could not be obtained by ground survey except at prohibitive cost. Government aerial photographic maps are available for most sections of the United States from the Cartographic Division, Soil Conservation Service, Federal Center Building, Hyattsville, Maryland 20782. Prints are available in sizes up to 40 by 40 in. for a 9-square-mile area. Local offices may have prints that can be enlarged to 1 in. = 200 ft and retain good detail. When ordering prints, include the identification, such as that shown at the top of the photo in Fig. 4.4. Each negative and print will have such data as "6-30-49; DE-1F-140," indicating the date flown; the photograph identifying symbol (DE); the photograph roll number (1F); and the exposure number in the roll (140).

Aerial photographs may be obtained from the State Department of Transportation, other state and local agencies, and private surveying companies. For maps of drainage systems photographs may be taken with hand-held cameras from low flying light aircraft. Although the scale may be distorted, these maps are useful for the location of drains at a later time. Satellite and space photography may be obtained from the U.S. Geological Survey, EROS Data Center, Sioux Falls, SD 57198.

Another source of aerial photographs is the 2 × 2 in. color or infrared slides taken for the U.S. Department of Agriculture, Agricultural Stabilization and Conservation Service for checking the compliance and acreage of conservation practices on individual farms. These slides are taken once and sometimes twice each year and may be purchased from the local offices of that agency at nominal cost. Although prints from slides may not be as accurate or as high quality as those that are obtained from high altitude black and white aerial photos, they are likely to be more up to date and may be adequate for many purposes.

By viewing adjacent pairs of photographs of the same object, a three-dimensional image may be seen with a stereoscope. These devices range from simple lens or mirror stereoscopes to large complex stereoscopic plotting instruments, from which contour maps may be prepared. Unless the area is greater than a few hundred acres, land surveys and hand plotting is more economical than stereophotographic mapping techniques.

4.4 Soil Survey Maps

Most counties have soil survey maps, which have been prepared by soil scientists of the Soil Conservation Service and the state experiment stations. These have been prepared from aerial photographs, usually with a scale of 1667 ft per in., on which the following information may be added.

1. Soil series, type and phase, including land slope classes and extent of erosion. (Slope classes and erosion are not shown in Fig. 4.5.)
2. Roads, streams, houses, and cities.
3. Section numbers, legal description, and township names.

These maps do not show contour lines. An example of such a map is shown in Fig. 4.5. Each soil mapping unit is indicated by a designation, such as BfA2, in which Bf is the soil series and type, A is the slope class, and 2 is the degree of erosion. Several mapping units may be grouped together and called a "Soil Management Group." These are useful for making general recommendations for lime and fertilizer, for forestry plantings, for septic disposal systems, and for the design of irrigation and drainage systems. Further grouping into eight land capability classes is discussed in Chapter 19.

These maps are available from the Superintendent of Documents, U.S. Government Printing Office, Washington, D.C., and other local sources. Some maps may not have all of the information listed above, and in some counties maps have not yet been prepared.

4.5 Topographic Maps

A topographic map is one that shows the relief or the topography of an area. Relief is usually shown by a contour line, which passes through all points having the same elevation. It is the best and simplest method of accurately representing the three-dimensional surface of an area on a two-dimensional drawing. The irregular lines shown on the farm map in Fig. 4.6 are contour lines. Such a map is useful for planning erosion control, water supply, drainage, irrigation, and other conservation systems.

Large-scale quadrangle topographic maps for 79 percent of the contiguous United States have been prepared by the U.S. Geological Survey (Fig. 4.7). These usually have a scale of 2000 ft per in. (1 : 24,000) and a contour interval of 5 to 20 ft. Location is by latitude and longitude rather than by counties. Roads, buildings, streams, water areas, forestland, and many other features are shown. They are multicolored for easy identification. Although quadrangle maps are valuable for large-scale applications, the contour interval is too wide for most on-farm conservation practices. A portion of such a map as shown in Fig. 4.7 includes the same area as in Fig. 4.5, the soil survey map. Some property lines are shown, but green forested areas are not reproduced. This area in Ohio was surveyed by metes and bounds.

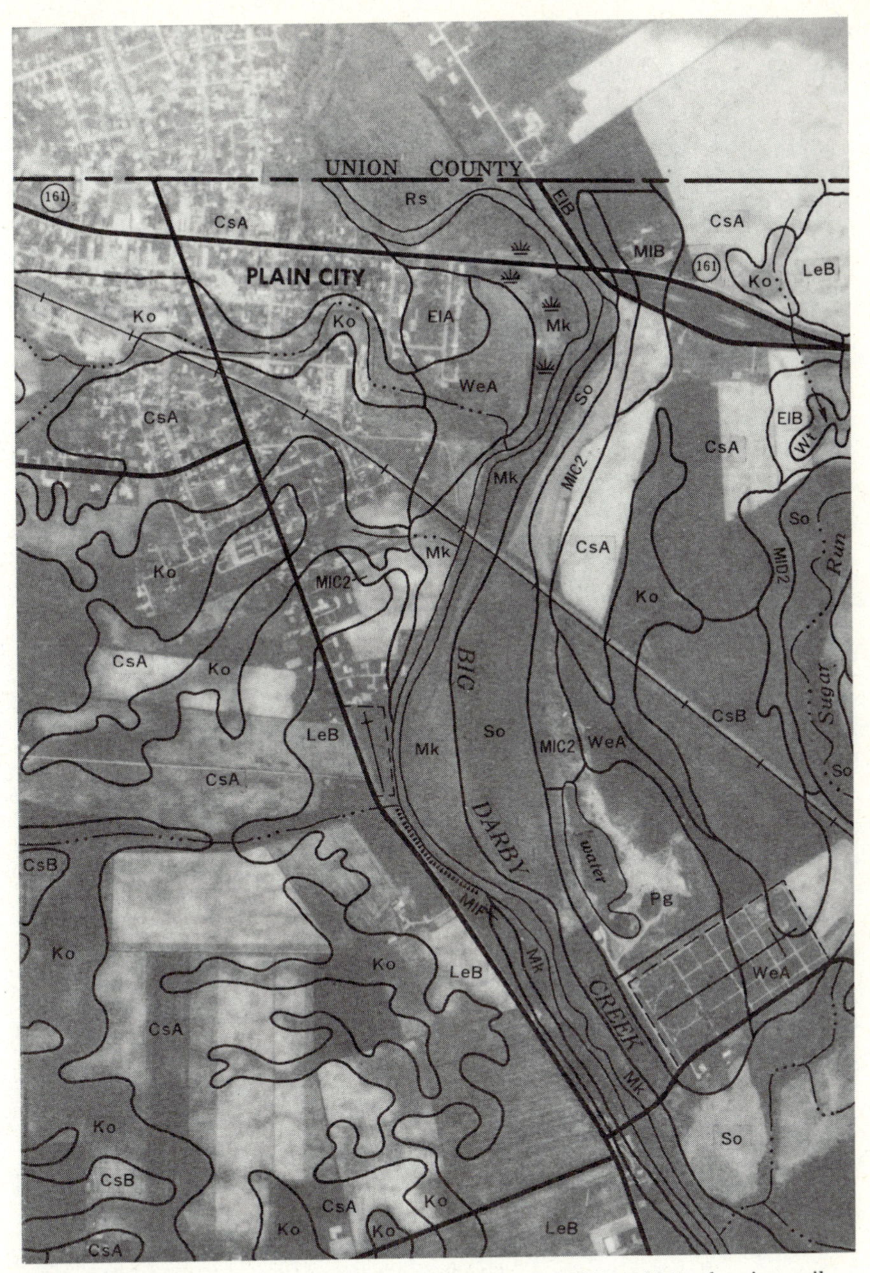

Figure 4.5. Soil survey map from Madison County, Ohio (1981), showing soil series and types. (*Courtesy Soil Conservation Service.*)

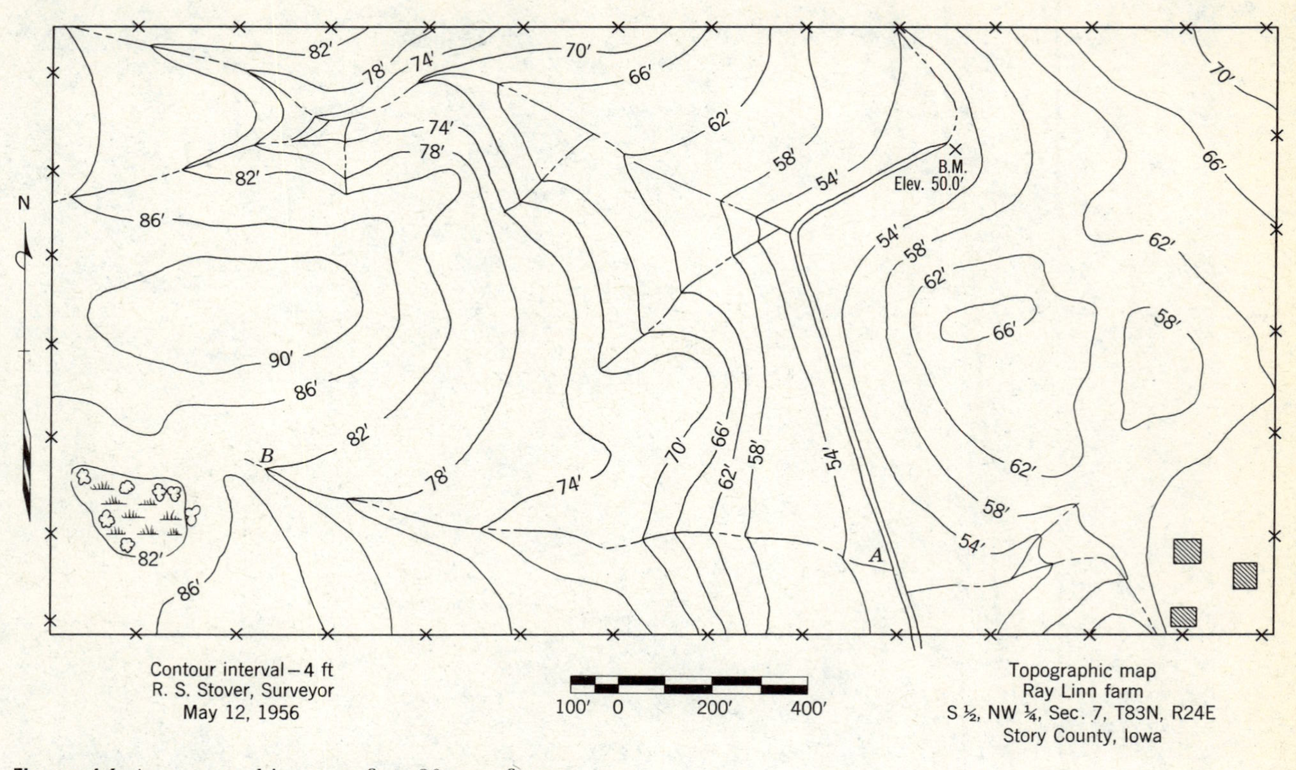

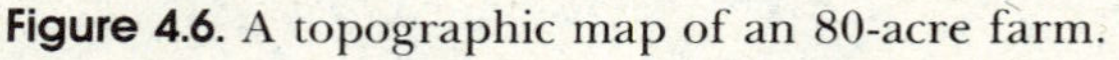

Figure 4.6. A topographic map of an 80-acre farm.

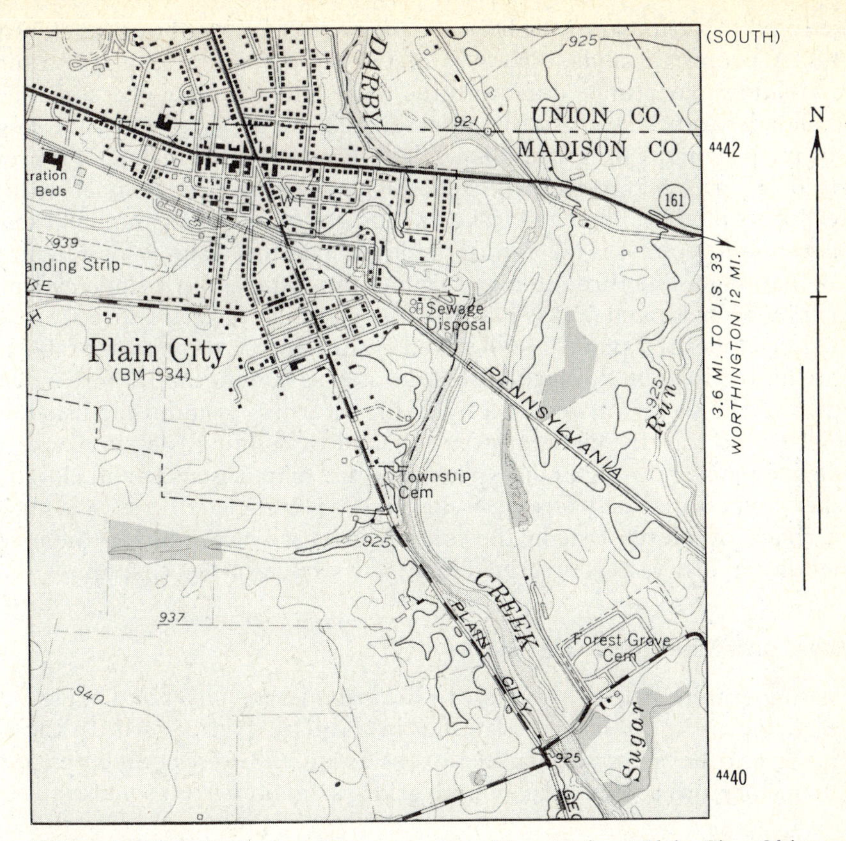

Figure 4.7. U.S. Geological Survey topographic map from Plain City, Ohio, quadrangle with a 5-ft contour interval. (Covers about the same area as soil survey map in Fig. 4.5.)

U.S. Geological Survey quadrangle maps are available from this agency and more than 2500 private retailers. For maps east of the Mississippi River, order from 1200 S. Eads St., Arlington, Virginia 22202 and for maps west of the Mississippi River from Box 25286, Building 41, Federal Center, Denver, Colorado, 80225.

MAP SURVEYS

Several types of map surveys are described, the selection of which will depend on the equipment available and the experience of the surveyor.

4.6 Traverse Table and Level

The traverse table for horizontal control and the level for vertical control are set up close together in the field. The traverse table is a board about 15 in.

square fitted to a tripod on which an alidade is placed for sighting, as shown in Fig. 4.8. A peep-sight alidade consists of two sight vanes attached to a metal ruler. Sighting may also be done with the top edge of a triangular scale. Telescopic alidades with stadia hairs are also available, but they are expensive. Levels for this purpose must have stadia hairs. The procedure for making a survey is shown in Fig. 4.9. A traverse table and level are set up over point *A* within a few feet of each other. Point *a* is arbitrarily located on the map, and the distance and elevation to point 1 is taken with the level. The first point is located on the map by sighting a line through *a* and scaling the distance. In a similar manner any point may be located on the map and the elevation recorded directly on the map. After all desired points are obtained from setup *A*, the traverse table and level are moved to point *B*, which must be located as *b* on the map before moving. Elevation of the level is maintained by taking a turning point in the usual way. At point *B*, the traverse table must be oriented by sighting back to *A* with the alidade line passing through points *a* and *b* on the map. Distances and elevations are taken and plotted as before until the survey is completed. A transit may be used in place of a level. Usually the contour lines are drawn in and the map is finished in the office, as is described in a later section of this chapter.

4.7 Level and Steel Tape Grid Survey

A good topographic map can be made by the grid method illustrated in Fig. 4.10. The field shown in the sketch is laid out in a square grid pattern. In the case shown the grid lines are run parallel to the south and west boundaries of the field. Some persons prefer to lay out the grid system on perfect squares for easy

Figure 4.8. Traverse table and peep-sight alidade. (*Courtesy W. & L. E. Gurley Co.*)

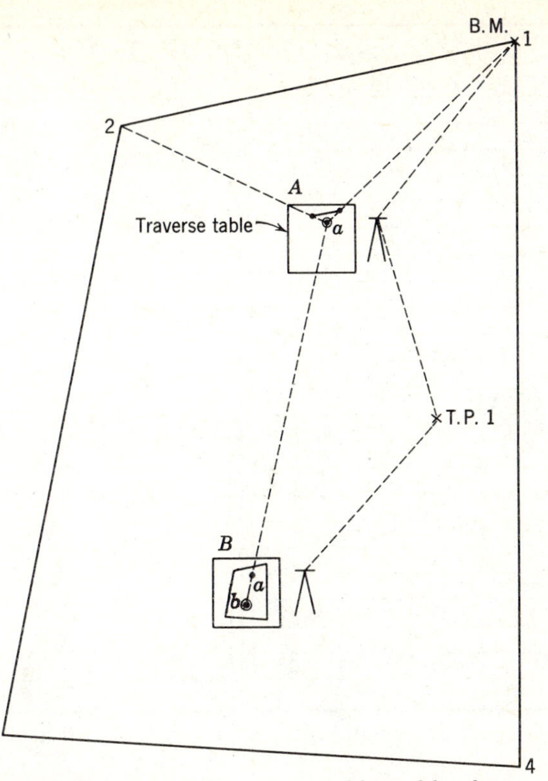

Figure 4.9. Survey with traverse table and level (or transit).

plotting on graph paper. Grid points may be readily identified by the scheme noted in the sketch. For example, point *E* 5 is the intersection of the east–west line through 5 with the north–south line through *E*. It is not necessary to stake each grid point. If a tall stake is placed at each point of the east–west lines through 0 and through 1 and at each point of the north–south lines through *A* and through *B*, it is possible to locate oneself at any other grid point in the field by lining up on the appropriate pairs of tall stakes.

The dimensions of the grid will depend upon the topography and the detail desired. For gentle uniform slopes a 100-ft grid pattern is usually satisfactory. The sides of the field are chained and the internal angles at the corners are measured. A check on the accuracy of the measurement of angles may be made by applying the rule that the sum of the internal angles of any closed figure bounded by straight lines is $(n-2)180°$ where n is the number of sides bounding the figure. If equipment is not available for measuring angles, the shape of the field may be determined by chaining the diagonals in addition to the sides.

With the grid system located on the ground, the elevations of the various grid points are determined by profile leveling. That is, a foresight is taken to each grid point, and the elevations are calculated as indicated in Fig. 4.10. In addition

JOHN DOW FARM
SE 1/4, SE 1/4, SEC. 32, T84N, R24W
TOPOGRAPHIC LEVELING

STA.					
X	Y	B.S.	H.I.	F.S.	ELEV.
B.M.1		4.32	94.94		90.62
A	6			7.7	87.2
A	5			5.2	89.7
A	4			0.4	94.5
A	3			5.8	89.1
A	2			7.8	87.1
A	1			8.8	86.1
A	0			9.9	85.0
B	0			10.6	84.3
CONTINUE	LINES	B, C, AND D.			
D	2	7.37	90.88	11.43	83.51
E	0			7.8	83.1
E	1			9.9	81.0
D+70	1			10.1	80.8
E	1+70			10.6	80.3
E+50	2+30			10.7	80.2
CONTINUE	LINES F,	G, H, AND I			
T.P.2		5.93	92.25	4.56	86.32
B.M.1				1.60	90.65
			ERROR		0.03

MAY 20, 1984
WARM, CLEAR
J. JONES ⊼ 12
R. SMITH φ

TOP 3' BOULDER ON FIELD BOUNDARY

B.M.1 90.62
T.P.1
T.P.2
N
Y
A B C D E F G H I

Figure 4.10. Field notes for topographic survey using the grid method.

to all grid intersections, intermediate points should be taken at high or low elevations and where slope changes occur. For example, a point 70 ft from *D* to *E* on line 1 is shown as (*D* + 70)1 in Fig. 4.10. Likewise (*E* + 50)(2 + 30) is 50 ft from *E* and 30 ft from line 2. This system of identifying points is also convenient when the digital computer is used to make a contour map.

4.8 Transit Surveys

As discussed in Chapter 3, a transit can measure accurately both vertical and horizontal angles as well as stadia distances. Points in the field can be located by taking azimuth angles and stadia distances. Elevations, other than level readings, can be obtained by calculating vertical distances from vertical angles and stadia distances. Some transits have a special (Beaman) arc scale which simplifies these conversions. Format for keeping field notes can be found in surveying tests, such as Davis et al. (1981).

4.9 Rotating Beam Level Surveys

As described in Chapter 3, the electronic or laser beam level establishes a horizontal plane from which rod readings are taken to obtain elevations. The elevation of the beam is obtained from a bench mark as with a level. The Lenker rod with a light beam detection unit allows reading the elevations directly. When the detector unit is on the beam, a signal is given. Using an aerial photograph or suitable map, points are located and the rodman records the elevation directly on the map. Only one person is needed to make a laser survey. A special detector used on earth-moving equipment can also be mounted on a vehicle for surveying. With a suitable recorder a tape printout of distance and elevation can be obtained. Such a detector follows the beam automatically. Some contractors survey the ground surface prior to drain installation with this type of equipment. Land-grading machines equipped with a laser can make surveys of the land surface prior to leveling or grading to determine the cut or fill.

PREPARATION OF TOPOGRAPHIC MAPS

4.10 Characteristics of Contour Maps

A contour is an imaginary line of constant elevation on the surface of the ground. The shoreline of a lake is a contour readily seen in nature. A contour line on a map is a line connecting points on the map that represent points on the surface of the ground having the same elevation. The elevation of a contour line is given by a number that appears on the contour line. The following characteristics of contour lines are useful guides in drawing and interpreting contour maps:

1. The horizontal distance between contour lines is inversely proportional to the slope. Hence, on steep slopes the contour lines are spaced closely.
2. On uniform slopes the contour lines are spaced uniformly.
3. On plane surfaces the contour lines are straight and parallel to one another.
4. As contour lines represent level lines, they are perpendicular to the lines of steepest slope. They are perpendicular to ridge and valley lines where they cross such lines.
5. All contour lines must close upon themselves either within or without the borders of the map.
6. Contour lines cannot merge or cross one another except in the rare cases of vertical or overhanging cliffs.
7. A single contour line cannot lie between two contour lines of higher or lower elevations except in very rare instances.

A profile of the soil surface (elevations along a given line) may be plotted from a contour map as illustrated in Fig. 4.11. This profile along waterway *AB* in the south portion of the farm (Fig. 4.6) shows graphically the slope in the channel. Such information could be used for waterway design as discussed in Chapter 7.

4.11 Map Constructions

Maps drawn from traverse table data are usually completed in rough form in the field, the various features of the landscape being drawn in as the work

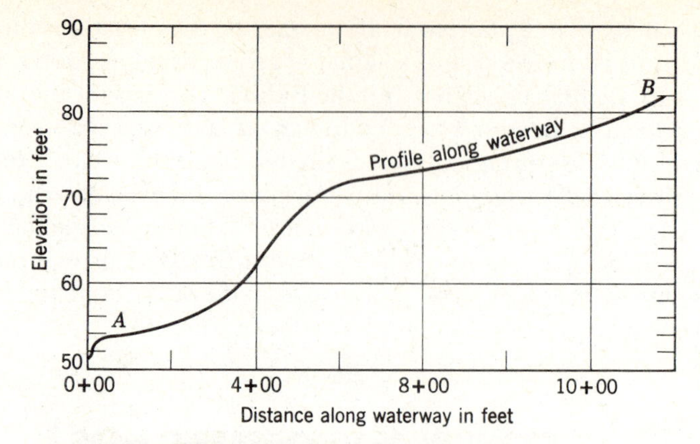

Figure 4.11. Profile along line *AB* in Fig. 4.6.

progresses. Where the grid system of topographic mapping is used, the field boundaries and the grid are reproduced on paper. In any system of topographic mapping the observed elevations are noted on the preliminary drawing as shown in Fig. 4.12.

Map Symbols. Map data gathered from a survey do not attain full usefulness until they are clearly presented on paper to form the finished map. Some features of the landscape occur so frequently that standard symbols have been adopted to facilitate their representation on maps. Figure 4.13 gives a number of common map symbols.

Scale. The relationship between distances on the map and distances on the ground is the scale of the map. If 1 in. on the map represents 100 ft on the ground, then the scale will appear on the map as 1 in. = 100 ft. This is often conveniently shown on the map in graphic form as indicated in the sample title block of Fig. 4.14. The graphic presentation of the scale is particularly desirable in that if the map is reduced or enlarged the graphic presentation of the scale is changed accordingly, whereas the scale as written out becomes incorrect. A scale of 1 in. = 100 ft may also be written 1 : 1200, a ratio of map to ground distance in any units.

Selection of the scale is based on the use of the map, the area covered, and the detail desired. In general the scale should conform to one of those found on a standard engineer's scale. These are 1 in. equals 10, 20, 30, 40, 50, or 60 ft, or any multiple of 10 of these values. A 160-acre farm can be conveniently represented on a 17 by 22-in. sheet to a scale of 1 in. = 200 ft.

Title Block. The following information should appear on the map or on a neatly arranged title block such as Fig. 4.14.

1. An arrow indicating true north, magnetic north, or both.

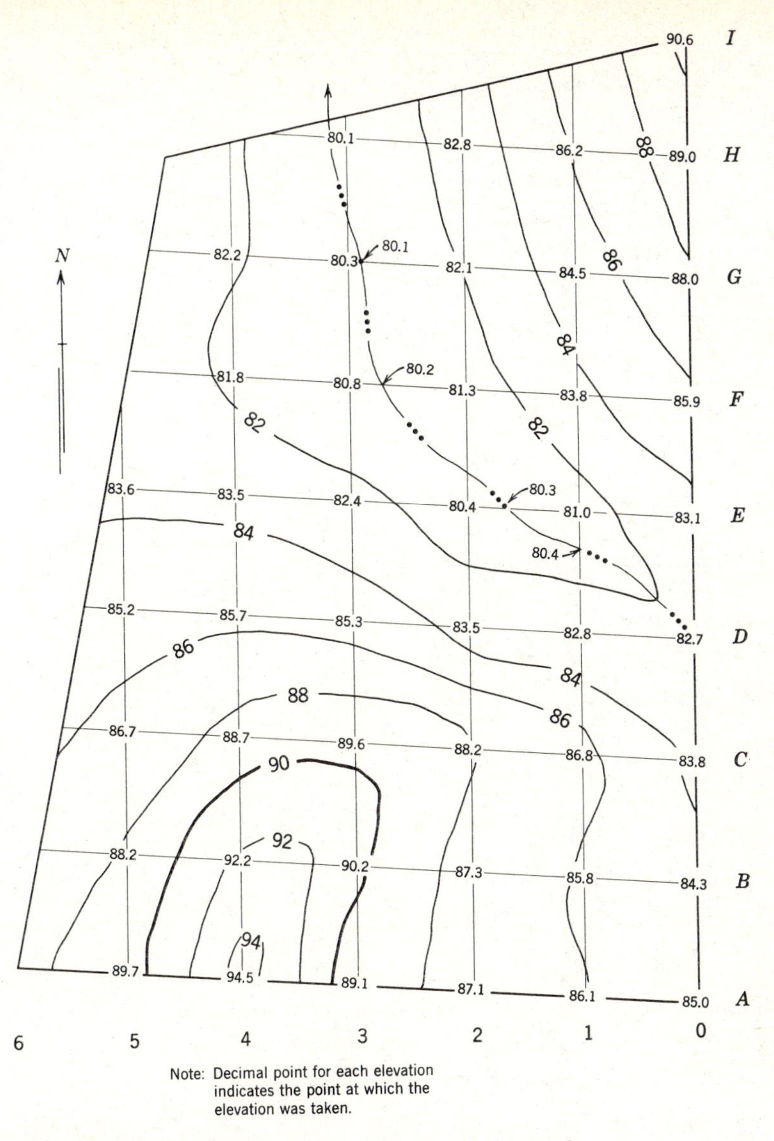

Figure 4.12. Contour map with a 2-ft contour interval. (Construction details included.)

2. A legend or key to symbols used if they are other than conventional symbols. (See Fig. 4.13.)
3. A graphic scale.
4. A statement of the kind or purpose of the map.
5. The name of the tract mapped.
6. The legal description of the tract mapped.

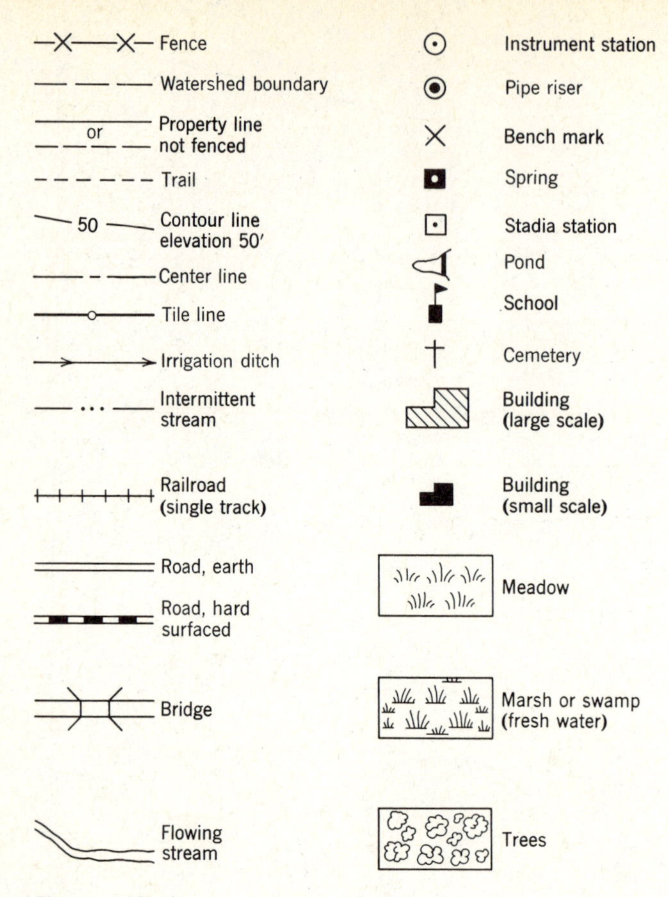

Figure 4.13. Common map symbols.

7. The name of the project or purpose for which the map is to be used. (This may be combined with item 4 in many instances.)
8. The name of the surveyor and draftsman.

Contour Lines. On a contour map the vertical distance represented by the spacing between adjacent contour lines is the contour interval. The same contour interval should be maintained throughout a given map to avoid confusion and misinterpretation. Selection of the contour interval will depend upon the slope of the ground, the scale of the map, the detail of the survey, and the purpose of the map. The contour interval together with the scale of the map should be selected to avoid excessive crowding of the contour lines.

Seldom does a contour line pass directly through a point of observed elevation. The location of the contour lines is determined by interpolation between two adjacent points of known elevation. It is assumed that the ground has a uniform slope between such points. For example, in Fig. 4.12 the difference in elevation

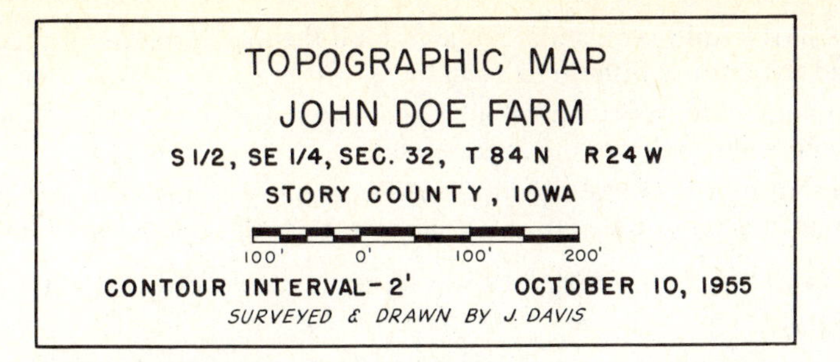

Figure 4.14. Example title block for a map.

between points H1 and $H2$ is 3.4 ft. From $H1$ to the 86-ft contour line there is a difference in elevation of 0.2 ft. Thus, the 86-ft contour line will fall 0.2/3.4 or $\frac{1}{17}$ of the horizontal distance from $H1$ to $H2$ away from $H1$. The 84-ft contour is 1.2 ft above $H2$; it will fall 1.2/3.4 or $\frac{6}{17}$ of the horizontal distance from $H2$ to $H1$ away from $H2$. This type of reasoning is applied to find the location of contour lines between each two adjacent points of known elevation.

Since the slope is assumed to be uniform between points, the rodman should keep this in mind when the field survey is made. Elevations should always be taken at the low points and at the high points. For example, in Fig. 4.12 the point F (2 + 70) is a low point and is necessary to locate the flow channel on the map.

The contour lines are drawn in as smooth freehand lines of uniform weight. Usually each 10-ft, etc. contour line (see 90-ft contour in Fig. 4.12) will be drawn in heavier than the intermediate lines. This enhances the readability of the map.

REFERENCES

Davis, R. E. et al. (1981) *Surveying Theory and Practice*, 6th ed., McGraw-Hill Book Co., New York.

Kissam, P. (1978) *Surveying Practice*, 3rd ed., McGraw-Hill Book Co., New York.

Moffitt, F. H., and H. Bouchard (1982) *Surveying*, 7th ed., Harper & Row, New York.

PROBLEMS

4.1 Give the complete legal description of 40 acres located in the extreme northwest corner of a section $23\frac{3}{4}$ miles north of the base line and $12\frac{1}{4}$ miles east of the principal meridian.

4.2 What should be the sum of all the internal angles of a field with five straight sides?

4.3 Determine the average land slope (in percent) between points A and B, which are 300 ft apart with each point on different contour lines. The contour interval is 2 ft and two other contour lines lie between points A and B.

4.4 What is the scale of an aerial photograph if the map distance between two points is 2.5 in. and the ground distance is 1650 ft?

4.5 Two points with elevations of 43.2 and 45.6 ft are 100 ft apart. How far should the 44-ft contour line be from the lowest point?

4.6 The shortest horizontal distance between two consecutive contour lines is 40 ft. If the contour interval of the map is 2 ft, what is the land slope in percent?

4.7 On a sketch of Section 18, T10N, R4E locate the $E\frac{1}{2}$, $SE\frac{1}{4}$, $NE\frac{1}{4}$. How many acres does it contain?

4.8 Determine the distance in miles between the center of Section 36, T84N, R14W and the center of Section 1, T5S, R14W.

4.9 A topographic map having a scale of 200 ft per in. is reduced to 75 percent of its original size by photographic processes. What is the scale of the reduced map in feet per inch? What is the advantage of a graphical scale?

4.10 The map distance between the 40- and the 60-ft contour lines is 1.5 in. If the scale of the map is 1 in. = 100 ft, compute the land slope in percent.

4.11 In a field with a 1.6 percent slope, how far is the 42-ft contour line from a point with an elevation of 41.52 ft?

CHAPTER 5

RAINFALL AND RUNOFF

Moisture, whether too much or too little, or poorly distributed, is one of the major limitations in agricultural production. The problems of preventing excessive movement of soil, retaining needed moisture, increasing the intake of surface water, adding needed water by irrigation, and removing excess water by drainage cannot be solved without consideration of rainfall–runoff relationships.

5.1 The Hydrologic Cycle

The water circulation on the earth consists of a continual movement of moisture over, through, and beneath the surface. This pattern, depicted in Fig. 5.1, is commonly referred to as the hydrologic cycle.

Precipitation may occur in many forms and may change from one form to another during its descent; or it may occur as frozen water particles, such as snow, sleet, or hail. Some of this precipitation evaporates partially or completely before reaching the ground; some precipitation changes from one form to another before reaching the earth's surface. Precipitation reaching the earth's surface may be intercepted by vegetation, it may infiltrate into the ground, or it may evaporate. Evaporation may be from the surface of the ground, from free water surfaces, or from the leaves of plants through transpiration. A portion of the total rainfall moves over the earth's surface as runoff while another portion moves into the soil surface, is used by vegetation, becomes part of the deep ground water supply, or seeps slowly to streams and to the ocean.

5.2 Air Masses

The characteristics of air masses are controlling factors in development of precipitation. Air masses are formed by continued association with specific surface and radiation conditions. Figure 5.2 gives the predominant air masses affecting

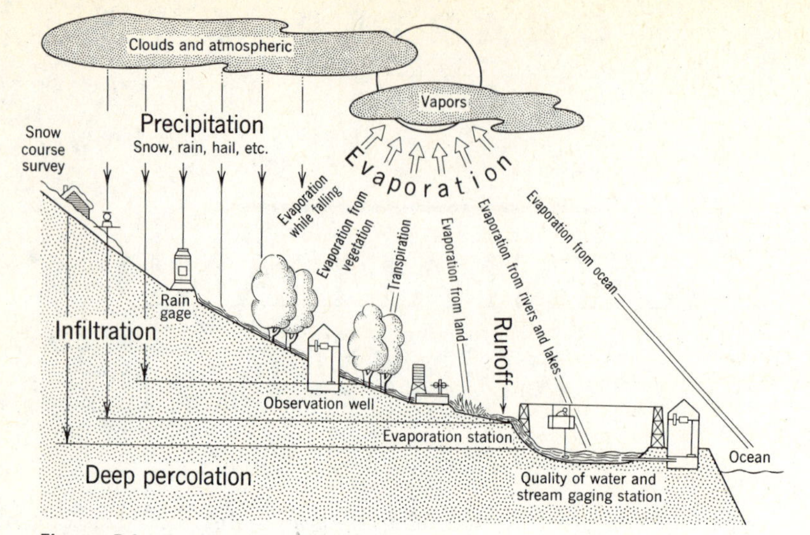

Figure 5.1. The hydrologic cycle.

the North American continent. The tropical maritime (mT) air formed over the Gulf of Mexico is subjected to considerable heating by the sun. As a result of long association with the water surface, it takes up much moisture and provides the moist warm air characteristic of the southerly winds in the central and eastern portions of the United States. By contrast the cold air masses known as polar continental (cP) air are generally formed over north-central Canada. Gigantic high-pressure centers may lie here for as long as several weeks. As these air masses have low moisture and often rest on large snow-covered areas, they have considerable negative radiation from the surface of the earth and become very cold.

Other important air masses shown in Fig. 5.2 are those denoted by S, the hot continental air formed in summer over the southwest desert areas of this country, and by mP, the polar maritime air mass, such as is formed over the northern parts of the Atlantic and Pacific oceans.

5.3 Sources of Moisture in Precipitation

The air mass that contributes the largest amounts of moisture to the central and eastern portions of the United States is the tropical maritime (mT), which moves warm moist air in from the Gulf of Mexico. On the west coast most of the precipitation comes from the polar maritime (mP) air, which carries moisture in from the Pacific Ocean. The portion of precipitation originating from continental evaporation is very small. This evaporation cannot increase local moisture. For local evaporation to have this effect would require complete stagnation of air, something that seldom, if ever, occurs. Instead, most of the moisture removed by evaporation is carried away by cool, dry continental air masses.

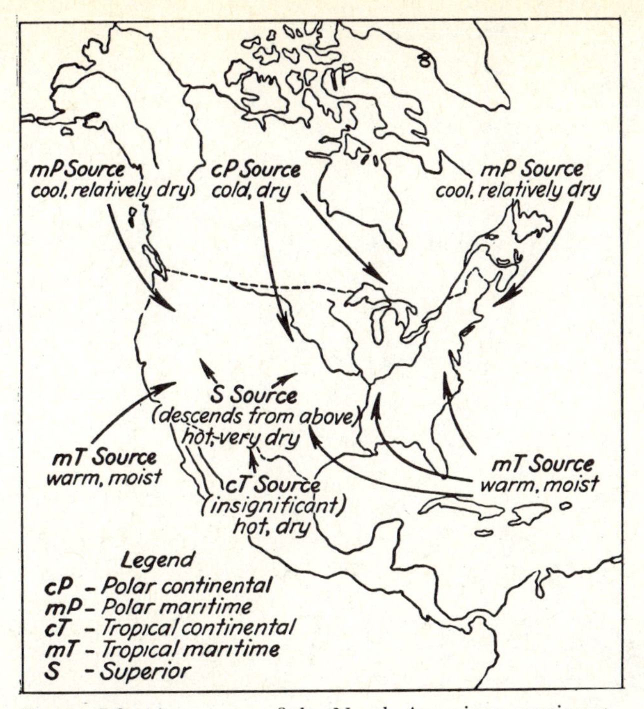

Figure 5.2. Air masses of the North American continent. *Source:* Rouse (1950).

5.4 Influence of Land and Water Masses

Land and water masses are important in their effects on weather. Since land cools much more rapidly than water, the air masses passing over water are not so drastically changed. Also in winter the formation of snow provides an excellent opportunity for negative radiation and the resultant formation of very cold air masses. Uneven distribution of land and water areas on the earth's surface also causes deviations from the expected circulation patterns because water surfaces offer less resistance to air movement than do the generally rougher land surfaces.

5.5 High and Low Pressure Areas

The distribution of atmospheric pressure on the earth's surface is indicated on weather maps (Fig. 5.3*a*) by means of lines of constant pressure called isobars. In general the wind blows nearly parallel to the isobars with the friction between the moving air and the earth's surface giving the surface winds a slight component toward the area of low pressure. Thus high- and low-pressure centers have wind directions as shown in Fig. 5.3*b*. The wind blows counterclockwise and slightly toward the center of low-pressure areas, giving an upward movement

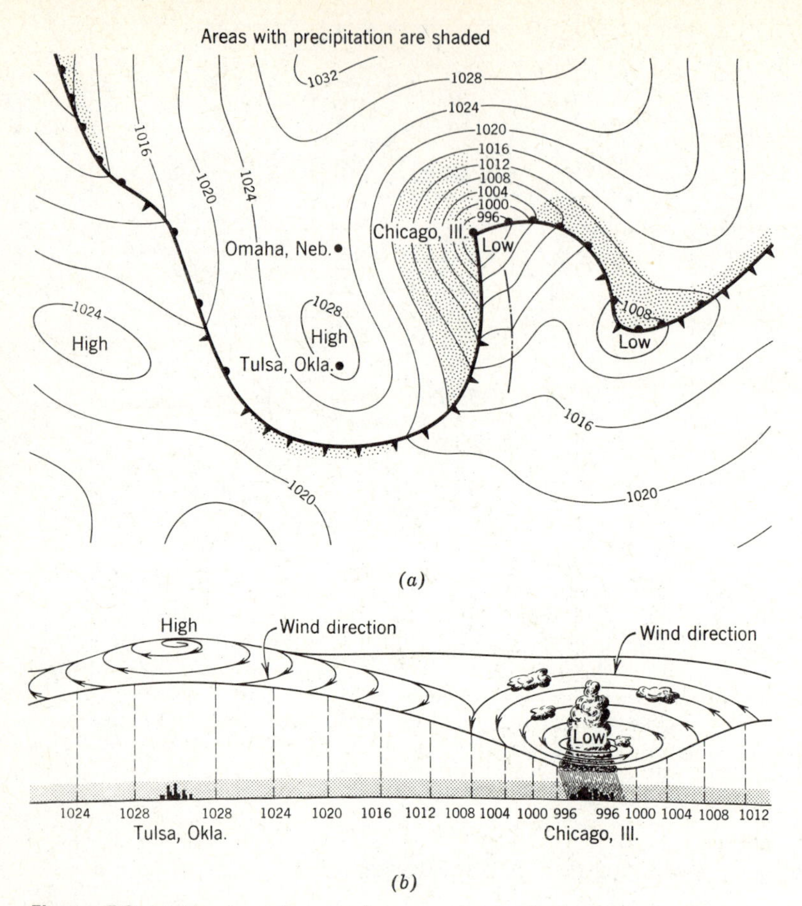

Figure 5.3. (*a*) Portion of a weather map in April showing cloudy weather in the East, rain in the Middle West, and clear skies in the Southwest. (*b*) Wind circulation around a high-pressure center at Tulsa and a low center at Chicago. (*Courtesy World Book Encyclopedia, 1961, and copyright by Rand McNally & Co., R.L. 70-S-35.*)

near the center which cools the air and increases the tendency for precipitation in these low-pressure centers. Conversely the air blows clockwise and slightly out of the center of high-pressure areas, giving a downward movement in the center of the high which warms the air and tends to cause clearer skies near the high-pressure center.

5.6 Weather Maps and Forecasting

Current weather conditions and forecasts are commonly depicted by weather maps that show the position of isobars, the ground position of fronts, temperature, location of high- or low-pressure centers, and areas of precipitation. A portion of such a weather map is shown in Fig. 5.3*a*. Official maps produced by NOAA (National Oceanic and Atmospheric Administration) show additional data, such as dew-point temperature, wind direction and velocity, cloud cover, and changes in barometric pressure.

Weather forecasts depend on the observations made at stations throughout the world and from space by photographic meteorological satellites. A forecast may be highly accurate for a short time, but the accuracy decreases as the prediction time increases. The function of the weather forecaster is to prognosticate the movement of frontal storms, their development, and the probable precipitation. As weather maps and satellite photos are now commonly shown in the news media, the individual is provided with an opportunity to practice forecasting, and to compare predictions with those of the National Weather Service. Long-range forecasts by the Weather Service cover periods of 5 days and 30 days.

5.7 Weather Modification

Drought and water shortages occur almost every year somewhere in the country. Such conditions create an interest in cloud seeding, one form of weather modification. Obviously, these practices cause legal problems because of conflict of interests and possible damages. Many states have enacted regulatory statutes or other measures that apply to weather modification, but there are few legal precedents.

Weather modification studies have been made for many years in many places. The effects of cloud seeding are extremely difficult to evaluate. Research in the Midwest has shown that cloud seeding can be expected to increase daily rainfall only by 10 percent, but hail may be reduced by 20 to 40 percent. More studies are needed before weather modification can be widely recommended.

STORM PATTERNS

In general, there are three major types of storm precipitation. The first is known as a frontal storm and is associated with the conflict between cold and warm air masses. The second is the air-mass thunderstorm, which is a result of convection over either land or water surfaces. The third is the orographic storm, in which precipitation results from the cooling of the air due to the upward movement of the air masses, especially over a range of mountains.

5.8 Frontal Storms

Frontal storms occur at boundaries of warm moist air and relatively dry cold air. The most common such combination in central and eastern United States is between tropical maritime air from the Gulf of Mexico and polar continental air from central Canada. During the fall, winter, and spring months, frontal storms are more likely to occur because the temperature variations across frontal boundaries are 10° to 30° Fahrenheit compared to differences of only a few degrees during the summer.

The greatest precipitation occurs in low-pressure centers which move along the air-mass boundary. The formation of these waves or storm centers is a gradual process, and is followed by the eventual disappearance of the low-pressure area. These storm centers often develop over continental North America and frequently do not disappear until they have passed far into continental Europe.

The process of the development and eventual disappearance of a frontal storm center is illustrated diagrammatically in Fig. 5.4. The opposite wind direction at the air-mass boundary in Fig. 5.4*a* often results in a circular motion, shown in Fig. 5.4*b*. Gradually this circulation increases, and the warm moist air rides over the heavier colder air. Gradually the warm and cold fronts develop as in Fig. 5.4*c*. The cold front occurs where the cold air is able to push the warm air back, and the warm front occurs where the warm air pushes the cold air before it. Since the cold front moves more rapidly than the warm front, it overtakes the warm front (Fig. 5.4*d*), trapping the warm air aloft and forming the occlusion

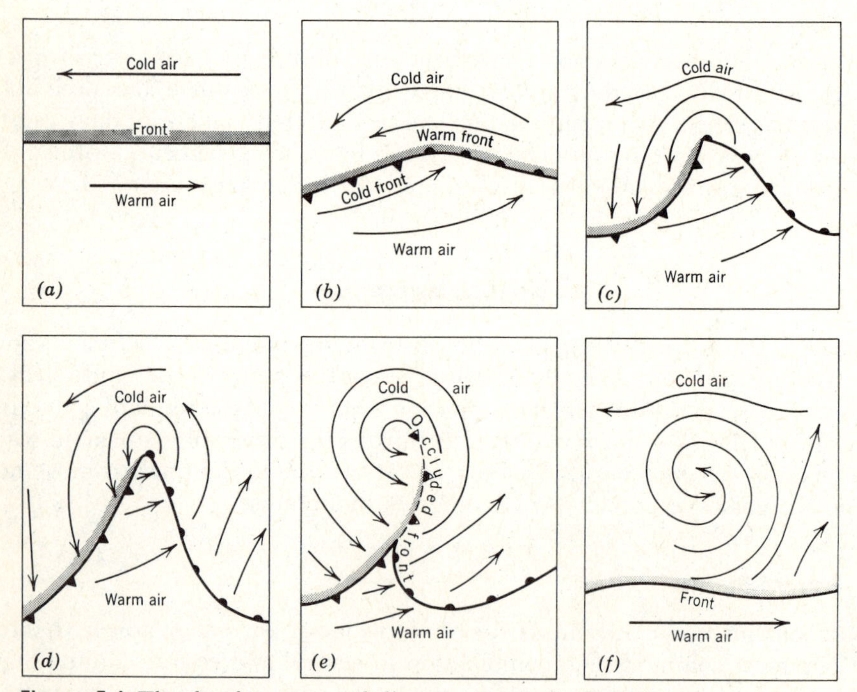

Figure 5.4. The development and disappearance of a frontal storm center. *Source:* U.S. Weather Bureau.

illustrated in Fig. 5.4*e*. Finally, as the wave dies, merely a swirl of air (Fig. 5.4*f*) persists where the wave occurred.

Cross sections of cold, warm, and occluded fronts are shown in Figs. 5.5, 5.6, and 5.7, respectively. The slope of the cold front is steeper than that of the warm front. Thus, the cold front causes a more rapid rise of the warm moist air and develops the violent thunderstorm type of precipitation. The more gentle slope of the warm front results in a more uniform precipitation, varying from light drizzles to heavy rains. This warm-front precipitation is generally of longer duration and covers larger areas than does cold-front precipitation.

Sometimes an additional type of front, known as the stationary front, forms, allowing the warm air to push continually over the cold air and causing continued precipitation. Often waves, causing large areas of precipitation, form and move along stationary fronts.

Both the paths followed and the velocity of movement of these storm centers are variable. Some of the paths commonly followed by these centers of low-pressure areas are given in Fig. 5.8. In general, they move from 200 to 500 miles in a 24-hr period, or from 8 to 20 mph.

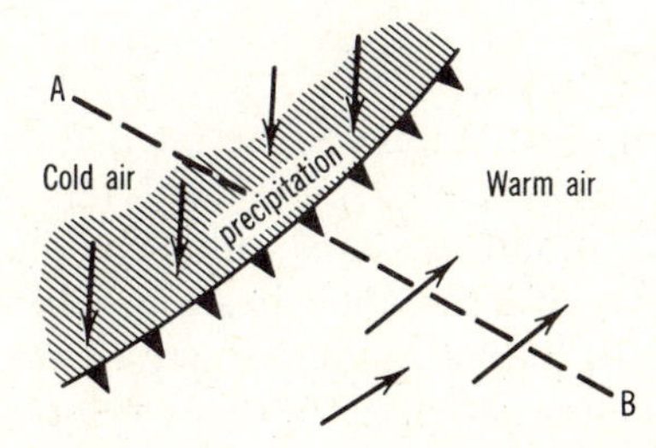

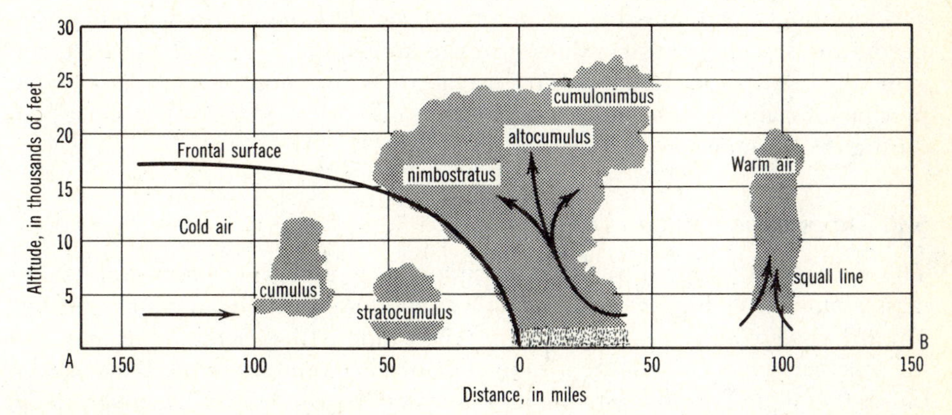

Figure 5.5. Cross section and clouds associated with a cold front. *Source:* Shaw and Elford (1958).

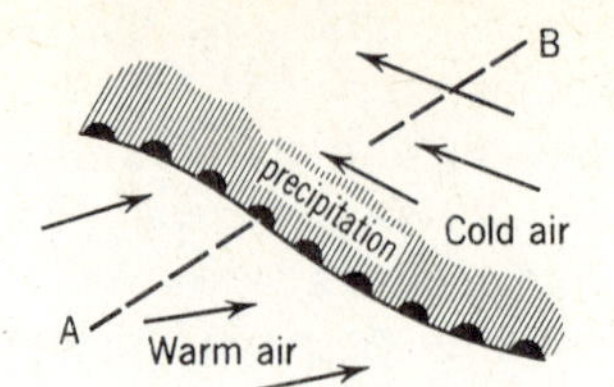

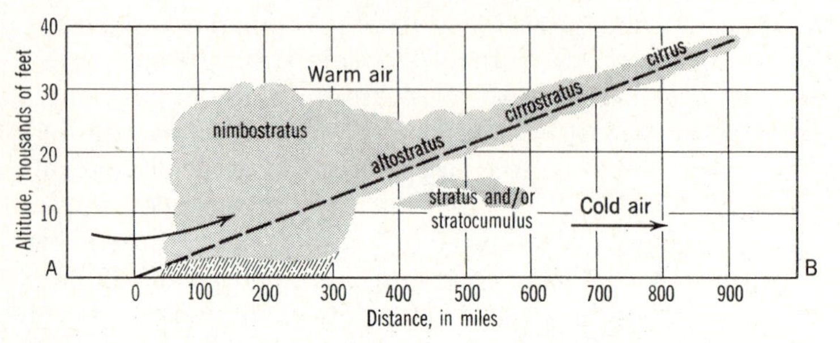

Figure 5.6. Cross section and clouds associated with a warm front. *Source:* Shaw and Elford (1958).

5.9 Convective Storms

During the summertime the contrast between the polar continental and the tropical maritime air is not so striking. During this period the air in the central and eastern states has a high moisture content and, especially near the surface, is subjected to considerable radiation heating. The heated air moves upward, being cooled both by the surrounding air and by the expansion process. When it is cooled to its condensation point, it forms a cloud of the convective type that may develop into a thunderstorm. Any source of ground heating, even a large fire, can set off circulation of this type. Whether such a convective circulation develops into a thunderstorm depends on the variation of temperature with height and the moisture conditions in the atmosphere. Though these storms generally cause precipitation only over small areas, they may result in very intense precipitation and are particularly important causes of floods on small watersheds during the summer.

5.10 Orographic Storms

The influence of topography on precipitation is especially important. As air masses move over high elevations, such as mountain ranges, the air is pushed upward, cooled, and oftentimes reaches the condensation point. Thus, much of the precipitation occurs on the upslope side of the mountain range. This process causes the highest annual precipitation found in North America. Conversely, as the air moves downslope it is warmed and, having had most of the moisture squeezed from it, deposits little precipitation. Thus, the mountain ranges along

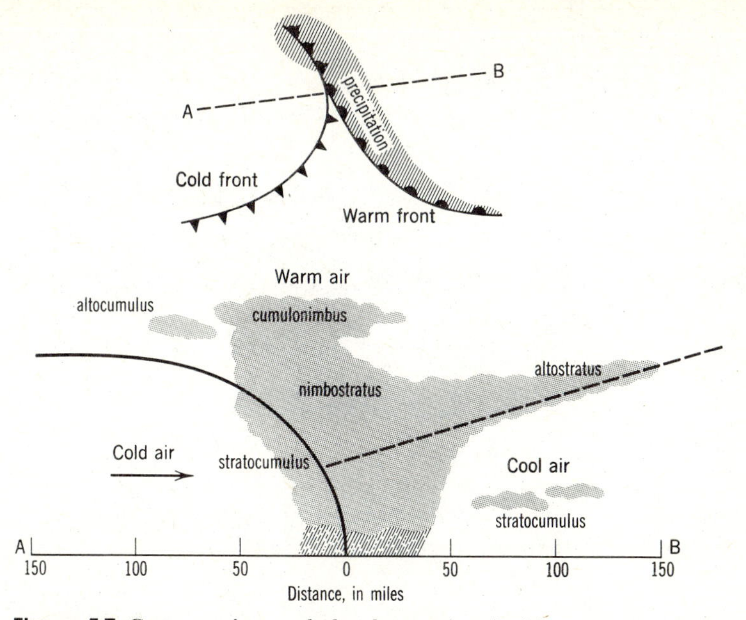

Figure 5.7. Cross section and clouds associated with an occluded front. *Source:* Shaw and Elford (1958).

the western coast are largely responsible for the arid areas further inland. As the central states are reached, this lack of moisture is compensated by the air movement from the Gulf of Mexico, which gradually builds up the precipitation as the east coast is approached.

Air temperatures normally decrease with height above the ground. This decrease called the lapse rate, averages about 4° F per 1000 ft of increase in altitude. Conversely, when air moves downward the temperature increases by the same amount. Temporary increases in temperature with increasing height, called inversions, are most apt to occur when warm and cold air masses come together. The moist warmer air overrides the heavier dry cold air. Radiaton cooling on clear, calm nights can also cause an inversion, a common cause of frosts.

RAINFALL

Design procedures for soil-and-water-conservation structures often require the rate and the depth of rainfall to be expected, as well as the frequency of occurrence. Since precipitation occurs randomly with respect to time and amounts, predicted values must be based on a statistical analysis of past records. In the United States, recording rain gages, which give the amount and the rate of rainfall, have been available only since about 1890. Total precipitation records were kept in many areas several years before this time.

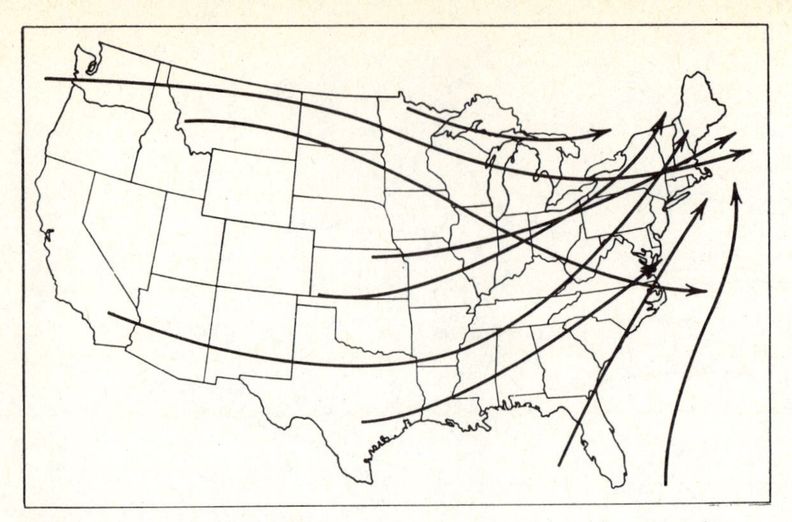

Figure 5.8. Paths followed by low-pressure centers moving over the continental United States.

5.11 Intensity, Duration, and Frequency of Rainfall

One of the most important rainfall characteristics is rainfall intensity, usually expressed in inches per hour. Very intense storms are not necessarily more frequent in areas having high total annual rainfall. In general, storms of high intensity last for fairly short periods and cover small areas. Storms covering large areas are seldom of high intensity, but may last for several days. The combination of relatively high intensity and long duration occurs infrequently, but when it does occur a large total amount of rainfall results. These infrequent storms cause much of our erosion damage and may result in devastating floods. In general, these unusually heavy storms are associated with warm-front precipitation. They are most apt to occur when the rate of front movement has decreased, when other fronts may pass by at close intervals, when stationary fronts persist in an area for a considerable period, or when tropical cyclones move into an area.

For a given frequency of occurrence, rainfall intensity decreases with duration of the storm, whereas the depth increases with duration. This relationship is shown in Table 5.1. Frequency of occurrence can be expressed as the return period, which is defined as the average number of years within which a given event will be equaled or exceeded. The data in Table 5.1 are for a 10-year return period. Rainfall intensity increases with the return period, as shown in Table 5.2. The actual intensities vary greatly with geographic location, but the relative intensities will be similar.

In designing any water-control facility, the desired return period storm is selected that will provide the lowest long-term cost. Designing for a lower intensity with a periodic failure may be more economical than designing for a high intensity storm with a corresponding high investment cost.

Table 5.1. Storm Duration and Rainfall Intensity and Depth at St. Louis, Missouri, for a Return Period of 10 Years

	Duration of Rainfall						
	10	*30*	*60*	*120*	*360*	*1440*	*min*
	(0.17)	*(0.5)*	*(1)*	*(2)*	*(6)*	*(24)*	*hr*
Intensity, iph	7.7	3.7	2.3	1.4	0.6	0.2	
Depth, inches	1.3	1.8	2.3	2.7	3.6	4.9	

A large amount of rainfall data is available from many stations in every state. Rainfall maps showing lines of equal rainfall have been prepared by the Weather Service for 1- and 24-hr durations and return periods of 2 and 100 years. By using these maps and a procedure developed by Weiss (1962), the curves shown in Fig. 5.9 were prepared for St. Louis, Missouri. Similar curves can be prepared for any location in the United States. An alternate procedure is to use the geographic rainfall factor from Fig. 5.10, which is illustrated in Example 5.1. It is an approximate method and should be used only where local data are not available.

Example 5.1. Determine the rainfall intensity and total rainfall for a 40-min duration storm that will occur once in 10 years at Raleigh, North Carolina.

Solution. From Fig. 5.9, read an intensity of 3.0 iph for a 40-min storm. By interpolation from Fig. 5.10, the geographic rainfall factor is 1.1 for Raleigh. The rainfall intensity,

$$i = 3.0 \times 1.1 = 3.3 \text{ iph} \quad (84 \text{ mm/h})$$

Rainfall depth in 40 min

$$I = 3.3 \times 40/60 = 2.2 \text{ in.} \quad (56 \text{ mm})$$

5.12 Average Rainfall Over a Watershed

The average rainfall depth over a watershed is less than the point rainfall given in Fig. 5.9. For areas less than a few square miles, the reduction is so small that it can be ignored. The correction is more for short duration storms of 30 min or less. Point rainfall is adequate for most small watersheds considered in this book.

Table 5.2. Rainfall Intensity and Frequency of Occurrence at St. Louis, Missouri, for a 30-min Duration Storm

	Return Period in Years					
	2	*5*	*10*	*25*	*50*	*100*
Intensity, iph	2.6	3.2	3.7	4.3	4.8	5.3
Relative intensity	0.7	0.9	1.0	1.2	1.3	1.4

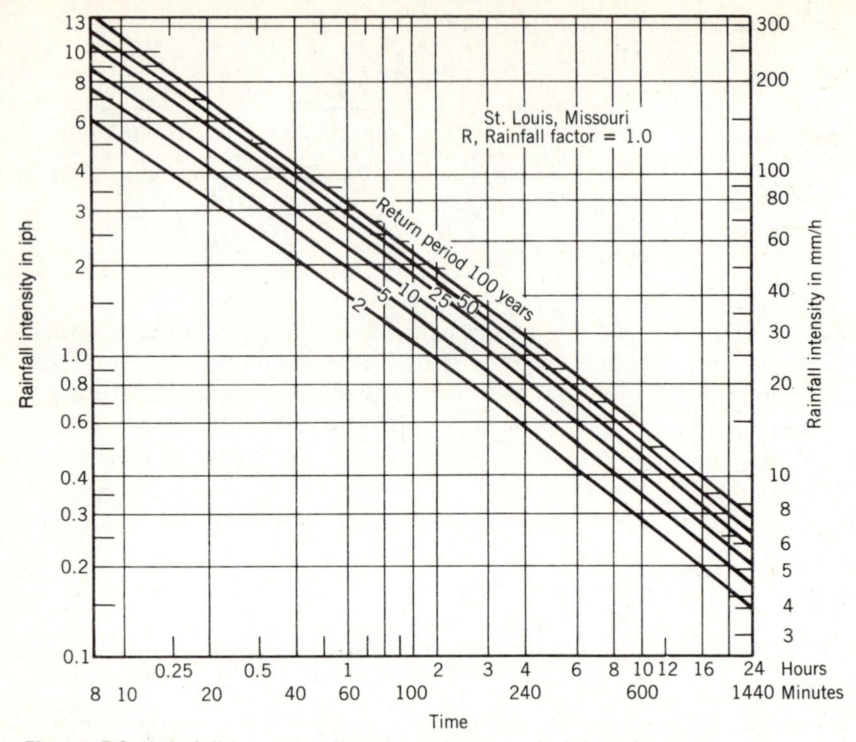

Figure 5.9. Rainfall intensity-duration-return period data for St. Louis, Missouri, with $R = 1.0$. *Source:* Hershfield (1961) and Weiss (1962).

RUNOFF

Surface runoff is that portion of precipitation that makes its way toward stream channels, lakes, or oceans as surface flow. The design of channels or structures to handle natural surface flows may involve the determination of peak rates of runoff, runoff volumes, and time distribution of runoff rates.

The factors affecting runoff may be divided into factors associated with precipitation and factors associated with the watershed. Precipitation factors include rainfall duration, intensity, and distribution of rainfall over an area. Watershed factors affecting runoff include size and shape of the watershed, orientation of the watershed, topography, geology, and surface culture of the watershed area.

5.13 Peak Runoff Rate by the Rational Method

With this method the design runoff rate is expressed by the equation,

$$q = CiA \tag{5.1}$$

where q is the design peak runoff rate in cubic feet per second (cfs), C is the runoff coefficient, i is the rainfall intensity in inches per hour (iph) for the design

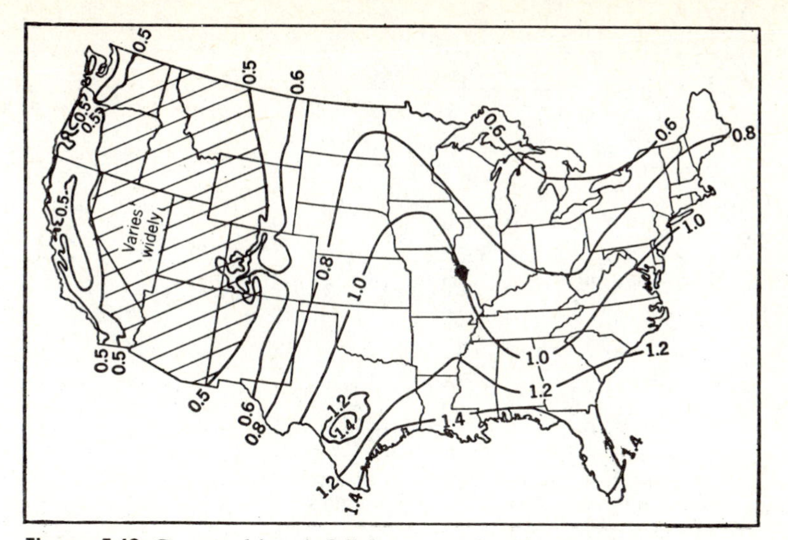

Figure 5.10. Geographic rainfall factor, R, for Fig. 5.9. *Source:* Hamilton and Jepson (1940).

return period and for a duration equal to the "time of concentration" of the watershed, and A is the watershed area in acres. The time of concentration of a watershed is the time required for water to flow from the most remote (in time of flow) point of the area to the outlet. It is assumed that when the duration of a storm equals the time of concentration, all parts of the watershed are contributing simultaneously to the discharge at the outlet. Table 5.3 provides a basis for estimating the time of concentration. The runoff coefficient C is defined as the ratio of the peak runoff rate to the rainfall intensity and is dimensionless. Table 5.4 provides estimates of the value of C for different soil and cover conditions with moderate slopes.

The use of the rational method is illustrated in Example 5.2.

Example 5.2. Determine the design peak runoff rate for a 50-year return period storm for the watershed in Fig. 5.11 at St. Louis, Missouri, containing 70 acres of row crop, *cp*, in cultivated clay loam soil and 40 acres of rolling sandy loam woodland. Both soils are in hydrologic group B. The maximum length of flow (*CDBA* in Fig. 5.11) is 2000 ft, and the difference in elevation of the most remote point C and the lowest point A is 10 ft.

Solution. The computed watershed gradient is $(10 \times 100/2000) = 0.5$ percent. From Table 5.3 read time of concentration of 20 minutes. From Fig. 5.9 read rainfall intensity of 6.3 iph for 50-year return period and 20 min duration (same as time of concentration). From Table 5.4 read runoff coefficients of 0.55 and 0.13 for the 70- and 40-acre conditions, respectively. The weighted runoff coefficient for the entire watershed is

$$C = \frac{(70 \times 0.55) + (40 \times 0.13)}{110} = 0.40$$

Table 5.3. Time of Concentration for Small Watersheds[a]

Maximum Length of Flow (ft)	*Time of Concentration (min)*					
	Watershed Gradient (percent)					
	0.05	*0.1*	*0.5*	*1.0*	*2.0*	*5.0*
500	18	13	7	6	4	3
1000	30	23	11	9	7	5
2000	51	39	20	16	12	9
4000	86	66	33	27	21	15
6000	119	91	46	37	29	20
8000	149	114	57	47	36	25
10,000	175	134	67	55	42	30
20,000	306	234	117	97	74	52

[a]Computed from $T_c = 0.0078\, L^{0.77} S^{-0.385}$, where L is the maximum length of flow in feet, S is the watershed gradient in feet per foot, and T_c is the time concentration in minutes.

Source: Kirpich (1940).

Substituting in Eq. 5.1,

$$q = 0.40 \times 6.3 \times 110 = 277 \text{ cfs} \quad (7.85 \text{ m}^3/\text{s})$$

Runoff coefficients are given for four hydrologic soil groups described in Table 5.4. More than 8000 soils listed in soil survey reports have been classified. These lists may be found in U.S. Soil Conservation Service (1979, 1972) and other references. The grouping is a simplified approach to identifying soils as to their intake capabilities. Runoff coefficients listed in Table 5.4 also increase with land slope and rainfall intensity. These factors must also be considered.

5.14 Peak Runoff Rate by the Soil Conservation Service (SCS) Method

The SCS method of estimating peak runoff rates is one of many that have been developed. Curve numbers are used to describe the soil-cover complex rather than the runoff coefficient as in the rational equation. A few curve numbers taken from a longer table are shown in Table 5.5 for each of the four hydrologic soil groups. Curve numbers and the runoff coefficient were developed independently and are not directly related. Both are inversely related to the intake rate of the soil.

To apply the SCS method a weighted curve number is computed for the watershed. Knowing the drainage area and the 24-hr rainfall, the peak runoff rate is selected from a graph for a specific curve number, type of storm, and slope. An example of such a graph is shown in Fig. 5.12 for the curve number 70, type II storm distribution, and for a moderate-slope watershed. The type II storm is a distribution pattern applicable to most areas in the contiguous United States. Separate graphs are available for flat and steep slopes and for

Table 5.4. Runoff Coefficients in the Rational Equation for Moderate Slopes

	Runoff Coefficient, C			
	Hydrologic Soil Group[a]			
Land Use, Crop, Management	*A*	*B*	*C*	*D*
Cultivated, with crop rotation				
Row crop, pp[b]	0.55	0.65	0.70	0.75
Row crop, cp[b]	0.50	0.55	0.65	0.70
Small grain, pp	0.35	0.40	0.45	0.50
Small grain, cp	0.20	0.22	0.25	0.30
Meadow	0.30	0.35	0.40	0.45
Pasture, permanent, moderate grazing	0.10	0.20	0.25	0.30
Woods, permanent, mature, no grazing	0.06	0.13	0.16	0.20
Urban residential				
30 percent of area impervious	0.30	0.40	0.45	0.50
70 percent of area impervious	0.50	0.60	0.70	0.80

[a]Hydrologic Soil Groups

A Well-drained sands and gravels with a high water transmission rate.

B Moderately to well-drained, moderately fine to moderately coarse texture with a moderate transmission rate.

C Poor to moderately well-drained, moderately fine to fine texture with a slow transmission rate.

D Poorly drained, clay soils with a high swelling potential, permanent water table, claypan, or shallow soils over nearly impervious material with a very slow transmission rate.

[b]cp, with good conservation practices.
pp, with poor conservation practices.

each five increments of curve numbers, such as 60, 65, 70, and so forth. The large number of graphs required precludes their inclusion in this book.

Example 5.3. Determine the peak runoff rate by the SCS method for the 110-acre watershed at St. Louis, Missouri, described in Example 5.2.

Solution. After selecting a curve number for B soils of 75 for row crops and 55 for woods from Table 5.5, compute the weighted curve number.

$$CN = \frac{70 \times 75}{110} + \frac{40 \times 55}{110} = 67.7$$

From Fig. 5.9 read a rainfall intensity of 0.26 iph for a 50-year storm of a 24-hr duration from which

$$I = 0.26 \times 24 = 6.24 \text{ in. in } 24 \text{ hr}$$

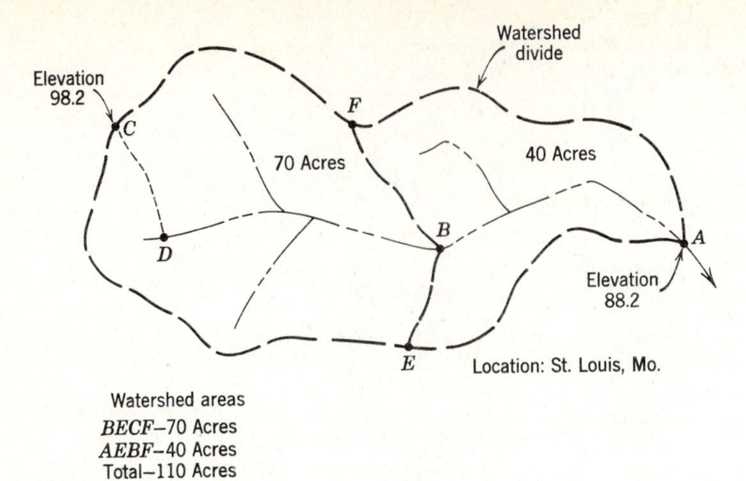

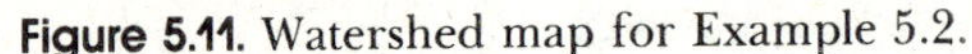
Figure 5.11. Watershed map for Example 5.2.

From Fig. 5.12, read q = 175 cfs (4.96 m³/s).

Peak flow rates computed by this method and the rational method given in Example 5.2 (277 cfs) should be the same. This wide difference illustrates the inadequacies of peak runoff methods, for which there is no simple explanation.

5.15 Runoff Hydrographs

A hydrograph is a graphical or tabular representation of the runoff rate (ordinate) from a watershed versus time with zero as the beginning of flow. For flow to begin the soil must be saturated, especially in small watersheds where stream flow is mostly surface runoff and not ground water flow. From uniform rainfall over the entire watershed, the flow rate will rise rapidly to the peak rate and then will fall to a low rate, which may continue for some time. All of the factors mentioned previously which affect peak rates influence the shape and size of the hydrograph.

Hydrographs are most useful for routing flood flows through reservoirs or through stream channels for the purpose of predicting flood heights. An outstanding example of a reservoir-regulated stream is the St. Lawrence River, the outlet for all the Great Lakes. Its maximum flow is not more than 20 percent greater than the minimum flow. In contrast, the maximum flow of the Missouri River is 2900 percent greater than the minimum. Farm ponds, because of their relatively small storage volumes, have no significant effect on flood flows below the dam. Hydrographs tell much about the hydrology of a watershed, especially peak flow rate and volume of flow (the area under the hydrograph).

Table 5.5. Runoff Curve Numbers for Average Antecedent Rainfall Conditions

Land Use, Crop, Management	*Runoff Curve Numbers (CN)* *Hydrologic Soil Group*[a] *A*	*B*	*C*	*D*
Cultivated, with crop rotation				
Row Crop, pp[b]	72	81	88	91
Row Crop, cp[b]	65	75	82	86
Small Grain, pp	65	76	84	88
Small Grain, cp	61	73	81	84
Meadow	55	69	78	83
Pasture, permanent, moderate grazing	39	61	74	80
Woods, permanent, mature, no grazing	25	55	70	77
Roads, hard surface (also roof areas)	74	84	90	92

[a]See description in Table 5.4.
[b]pp, with poor conservation practices.
cp, with good conservation practices.
Source: U.S. Soil Conservation Service, *National Engineering Handbook*, Hydrology, Section 4 (1972).

5.16 Runoff Volume

The runoff volume from one or more storms is of interest where flood-control reservoirs or other structures are to be designed, while the total annual runoff amount, often called water yield, is needed for designing water-supply structures, such as farm ponds. Water yields can be estimated from Chapter 12. More accurate estimates can be obtained from frequency analyses of stream-flow records. Minimum water yields are of most concern where water supply is the problem, since the driest years are most critical. For flood-control purposes, maximum storm or seasonal volumes are needed. Most methods of estimating storm volume use rainfall, from which infiltration, interception, surface storage, and other losses are subtracted to obtain the runoff.

A widely accepted method for estimating storm runoff volume was developed by the U.S. Soil Conservation Service (1972), which is

$$Q = \frac{(I - 0.2S)^2}{(I + 0.8S)} \tag{5.2}$$

where Q = direct storm runoff volume in inches,
I = storm rainfall depth in inches, and

S = maximum potential difference between rainfall and runoff in inches, starting from the storm's beginning, and further defined as

$$S = \left(\frac{1000}{CN}\right) - 10 \tag{5.3}$$

where CN = runoff curve number, which varies from 0 to 100.

If $CN = 100$, $S = Q$ and $I = Q$, or 100 percent rainfall is runoff. Curve numbers can be obtained from Table 5.5. For small watersheds the duration of the storm may vary from about 3 to 24 hr. The duration may be determined by selecting the duration that gives the highest volume of runoff.

Example 5.4. Determine the runoff volume in acre-feet for designing a storm water detention reservoir. Assume a 25-year return period, 24-hr storm near St. Louis, Missouri. The soil is in hydrologic soil group C. Assume average antecedent soil moisture conditions for the 100-acre urban watershed in which 50 percent is grass and 50 percent is roads and roof areas.

Solution. From Fig. 5.9, read a rainfall intensity of 0.24 iph for which the 24-hr rainfall, $I = 0.24 \times 24 = 5.8$ in. (147 mm).

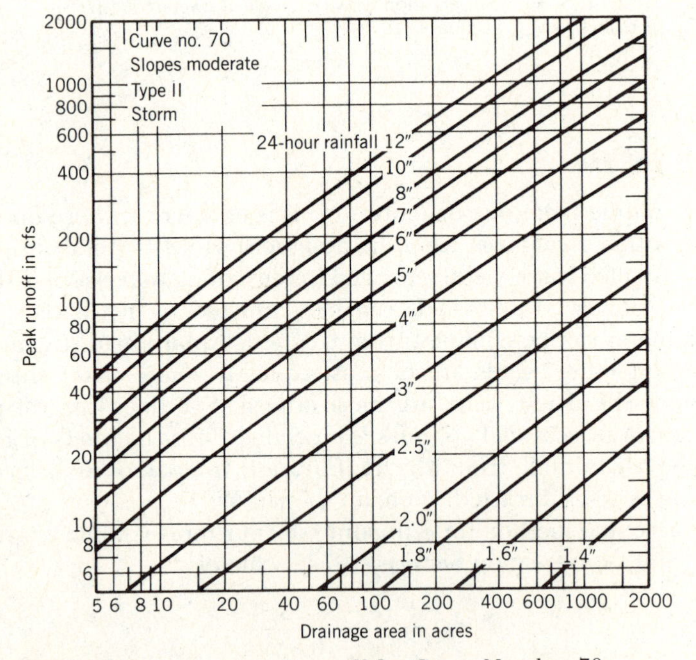

Figure 5.12. Peak rates of runoff for Curve Number 70, moderate slopes, and type II storm distribution. *Source:* U.S. Soil Conservation Service (1976).

From Table 5.5 for soil group C, read $CN = 74$ for pasture (grass) and $CN = 90$ for roads. The weighted average $CN = (74 \times 0.5) + (90 \times 0.5) = 82$.

Substituting in Eq. 5.3,

$$S = \left(\frac{1000}{82}\right) - 10 = 2.2 \text{ in.} \quad (56 \text{ mm})$$

and in Eq. 5.4,

$$Q = \frac{(5.8 - 0.2 \times 2.2)^2}{(5.8 + 0.8 \times 2.2)}$$

$$= \frac{5.4^2}{7.6} = 3.8 \text{ in.} \quad (97 \text{ mm})$$

By interpolation, Q may be read from Table 5.6.

$$\text{Runoff volume} = 3.8 \times 100 \times \left(\frac{1}{12}\right) = 31.7 \text{ ac ft}$$

Table 5.6. Storm Runoff Depth in Inches

Rainfall	*Curve Number, CN*						
(in inches)	*60*	*65*	*70*	*75*	*80*	*85*	*90*
1.0	0	0	0	0.03	0.08	0.17	0.32
1.2	0	0	0.03	0.07	0.15	0.28	0.46
1.4	0	0.02	0.06	0.13	0.24	0.39	0.61
1.6	0.01	0.05	0.11	0.20	0.34	0.52	0.76
1.8	0.03	0.09	0.17	0.29	0.44	0.65	0.93
2.0	0.06	0.14	0.24	0.38	0.56	0.80	1.09
2.5	0.17	0.30	0.46	0.65	0.89	1.18	1.53
3.0	0.33	0.51	0.72	0.96	1.25	1.59	1.98
4.0	0.76	1.03	1.33	1.67	2.04	2.46	2.92
5.0	1.30	1.65	2.04	2.45	2.89	3.37	3.88
6.0	1.92	2.35	2.80	3.28	3.78	4.31	4.85
7.0	2.60	3.10	3.62	4.15	4.69	5.26	5.82
8.0	3.33	3.90	4.47	5.04	5.62	6.22	6.81
9.0	4.10	4.72	5.34	5.95	6.57	7.19	7.79
10.0	4.90	5.57	6.23	6.88	7.52	8.16	8.78

Source: U.S. Soil Conservation Service (1979).

REFERENCES

Hamilton, C. L., and H. G. Jepson (1940) "Stock-water Developments; Wells, Springs, and Ponds," *U.S. Dept. Agr. Farmers' Bull. 1859.*

Harrold, L. L., G. O. Schwab, and B. L. Bondurant (1978) *Agricultural and Forest Hydrology*, Ohio State University Bookstore, Columbus, Ohio.

Hershfield, D. M. (1961) "Rainfall Frequency Atlas of the United States," U.S. Weather Bureau, Tech. Paper 40, May.

Kirpich, P. Z. (1940) "Time of Concentration of Small Agricultural Watersheds," *Civil Eng.* **10**:362.

Rouse, H. (1950) *Engineering Hydraulics*, John Wiley & Sons, New York.

Schwab, G. O., R. K. Frevert, T. W. Edminster, and K. K. Barnes (1981) *Soil and Water Conservation Engineering*, 3rd ed., John Wiley & Sons, New York.

Shaw, R. H., and C. R. Elford (1958) "Why Our Weather Changes," Reprint FS–763, *Iowa Farm Science* **13**(1):5–8.

U.S. Soil Conservation Service (1979) *Engineering Field Manual for Conservation Practices*, Washington, D.C.

U.S. Soil Conservation Service (1972) "Hydrology," *Natl. Eng. Handb., Section 4, Part I, Watershed Planning* (Lithographed), Washington, D.C.

Weiss, L. L. (1962) "A General Relation between Frequency and Duration of Precipitation," *Monthly Weather Rev.*, pp. 87–88, March.

PROBLEMS

5.1 Using the rational method, determine the peak runoff rate for a 10-year return period for a 100-acre (40.5 ha) watershed near St. Louis, Missouri. The watershed area is relatively flat, row crop, cp, consisting of clay and silt loam soil (B group). The watershed gradient is 0.5 percent and the maximum length of flow is 3540 ft (1079 m).

5.2 Solve Problem 5.1 if the watershed was at your present location.

5.3 Determine the peak runoff rate at point *B* of the watershed shown in Fig. 5.11. The elevation of *B* is 91.0 ft and the distance *BDC* is 1200 ft. The watershed area contains 30 acres of hilly pasture with clay soil (D) and 40 acres of row crop, cp, with sandy loam soil (A). Assume a return period of 10 years. Use the rational method.

5.4 Determine the peak runoff rate at point *A* in Fig. 5.11 if the runoff coefficients for areas *AEBF* and *BECF* are 0.50 and 0.28, respectively. The maximum length of flow *ABDC* is 2000 ft (610 m). Assume a return period of 10 years.

5.5 Determine the weighted runoff coefficient in the rational equation for a 120-acre (48.6 ha) watershed having 40 acres (16.2 ha) of woodland on flat, clay loam soil (C) and 80 acres of rolling land small grain, cp, with tight clay soil (D).

5.6 For your present location determine the runoff rate (rational method) for a grassed waterway just below the outlet of five parallel terraces spaced 90 ft (27 m) apart. Each terrace is 484 ft (148 m) in length. The field is rolling with silt loam soil (B). Time of concentration is 8 minutes and the design return period is 10 years. Soybeans, *cp*, is the row crop in the field.

5.7 For your present location determine the peak runoff rate for a 25-year return period storm for a 50-acre (20.2 ha) watershed in permanent woods with hydrologic soil group C. Slopes are moderate. Use the SCS method.

5.8 Assuming a time of concentration of 15 min determine the peak runoff rate for Problem 5.7 using the rational method.

5.9 For a 24.7-acre (10 ha) watershed having a B soil and for a 5-year return period rainfall rate of 3.0 iph (76 mm/h) compare the runoff rate from row crop, pp, to a year when the field is in meadow crop.

5.10 Solve Problem 5.7 if the same runoff characteristics are applied to a watershed in southern Mississippi.

5.11 Determine the rainfall intensity and depth for a 10-min and a 24-hour storm at St. Louis, Missouri, having a return period of 100 years. Compute the change from the 10-year return period values given in Table 5.1 as a percentage of the 10-year values.

5.12 For your present location determine the rainfall intensity for a 25-year return period storm with durations of 60 min and 24 hr, using values from Figs. 5.9 and 5.11. Compute the total rainfall that will fall for each of these durations.

5.13 Compare the rainfall amounts obtained in Problem 5.12 to the actual records published by the Weather Service or from other local records. Compute the percentage error for each of the two storm durations.

5.14 Determine the storm runoff volume in inches depth from a 50-year return period storm at your present location. Assume average antecedent soil moisture conditions and the duration of the storm as 3 hr. One third of the watershed is in woods and two thirds in row crop, cp, and all soils are classed as hydrologic soil group B.

CHAPTER 6

SOIL EROSION BY WATER

Soil erosion caused by man's activities is sometimes called accelerated erosion to distinguish it from geological erosion. Geological erosion, which occurs where soil exists in its natural environment under native vegetation, is important to agriculture because it includes soil-forming processes as well as eroding processes. These processes, which have occurred for millions of years, develop and maintain the soil in a favorable balance, suitable for plant growth. Soil erosion includes not only the removal of individual soil particles, but also the loss of organic matter and plant nutrients as well. Plant nutrients may be attached to the particles or removed in a soluble form by water.

Erosion is one of the most important problems in the United States and in the world. It reduces the productivity of agricultural land, and the resulting sediment causes stream pollution and reduces the capacity of reservoirs.

Development of urban land often exposes bare soil subject to severe erosion for several years. During the construction period steep unstable slopes are created that cannot be protected by vegetation. Such construction areas can produce severe sedimentation and pollution of streams and downstream reservoirs.

In humid regions water is the primary agent causing erosion, and wind is important in some arid areas. Wind erosion may be serious on sandy soils and on organic soils in humid areas, particularly during dry, windy seasons. Wind erosion will be discussed further in Chapter 9.

6.1 Factors Affecting Erosion by Water

The major variables affecting soil erosion are climate, soil, vegetation, and topography. Of these, vegetation and to some extent the soil may be controlled. Climatic factors and topographic factors other than slope length, are beyond the ability of man to control.

The factors that influence water erosion are the same as those that affect runoff (see Chapter 5). Erosion studies in the Midwest have shown that rainfall energy (which varies as the logarithm of the rainfall intensity); soil moisture prior to rainfall; soil chemical and physical properties; vegetative cover; degree of land slope; and length of slope are the most important. Forces involved in soil erosion are either (1) attacking forces that remove and transport the soil particles or (2) resisting forces that retard erosion, such as the resistance of soil to dispersion and movement, and the effect of vegetation.

The effect of rainfall on soil and water losses is shown in Fig. 6.1. Soil and runoff are compared for different amounts and intensities of rain. First, considering the rainfall amount groups, small and medium storms of less than 3 in. accounted for 90 percent of the total rainfall and 95 percent of the soil loss. Heavier storms exceeding 3 in. accounted for 10 percent of the rainfall and 5 percent of the soil loss. Total soil loss varies with the total runoff for the different rainfall amount groups. The soil loss per inch of runoff, however, decreased as the size of the storms increased. When storms are grouped by rainfall intensities, entirely different results are obtained. Rains of less than 1.5 iph intensity represented about 44 percent of the total rainfall and produced only 11 percent of the soil loss. At the other extreme, intensities greater than 4.5 iph represented only 7 percent of the total rainfall but produced 18 percent of the soil loss. The soil loss per inch of runoff increased as the intensity increased, even though the

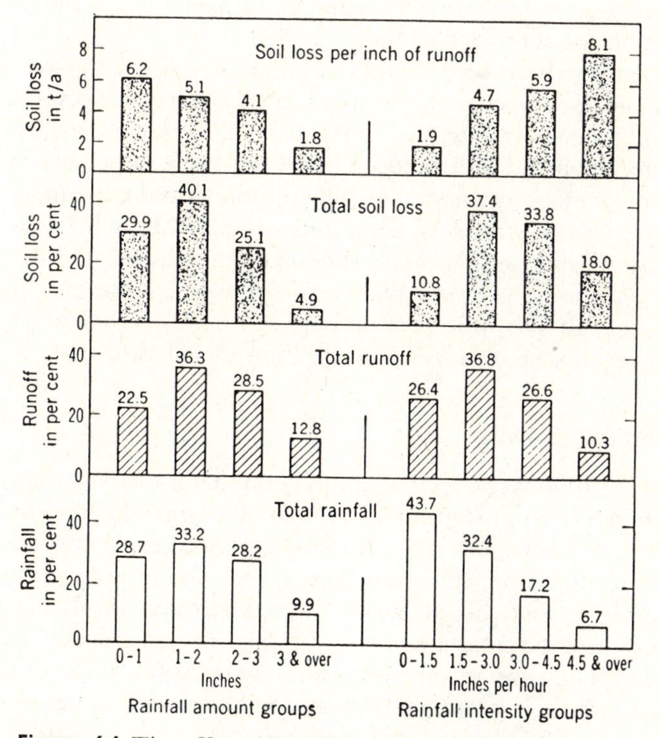

Figure 6.1. The effect of rainfall and runoff on soil loss from a sandy clay loam soil in North Carolina. *Source:* Copley et al. (1944).

total rainfall decreased. One or two high-intensity rains frequently cause as much soil loss as all other storms during a season or during several seasons.

6.2 Raindrop Erosion

Raindrop erosion is soil splash resulting from the impact of water drops directly on soil particles or on thin water surfaces. As shown in Fig. 6.2, raindrops strike bare soil with considerable impact and splash over 2 ft high and 5 ft laterally on level surfaces. Some raindrops may fall as fast as 30 mph. In addition to splashing soil, falling raindrops keep fine material and plant nutrients in suspension to be easily removed in the runoff, and they may wash out seeds.

Tremendous quantities of soil are splashed into the air, most of it more than once. The amount of soil splashed into the air as indicated by the splash losses from small elevated pans was found to be fifty to ninety times greater than the washoff losses. On bare soil it is estimated that as much as 100 tons of soil per acre are splashed into the air by heavy rains. On level land raindrop splash is not serious, but on sloping fields considerably more soil is splashed downhill than uphill. This effect may account to a large extent for serious erosion on short, steep slopes. Erosion by small rills or channels increases with the length of slope and is greater on steeper areas and at the lower end of the field, whereas raindrop erosion occurs over the entire field.

Factors affecting the direction and distance of soil splash are slope, wind, surface condition, and impediments to splash such as vegetative cover and mulches. On sloping land, the angle of impact causes the splash reaction to be in a downhill direction. Components of wind velocity up or down the slope have an important effect on soil movement by splash. Surface roughness and impediments to splash tend to counteract the effects of slope and wind. Contour furrows and ridges break up the slope and cause more of the soil to be splashed uphill. If raindrops fall on crop residue or growing plants, the energy is absorbed and thus soil splash is reduced. Raindrop impact on bare soil not only causes splash but also decreases aggregation and causes deterioration of soil structure.

6.3 Sheet and Rill Erosion

The idealized concept of sheet erosion has been that it was the uniform removal of soil in thin layers from sloping land, resulting from sheet or overland flow occurring in thin layers. Current fundamental studies of the mechanism of erosion, in which both time lapse and high-speed photographic techniques have been used, indicate that this idealized form of erosion rarely occurs. Minute rilling takes place almost simultaneously with the first detachment and movement of soil particles. The constant meander and change of position of these microscopic rills obscure their presence from normal observation, hence establishing the false concept of sheet erosion.

The beating action of raindrops combined with surface flow causes this initial microscopic rilling. From an energy standpoint raindrop erosion is far more

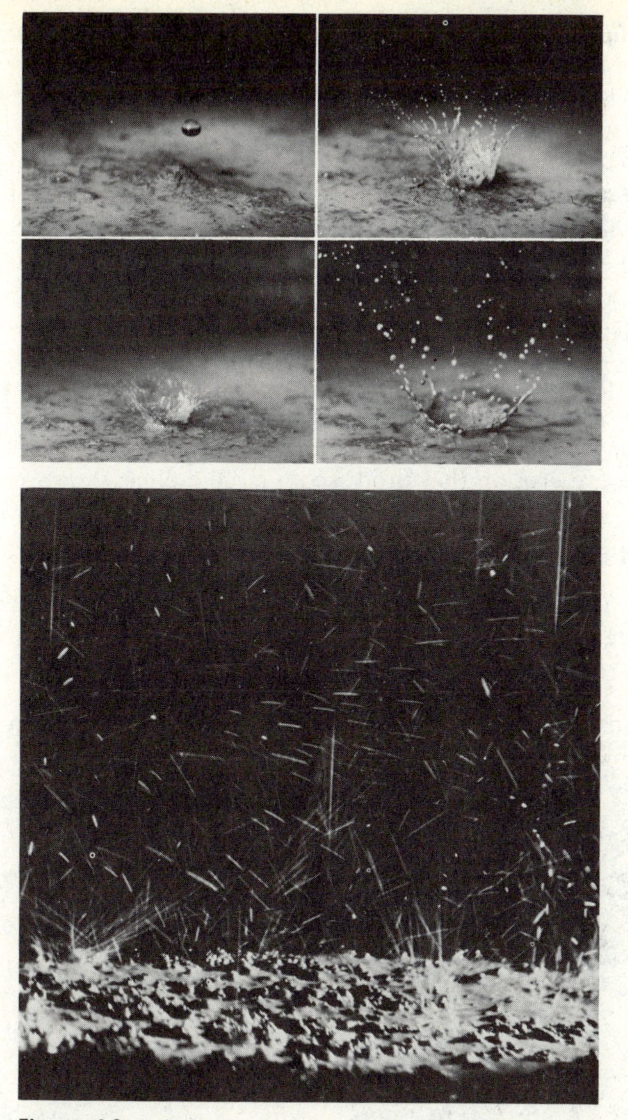

Figure 6.2. Raindrop splash. Top: impact from a single drop. Bottom: splash pattern during a storm. (*Courtesy U.S. Soil Conservation Service.*)

important because raindrops have velocities of about 20 to 30 fps, whereas overland flow velocities are about 1 to 2 fps. Raindrops cause the soil particles to be detached and the increased sediment reduces the infiltration rate by sealing the soil pores. Areas where loose, shallow topsoil overlies a tight subsoil are most susceptible to erosion. The eroding and transporting power of sheet flow are functions of the depth and velocity of runoff for a given size, shape, and density of soil particle or aggregate.

Rill erosion is the removal of soil by water from small but well-defined channels or streamlets when there is a concentration of overland flow. Conventionally, rill erosion occurs when these channels have become sufficiently large and stable to be readily seen. Rills are small enough to be easily removed by normal tillage operations. Although rill erosion is often overlooked, it is the form of erosion in which most soil erosion occurs.

6.4 Gully Erosion

Gully erosion produces channels larger than rills. As distinguished from rills, gullies cannot be obliterated by tillage. Thus, gully erosion is an advanced stage of rill erosion, much as rill erosion is an advanced stage of sheet erosion. An example of severe erosion that produced a large gully is shown in Fig. 6.3.

6.5 Universal Soil Loss Equation (USLE)

This equation was developed empirically from soil losses measured from small plots beginning in the 1930s. Data were collected in many states, primarily those

Figure 6.3. Severe gully erosion. (*Courtesy U.S. Soil Conservation Service.*)

in the eastern humid region. The USLE has been applied to planning conservation measures for farms and to predicting nonpoint sediment losses for pollution control programs. Although the USLE is a simplified expression of a complex set of interacting variables, it is the most widely accepted method of estimating sediment loss.

The USLE for estimating the average annual soil loss is

$$A = RKLSCP \tag{6.1}$$

where A = average annual soil loss in t/a (tons per acre)

R = rainfall and runoff erosivity index by geographic location (Fig. 6.4)

K = soil-erodibility factor in t/a, which is the average soil loss per unit of erosion index for a soil in cultivated continuous fallow with a slope length of 72.6 ft and a slope of 9 percent (Table 6.1)

LS = topographic factor, L for slope length and S for percent slope (Fig. 6.5)

C = cropping-management factor, the ratio of soil loss for given conditions to that from cultivated continuous fallow (Table 6.2)

P = conservation practice factor, the ratio of soil loss for a given practice to that for up and down the slope farming (Table 6.3)

The index factor R is a relative value closely correlated to the product of the kinetic energy of a storm and the maximum 30-min intensity. Average annual values of R (2-year return period) given in Fig. 6.4 are generally the most useful. However, losses for 181 specific locations for 5-, 10-, and 20-year return periods for annual and single-storm erosion have been developed by Wischmeier and Smith (1965). Average R values for single storms are also given.

The soil-erodibility factor K was developed for several bench mark soils by direct measurement from small plots. These factors are available by soil types in each state. Since it is impractical to list several thousand soils, Table 6.1 can serve as a general guide. Correlation of soil loss with physical and chemical properties is not yet fully known. The K factor in Eq. 6.1 has dimensions of t/a. Conversion of the equation to metric units can be done easily by changing K to metric tons per hectare (Mg/ha).

The topographic factor LS adjusts the soil loss from the standard length of 72.6 ft and 9 percent slope, respectively (values arbitrarily selected). For these values $L = 1$ and $S = 1$. These factors can be calculated separately from the equations

$$L = (y/72.6)^m \tag{6.2}$$

$$S = 65.41 \sin^2 \theta + 4.56 \sin \theta + 0.065 \tag{6.3}$$

where L = slope length factor

y = slope length in feet

m = a constant, 0.5 for slopes 5 percent and steeper, 0.4 for 3.5 to 4.5 percent slopes, 0.3 for 1 to 3 percent slopes, and 0.2 for <1 percent slopes

S = percent slope factor

θ = field slope in degrees (for <20 percent slopes, percent slope = 100 sin θ, approximately)

The product LS can be read directly from Fig. 6.5. The slope length is measured from the point where surface flow originates (usually the top of the ridge) to the outlet channel or a point down slope where deposition begins.

The cropping-management factor C includes the effects of cover, crop sequence, productivity level, length of growing season, tillage practices, residue management, and time distribution of erosive rainstorms. Values for C in the USLE are expressed as a ratio of the soil loss for crops to that for continuous fallow. They are given as a percentage in Table 6.2 for a few conditions. Wischmeier and Smith (1978) give many more conditions and discuss detailed procedures for computing this factor. The time distribution of the rainfall-erosion index will vary with geographic location. The monthly distribution for a few locations is given in Table 6.4. These percentages can be applied to the crop, tillage, and residue conditions for any time period of the year to give a weighted value of C for the year. Where a crop rotation is followed, the soil loss can be estimated for all crops in the rotation and an average annual soil loss computed (see Example 6.3). In many states average C values for common rotations have been developed.

The conservation practice factor P is given in Table 6.3 for contouring, strip cropping, and terracing. The reduction in soil loss at a given slope is about 50 percent for the next more intensive practice in the order given above. In humid regions strip cropping and terracing by definition include contouring. As will be discussed in Chapter 8, terracing affects the slope length; thus, the horizontal terrace spacing is the slope length in the USLE. The terracing factor shown will give the soil loss from the field rather than the loss to the terrace channel. To obtain the soil loss to the channel, the contouring factor is used in the USLE. Terraces do not control the soil loss from the slope; this control must be obtained by contouring or good crop and soil management. The maximum slope lengths for which the contouring and strip cropping factors apply are shown in Table 6.3. For slopes less than 12 percent the P factor for farming parallel to the field boundaries may be taken as 0.8. This practice may be desirable where contours are nearly parallel to a boundary or the field has short irregular slopes.

By selecting appropriate factors for Eq. 6.1, the estimated soil loss can be determined for a given set of conditions. If the soil loss is greater than the minimum required to maintain a high level of crop productivity, it may be reduced by changing cropping-management practices or conservation practices or both. In addition to the physical factors, economic, social, and others need to be considered in establishing soil-loss tolerances, called T values. In the United States, research has shown that T values vary with topsoil depth within the range of 2 to 5 t/a, annually. These maximum rates have been determined for each

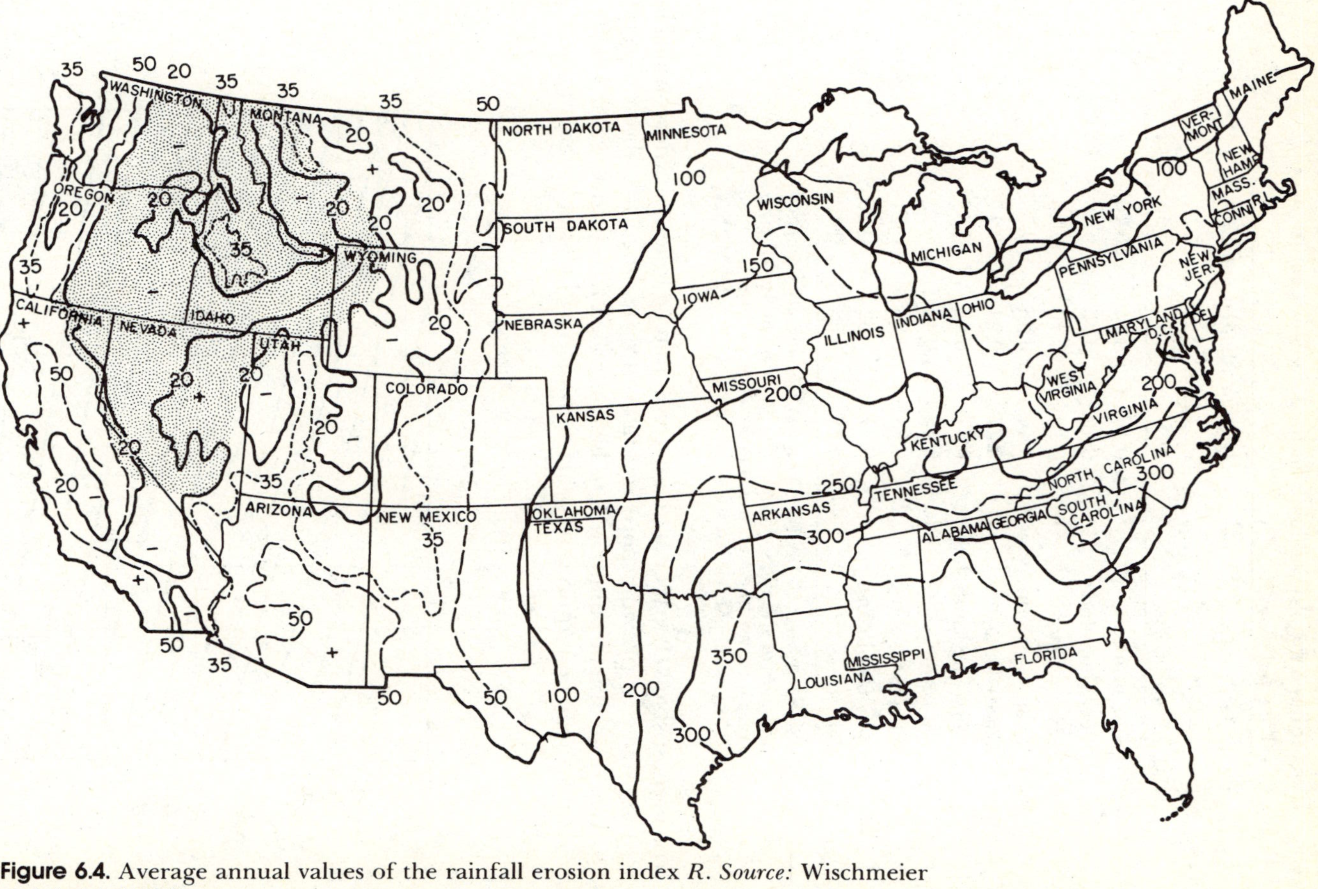

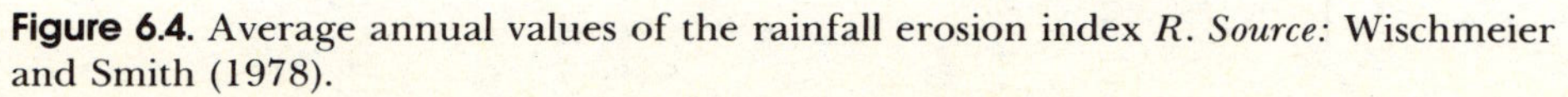

Figure 6.4. Average annual values of the rainfall erosion index *R*. *Source:* Wischmeier and Smith (1978).

Table 6.1. Soil-Erodibility Factor K by Soil Texture in Tons per Acre[a]

	Organic Matter Content in Percent		
Textural Class	*0.5*	2	4
Fine sand	0.16	0.14	0.10
Very fine sand	0.42	0.36	0.28
Loamy sand	0.12	0.10	0.08
Loamy very fine sand	0.44	0.38	0.30
Sandy loam	0.27	0.24	0.19
Very fine sandy loam	0.47	0.41	0.33
Silt loam	0.48	0.42	0.33
Clay loam	0.28	0.25	0.21
Silty clay loam	0.37	0.32	0.26
Silty clay	0.25	0.23	0.19

[a]Selected from USDA-EPA Vol. I (1975) and are estimated averages of specific soil values. For more accurate values by soil types, use the local recommendations of the U.S. Soil Conservation Service or state agencies (1 t/a = 2.24 Mg/ha).

soil type in most states. Local or state guides should be followed where available as these are likely to provide more accurate estimates than the above general factors. Where sediment pollution is to be controlled to improve water quality, lower T values may be required or other criteria may be established.

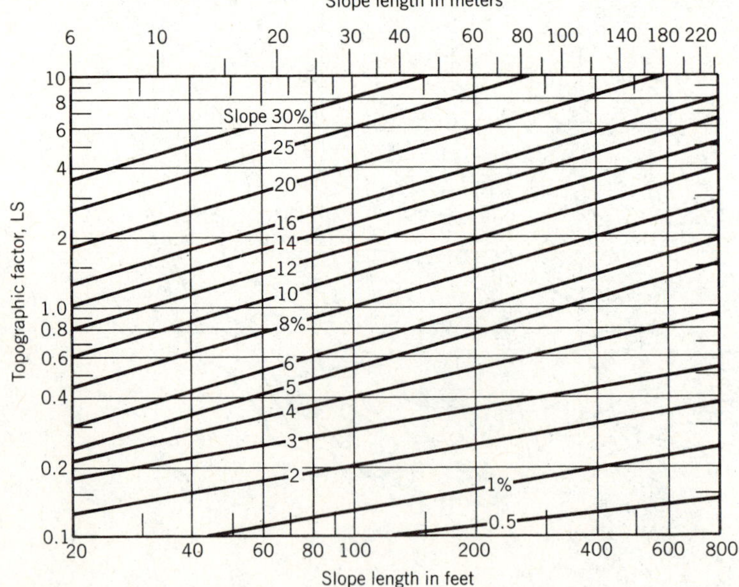

Figure 6.5. Topographic factor LS. *Source:* Wischmeier and Smith (1978).

Table 6.2. Ratio of Soil Loss (Percentage) from Cropland to Corresponding Loss from Continuous Fallow—Factor *C*

Line[a] No.	Cover, Crop Sequence, and Management	Spring Residue (lbs/A)	Crop-Stage Period[d] F (%)	SB (%)	1 (%)	2 (%)	3[b] (%)	4[c] (%)
	Corn After Corn, Grain Sorghum, Small Grain, or Cotton in Meadowless Systems							
Moldboard plow conv. till:								
2	RdL, spring TP	3400	36	60	52	41	24	30
6	RdL, fall TP	GP	49	70	57	41	24	—
10	RdR, spring TP	GP	67	75	66	47	27	62
14	RdR, fall TP	GP	77	83	71	50	27	—
28	No-till plant in crop residue	3400	—	8	8	8	8	19
	Corn in Sod-Based Systems							
2[e]	RdL, spring TP 1st year after sod (2–3 t/a)	3400	11	27	23	20	13	20
6[e]	RdL, fall TP 1st year after sod (2–3 t/a)	GP	15	32	26	20	13	—
10[e]	RdR, spring TP 1st year after sod (2–3 t/a)	GP	20	34	30	24	15	40
14[e]	RdR, fall TP 1st year after sod (2–3 t/a)	GP	23	37	32	25	15	—
6[e]	RdL, fall TP 2nd year after sod (2–3 t/a)	GP	37	60	48	37	23	—
14[e]	RdR, fall TP 2nd year after sod (2–3 t/a)	GP	58	71	60	45	26	—
104	No-till plant in killed sod, 1–2 t/a yield		—	2	2	2	2	2
107	Strip till, 1–2 t/a yield, 40% cover		—	4	4	4	4	6
	Corn After Soybeans							
110	Spring TP, conv. till	GP	47	78	65	51	30	37
113	Fall TP, conv. till	GP	53	81	65	51	30	—
121	No-till, plant in crop residue	GP	—	33	29	25	18	33
	Soybeans After Corn							
124	Spring TP, RdL, conv. till	GP	39	64	56	41	21	—
127	Fall TP, RdL, conv. till	GP	52	73	61	41	21	—
	Small Grain After Corn, Small Grain, Grain Sorghum or Cotton in Sod-Based Systems							
131[e]	Spring grain in disked residues, 50% cover, 2nd year after sod (2–3 t/a)	3400	15	18	14	12	4	22

Table 6.2. Ratio of Soil Loss (Percentage) from Cropland to Corresponding Loss from Continuous Fallow—Factor C

			Crop-Stage Period[d]					
Line[a] No.	*Cover, Crop Sequence, and Management*	*Spring Residue (lbs/A)*	*F (%)*	*SB (%)*	*1 (%)*	*2 (%)*	*3[b] (%)*	*4[c] (%)*
143[e]	Winter grain after fall TP, RdL, 50% mulch 2nd year after sod (2–3 t/a)	GP	—	39	34	25	7	22
	Small Grain After Summer Fallow							
149	With grain residues	1000	—	26	21	15	7	—
155	With row crop residues	1000	—	40	31	24	10	—
	Established Sod or Meadow							
	Grass or legume mix, 3–5 t/a		0.4 for entire year					
	2–3 t/a		0.6 for entire year					
	1 t/a		1.0 for entire year					

[a]Numbers from 160 line table in Wischmeier and Smith (1978).
[b]For 90% canopy cover at maturity.
[c]For all residues left on field.
[d]Crop-stage period
F (rough fallow), plowing to secondary tillage.
SB (seedbed), secondary tillage for seedbed preparation until crop has 10% canopy cover.
1 (establishment), end of SB until crop has 50% canopy cover.
2 (development), end of 1 until 75% canopy cover.
3 (maturing crop), end of 2 until crop harvest.
4 (residue or stubble), harvest to plowing or new seeding.
[e]Adjusted from Table 5–D for 2–3 t/a hay yield from Wischmeier and Smith (1978).
Symbols: GP, good productivity; M, grass and legume meadow; RdL, crop residues left on field; RdR, crop residues removed; TP, plowed with moldboard plow; t/a, tons per acre.
Source: Wischmeier and Smith (1978).

Example 6.1. Determine the soil loss for the following conditions: Location Memphis, Tennessee, K = 0.1 t/a, y = 400 ft (slope length), s = 10 percent slope, C = 0.18 (approximate for corn-corn-oats-meadow rotation with good management), and the field is to be contoured.

Solution. From Fig. 6.4 read R = 310 as the rainfall erosion index, from Fig. 6.5 read LS = 2.8, and from Table 6.3 read P_c = 0.6. Substituting in Eq. 6.1,

$$A = 310 \times 0.1 \times 2.8 \times 0.18 \times 0.6 = 9.4 \text{ t/a} \quad (21.1 \text{ Mg/ha})$$

Example 6.2. If the minimum soil loss to maintain productivity for the conditions in Example 6.1 is 3.0 t/a, what practices coud be adopted to accomplish this reduction?

Solution. Since C and P are the only practices that can be changed, the following combinations are possible by substituting the appropriate factors in Eq. 6.1:

Table 6.3. Recommended Conservation Practice Factors P

Percent Slope	P_c *Contouring (Maximum Slope Length)*	P_{sc}[2] *Strip Cropping (Maximum Strip Width)*	P_{tc}[3] *Terracing and Contouring (Graded Channels, Grass Outlets)*
1–2	0.6 (400)[1]	0.30 (130)	0.12
3–5	0.5 (300)	0.25 (100)	0.10
6–8	0.5 (200)	0.25 (100)	0.12
9–12	0.6 (120)	0.30 (80)	0.12
13–16	0.7 (80)	0.35 (80)	0.14
17–20	0.8 (60)	0.40 (60)	0.16
21–25	0.9 (50)	0.45 (50)	0.18

[1]Length may be increased with residue cover.

[2]For four-year rotation of a row crop, small grain with meadow seeding, and two years of meadow. Double these values for alternate strips of row crop and small grain. Adjust strip width downward to accommodate farm equipment width.

[3]Values for computing soil loss from the field. Use a value of 0.06 at all slopes for pipe outlet terraces. For soil loss to the terrace channel, use the contouring factor and the appropriate horizontal terrace spacing as the slope length.

Source: Wischmeier and Smith (1978).

Conservation Practice	*LS (Fig. 6.5)*	*C Factor*	*P (Table 6.3)*	*A*[a] *t/a*	*Remarks*
(1) Strip crop	2.8	0.18	0.3	4.7	Too high
(2) Strip crop	2.8	0.12 max	0.3	3.0	Satisfactory
(3) Terracing	1.1	0.18	0.12	0.7	Low loss
(4) Terracing	1.1	0.73 max	0.12	3.0	Satisfactory
(5) Terracing	1.1	0.15 max	0.6	3.0	Loss to terrace channel

[a]Computed using $RK = 31$ from Example 6.1.

Note that strip cropping (1) is not adequate; however, in line (2) the 3.0 t/a soil loss with strip cropping is possible by reducing the cropping-management factor C to 0.12 or less. In line (3), terracing with $C = 0.18$ would give a field soil loss lower than required. Thus, with terracing, the cropping-management factor may be increased to 0.73 without exceeding the permissible soil loss of 3.0 t/a from the field. In lines (3), (4), and (5) the terrace slope length is assumed to be 65 ft (see Chapter 8) which reduces the LS factor to 1.1. If the soil loss to the terrace channel is to be 3.0 t/a or less, C factor in line (5) should not be exceeded.

Example 6.3. If strip cropping is selected as the most desirable conservation practice in Example 6.2, what recommendations could be made for the cropping-management practice?

Solution. From line (2) in Example 6.2 the maximum C factor allowable to keep soil losses to 3.0 t/a is 0.12. By trial and error select a corn-small grain-meadow rotation, fall plowing,

Table 6.4. Monthly Distribution of the Rainfall and Runoff Erosivity Index at Selected Locations

	Percent of Annual Erosion at Locations					
Month	*A*	*B*	*C*	*D*	*E*	*F*
January	0	2	6	2	0	15
February	1	2	7	2	0	12
March	1	4	8	3	0	8
April	4	6	12	4	5	2
May	11	11	11	6	9	3
June	26	20	11	14	9	5
July	24	19	12	23	17	1
August	18	15	8	20	42	1
September	11	10	6	15	2	7
October	3	6	5	6	16	11
November	1	3	8	3	0	16
December	0	2	6	2	0	19

Locations:

A, Area 2—northwestern Iowa, northern Nebraska, and southeastern South Dakota.

B, Area 16—northern Missouri, central Illinois, Indiana, and Ohio.

C, Area 22—Louisiana, Mississippi, western Tennessee, and eastern Arkansas.

D, Area 29—Atlantic Coastal Plains of Georgia and the Carolinas.

E, Pueblo, Colorado.

F, Portland, Oregon.

Source: Wischmeier and Smith (1978).

and corn residue plowed under. Compute the weighted average *C* value for the three-year rotation as follows:

Crop (1)	*Months* (2)	*(Crop Stage) Percent C (Table 6.2)* (3)	*Percent Annual (Table 6.4)* (4)	*Weighted C Factor (3) × (4)* (5)
Corn, first year	Jan.–Mar.	(F) 15	21	0.032
	Apr.	(SB) 32	12	0.038
	May	(1) 26	11	0.029
	June	(2) 20	11	0.022
	Jul.–Sep.	(3) 13	26	0.034
Small grain, with meadow seeding	Oct.–Feb.	(F) 15	32	0.048
	Mar.	(SB) 18	8	0.014
	Apr.	(1) 14	12	0.017
	May	(3) 4	11	0.004

	Jun.–Sep.	(4)	22	37	0.081
Meadow	Oct.–Oct.	—	0.6	105	0.006
	Nov.–Dec.	(F)	15	14	0.021
				Three-year total	0.346

Average annual C = 0.346/3 = 0.115

Since 0.115 is less than 0.12, the three-year rotation and soil management practices are satisfactory.

Several other combinations of crops and soil management practices could be selected provided that the average annual C value is less than the maximum allowed.

6.6 Erosion from Construction Sites

The procedures and data for cropland erosion can be adapted to highway, residential, and commercial sites, which may have bare soil exposed during the construction period. The USLE will provide relative erosion rates for different development plans or show the amount of deposition to be trapped in sediment basins or both. For construction periods of less than one year, the appropriate R value for the period may be selected from Table 6.4. Erosion rates can be predicted for various subsoil horizons as they are exposed during the construction period. Some subsoils are substantially more erodible than the original topsoil, and others are less erodible. The LS factors in Fig. 6.5 apply within the limits given, but the values for embankment slopes of 2 : 1 and 3 : 1 are not available. When the slope is concave or convex, the LS values need to be adjusted. Diversions or stabilized waterways can be constructed to reduce the slope length. The cropping-management factor C and the conservation practice factor P are usually taken as 1.0. For long construction periods, buffer strips or grass or small grain, or anchored mulch may be applied for better control. For further information, consult Wischmeier and Smith (1978) and other references.

6.7 Erosion from Irrigated Lands

Erosion may occur on irrigated land, especially from surface methods, such as furrow, border, and flooding. It is less likely from sprinkler or subirrigation methods. Most erosion will be within the field rather than from the field. With surface methods of irrigation, stream flow is greatest where it enters the field at the upper end. As the water proceeds downstream, the flow is decreased by the amount that infiltrates into the soil. Erosion is usually greatest during the initial time of irrigation. The dry surface soil, sometimes loosened by cultivation, is readily moved by the flowing water. Clear water tends to pick up the finer soil particles carrying them downstream where they seal the soil at the lower end of the field. Removal of the fine particles at the upper end may increase the water intake rate.

Erosion cannot be eliminated, but it can be minimized. Observation of erosion with different size streams is about the only way to determine the permissible flow rate. With irrigation the most severe erosion will occur at the upper end of the field. With natural rainfall the reverse applies as erosion is greatest at the lower end of the field. By properly designing irrigation systems, erosion can be controlled.

6.8 Plant Nutrient and Organic Matter Losses

The loss of plant nutrients from the soil is at least as important as the loss of the soil itself. Since the erosion process causes the finer soil particles and organic matter to be removed, and since such material furnishes most of the base exchange capacity of the soil, providing storage for plant food, the removal of these smaller and lighter particles greatly decreases the fertility of the soil.

The effect of slope on annual soil and plant nutrient losses is shown in Fig. 6.6. The loss of nutrients and soil increases as the slope increases. Plant nutrients and organic matter losses are given as a percentage of the total in the top 6 in. of soil. Except for potassium, nutrient losses varied linearly with the slope and increased with the soil loss. The data are from plots about 60 ft in length and farmed on the contour. An average of 93 percent of the losses shown in Fig. 6.6 occurred during the years that the plots were in corn.

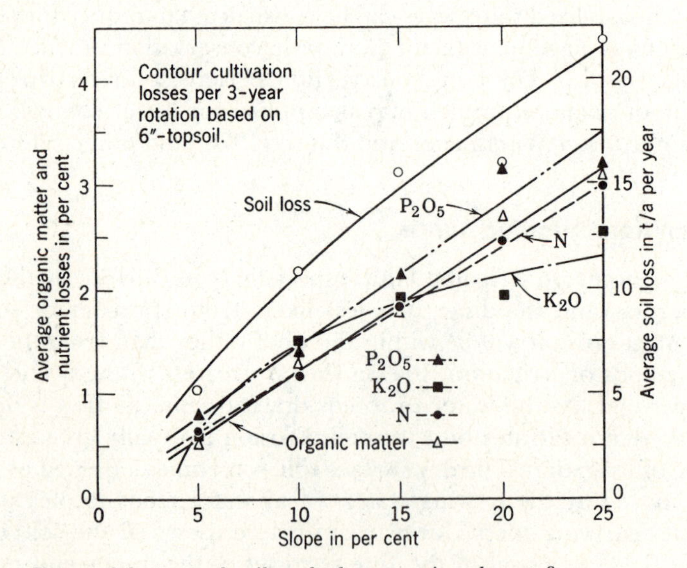

Figure 6.6. Annual soil and plant nutrient losses from topsoil versus slope. (*Data from Moody, 1948.*)

REFERENCES

Copley, T. L. et al. (1944) "Effects of Land Use and Season on Runoff and Soil Loss," *N. Carolina Agr. Exp. Sta. Bull. 347*.

Hudson, N. (1981) *Soil Conservation*, 2nd ed., Cornell University Press, Ithaca, N.Y.

Moody, J. E. (1948) "Plant Nutrients Erode Too" (unpublished research), Va. Agr. Exp. Sta.

Schwab, G. O., R. K. Frevert, T. W. Edminster, and K. K. Barnes (1981) *Soil and Water Conservation Engineering*, 3rd ed., John Wiley & Sons, New York.

Wischmeier, W. H., and D. D. Smith (1978) *Predicting Rainfall Erosion Losses—A Guide to Conservation Planning*, U.S. Dept. Agr. Handb. 537, U.S. Government Printing Office, Washington, D.C.

Wischmeier, W. H., and D. D. Smith (1965) *Predicting Rainfall-Erosion Losses from Cropland East of the Rocky Mountains*, USDA-ARS Agr. Handb. 282, U.S. Government Printing Office, Washington, D.C.

U.S. Department of Agriculture Research Service and Environmental Protection Agency (1975) *Control of Water Pollution from Cropland*, Vol. I; (1976) Vol. II, U.S. Government Printing Office, Washington, D.C.

PROBLEMS

6.1 If the average annual soil loss on a slope of 8 percent with up- and downhill farming is 10 t/a (22.4 Mg/ha), what is the soil loss for contouring, for strip cropping, and for terracing (including contouring) for the same conditions?

6.2 If the soil loss at Memphis, Tennessee, for a given set of conditions is 5 t/a (11.2 Mg/ha), what is the expected soil loss at your present location if all factors are the same except the rainfall factor?

6.3 If the degree of slope is increased from 2 percent to 10 percent, what is the relative increase in erosion? Assume other factors are constant.

6.4 If the soil loss for a 10 percent slope and a 200-ft (61 m) length of slope is 4 t/a (9 Mg/ha), what soil loss could be expected for a 400-ft (122 m) slope length?

6.5 Determine the soil loss in tons per acre for a field at your present location if $K = 0.15$ t/a, $y = 300$ ft (91 m), $s = 10$ percent, $C = 0.2$, and up and down the slope farming is practiced. What conservation practice should be adopted if the soil loss is to be reduced to 3.5 t/a (7.8 Mg/ha)?

6.6 If the soil loss for up and down the slope farming at your present location is 35 t/a (78 Mg/ha) from a field having a slope of 6 percent and slope length of 400 ft (122 m), what will be the soil loss if the field is terraced with a horizontal spacing of 85 ft (26 m)? Assume that the cropping-management conditions remain unchanged and soil is removed from the field.

6.7 Compute the soil loss in t/a from a completely terraced field at your present location assuming a slope of 6 percent, $C = 0.2$, $K = 0.3$, and a field slope length of 640 ft (195 m) with terraces at 80-ft (24-m) intervals. Determine the soil lost (1) at the terrace outlet, (2) to the terrace channel (use contouring factor), and (3) to the terrace channel if upslope plowing is practiced (use strip cropping factor).

6.8 The soil loss for a field with an 8 percent slope was 40 t/a (90 Mg/ha) for up- and downhill farming. Compute the expected average annual soil loss if the slope length is reduced from 400 to 100 ft (122 to 30 m) by terracing and the cropping-management factor is changed from 1.1 to 0.8.

6.9 Which field, A or B will produce the highest soil loss. Show your calculations.

Soil Loss Factor	*Field A*	*Field B*
Soil-erodibility K	0.1	0.3
Slope length-slope LS	6.0	2.0
Cropping-management C	0.15	0.2
Conservation practice P	0.6	0.3

If the fields are located in southeastern Iowa, compute the average annual soil loss in t/a for field B.

6.10 Compute the average annual (calendar year) cropping-management factor for a two-year corn-soybean rotation, conventional tillage, and crop residues left in the field. Assume fall plowing on November 1, crop stage SB in April and May, 1 in June, 2 in July, and 3 in August through October. The location is in northern Missouri (area 16).

6.11 If the allowable soil loss tolerance is 4 t/a (9 Mg/ha), the soil is clay loam with 2 percent organic matter located in southeastern Iowa, $LS = 2.0$, and $P_c = 0.6$, compute the maximum value for the cropping-management factor C.

CHAPTER 7

VEGETATED WATERWAYS

Wherever water flows over unprotected soil it may pick up and carry along soil particles. If such flows are concentrated by natural topography or works of man, gullies may develop that destroy valuable farmland. Fields are divided by gullies into smaller fields, rows are shortened, movement between fields is obstructed, and the farm value is decreased. Gully development jeopardizes roads, bridges, buildings, and fences; sediment from gullied areas causes local as well as downstream damage.

The hazards of gully formation can frequently be avoided by providing properly proportioned channels protected by vegetation which absorbs the energy of runoff without damage. Where slopes are steep or runoff volumes large, it may be necessary to utilize permanent control structures in addition to vegetal protection (Chapter 10).

7.1 Uses of Vegetated Outlets and Watercourses

Stabilization of waterways is necessary where runoff occurs from natural watersheds or from terrace systems, contour furrows, diversion channels, or for emergency spillways for farm ponds or other structures. Continuing low flows, such as discharge from tile drains, should not be directed down vegetated waterways. The prolonged wet condition in the waterway may result in poor vegetal protection. Under such condition it may be desirable to provide subsurface drainage under the waterway.

Vegetated watercourses are also constructed across the slope to serve as diversion channels, which are similar to terraces. The function of a diversion channel is comparable to that of the top terrace; however, the drainage area and the cross-sectional area of the diversion ditch are generally larger than for the terrace. Diversion ditches divert water away from active gullies and farm buildings, protect bottom land from overflow, and intercept runoff when it is not desirable or otherwise possible to control the runoff because of topography, land ownership, or other reasons.

Under some conditions trees, shrubs, and vines may be more suitable for gully control than grasses and legumes. Natural revegetation may be stimulated by diverting flow and fencing livestock out of the areas. A gradual succession of plant species native to the region may become established. Plantings of recommended species will speed up revegetation. Variety in plantings will reduce the danger of destruction by disease and climatic extremes. Such revegetated areas may serve as wildlife refuges or provide timber.

7.2 Channel Cross-Sectional Area

Vegetated waterways may be constructed to be parabolic, trapezoidal, or triangular in cross section. These cross sections, together with formulas for their areas, top widths, and wetted perimeters, are shown in Fig. 7.1. The top width of the channel together with its cross-sectional shape and depth determine its cross-sectional area of flow. The wetted perimeter is the length of contact of the water with the bottom and sides of the channel. The area and wetted perimeter together with the slope of the channel and the nature of the channel lining determine its flow capacity, as discussed in Section 7.5. The hydraulic radius R is the cross-sectional area divided by the wetted perimeter. In Fig. 7.1 the depths are dimensioned for each cross section; the smaller depth d is the actual design depth of flow for the channel and is used in calculations of channel capacity. The larger depth D is the total construction depth of the channel and should be used in calculations of the amount of earth to be removed in excavating and forming the channel. The difference $D - d$ is known as the freeboard. It is the clearance in the channel above the anticipated high water line and overflow condition. In vegetated waterways a freeboard should be provided to insure against overflow.

The parabolic channel is similar in cross section to a natural waterway. Under the normal action of channel flow, deposition, and bank erosion, trapezoidal and triangular sections tend to become parabolic. In some cases no earth work is necessary to form a parabolic channel. The natural cross section of a drainage way or meadow outlet is adequate, and only the boundaries need be defined and good vegetation established to produce a satisfactory vegetated watercourse. Parabolic channels should be selected for natural drainageways, and trapezoidal or triangular cross sections adopted where the entire channel must be excavated.

The selection of a channel cross section is also influenced by the type of excavation machinery. Blade-type machines can construct trapezoidal channels only if the bottom width of the channel is wider than the working width of the blade. Triangular channels are readily constructed with blade-type equipment. Plows, bulldozers, or scrapers are well adapted to the construction of parabolic cross sections.

Ease of maintenance and convenience in crossing with field machinery should be carefully considered in the selection of a channel cross section. Side slopes

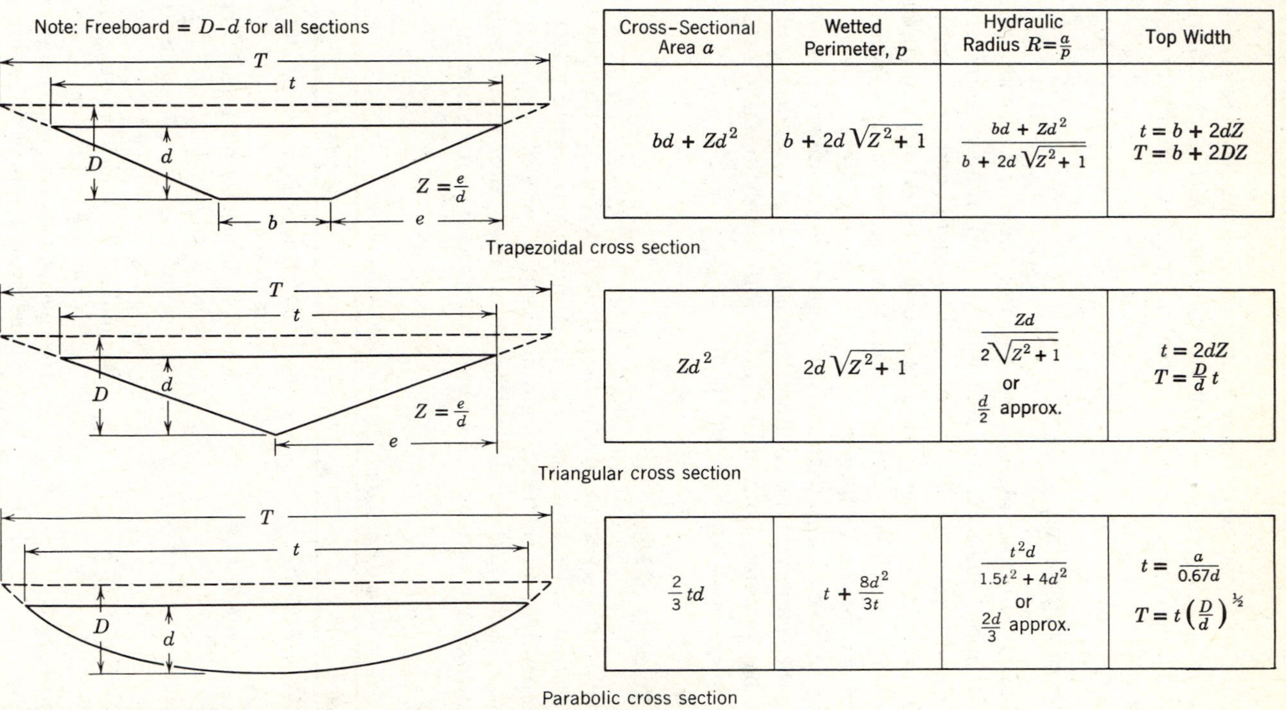

Cross-Sectional Area a	Wetted Perimeter, p	Hydraulic Radius $R=\frac{a}{p}$	Top Width
$bd + Zd^2$	$b + 2d\sqrt{Z^2+1}$	$\frac{bd + Zd^2}{b + 2d\sqrt{Z^2+1}}$	$t = b + 2dZ$ $T = b + 2DZ$
Zd^2	$2d\sqrt{Z^2+1}$	$\frac{Zd}{2\sqrt{Z^2+1}}$ or $\frac{d}{2}$ approx.	$t = 2dZ$ $T = \frac{D}{d}t$
$\frac{2}{3}td$	$t + \frac{8d^2}{3t}$	$\frac{t^2d}{1.5t^2 + 4d^2}$ or $\frac{2d}{3}$ approx.	$t = \frac{a}{0.67d}$ $T = t\left(\frac{D}{d}\right)^{½}$

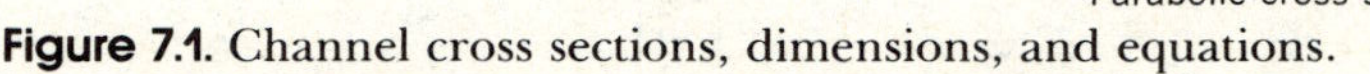
Figure 7.1. Channel cross sections, dimensions, and equations.

of 4 : 1 or flatter are desirable to facilitate maintenance by mowing and to provide easy crossing.

Low flows in wide flat bottoms of trapezoidal channels may give sediment deposits that result in meandering of higher flows with accompanying local damage to vegetation. Thus, a slight "V" bottom may be desirable in a nominally trapezoidal channel.

7.3 Selection of Vegetation

The vegetation selected for a given waterway is governed by soil and climatic conditions; duration, quantity, and velocity of runoff; time required to develop a good cover and ease of establishment; availability of seed or plant materials; suitability for utilization by the farmer as a seed or hay crop; and spreading of vegetation to adjoining fields. Table 7.1 lists some of the grasses recommended for use in waterways in various regions of the United States.

Uniformity of cover is important, as the stability of the most sparsely covered area controls the stability of the channel. Thus, bunch grasses do not offer good protection. They produce nonuniform flows with highly localized erosion, and their open roots do not bind the soil firmly.

7.4 Design Velocity of Flow

The design velocity of a vegetated waterway should be such that neither scouring nor sedimentation will take place. The maximum permissible velocity in the channel is dependent on the type, condition, and density of vegetation and the erosive characteristics of the soil. Sedimentation may occur in a channel if the velocity of the water is reduced. A change from steep to flat slope, an increase

Table 7.1. Vegetation Recommended for Grassed Waterways

Geographical Area of U.S.	*Vegetation*
Northeastern—*L, N, R, S*[a]	Kentucky bluegrass, red top, tall fescue, white clover
Southeastern—*N, O, P,T, U*	Kentucky bluegrass, tall fescue, Bermuda, brome, Reed canary
Upper Mississippi—*K, L, M, N*	Brome, Reed canary, tall fescue, Kentucky bluegrass
Western Gulf—*H, I, J, P, T*	Bermuda, King Ranch bluestem, native grass mixture, tall fescue
Southwestern—*C, D, G*	Intermediate wheatgrass, western wheatgrass, smooth brome, tall wheatgrass
Northern Great Plains—*G, F, H*	Smooth brome, western wheatgrass, red top switchgrass, native bluestem mixture

[a]Letters refer to land resource regions shown in Chapter 1, but recommended vegetation does not necessarily apply to all areas in the region.

in width of the channel without increase in slope, or an increase in the resistance of the vegetation to flow will cause the velocity of the water to decrease. Various channel cross sections of equal cross-sectional area are shown in Fig. 7.2 to illustrate the effect of shape on the velocity.

The velocity of flow in an open channel may be calculated by application of the Manning formula,

$$v = \frac{1.49}{n}\left(\frac{a}{p}\right)^{2/3} s^{1/2} \tag{7.1}$$

where v = velocity of flow in feet per second, fps
n = roughness coefficient (Table 7.2)
a = cross-sectional area of flow in square feet

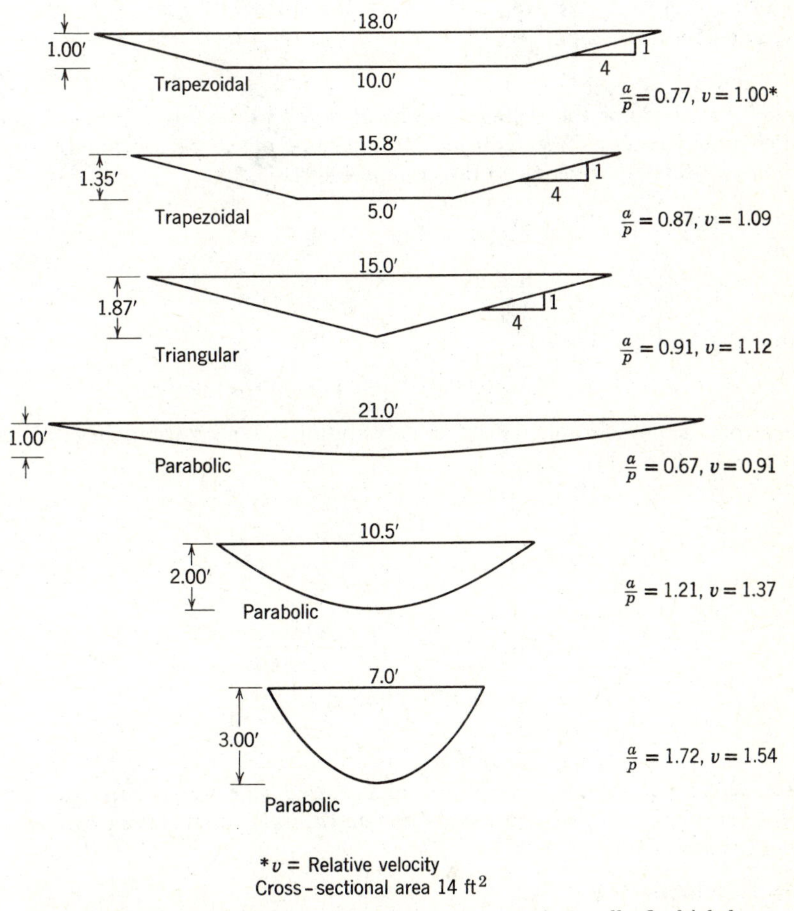

Figure 7.2. Relative velocity for various cross sections, all of which have the same cross-sectional area of 14 sq ft.

p = wetted perimeter, the length of the line of contact between the water and the bottom and sides of the channel around the cross section in feet (a/p is called the hydraulic radius)

s = slope of the channel in feet per foot

This equation can be solved with a nomograph given in Fig. 7.3. The Manning formula shows that the velocity increases as the ratio of the cross-sectional area to the wetted perimeter increases, and also as the slope increases. As the roughness of the channel lining increases, the velocity decreases. Suggested n values for several types of channels are given in Table 7.2.

The design velocity is an average velocity within the channel. The velocity in actual contact with the vegetation or within the channel bed is much lower than the average velocity. The distribution of velocity in a grass-lined channel is shown in Fig. 7.4. In this drawing the average velocity is about 2.5 fps, but the velocity in contact with the vegetation is less than 1 fps. Suggested design average velocities are given in Table 7.3. These vary for different grass species, soil erodibility, and channel slope.

Example 7.1. Determine the average velocity of flow and the top width for a parabolic-shaped channel having a flow depth of 3.9 ft, a slope of 0.8 percent, and a roughness coefficient $n = 0.03$. The required flow rate is 400 cfs.

Solution. From Fig. 7.1,

$$R = \frac{2d}{3} = 2 \times \frac{3.9}{3} = 2.6$$

Substitute R, s, and n in Eq. 7.1,

$$v = 1.49(2.6)^{2/3}\,(0.008)^{1/2}/0.03 = 8.40 \text{ fps } (2.56 \text{ m/s})$$

or read v from the nomograph in Fig. 7.3. Substituting in Eq. 7.2 using the cross-sectional area from Fig. 7.1, $a = 2td/3$,

$$q = av = \frac{2tdv}{3}$$

from which by solving for

$$t = \frac{3q}{2dv} = \frac{3 \times 400}{2 \times 3.9 \times 8.40}$$
$$= 18.3 \text{ ft } (5.6 \text{ m})$$

Example 7.2. The average velocity of flow in a grassed waterway is 5 fps with a channel slope of 4 percent. If all the other factors remain constant, what is the expected velocity if the slope is increased to 8 percent?

Solution. Substituting in Eq. 7.1,

$$v\ (4\%) = C\ (0.04)^{1/2} = C \times 0.20 = 5 \text{ fps}$$

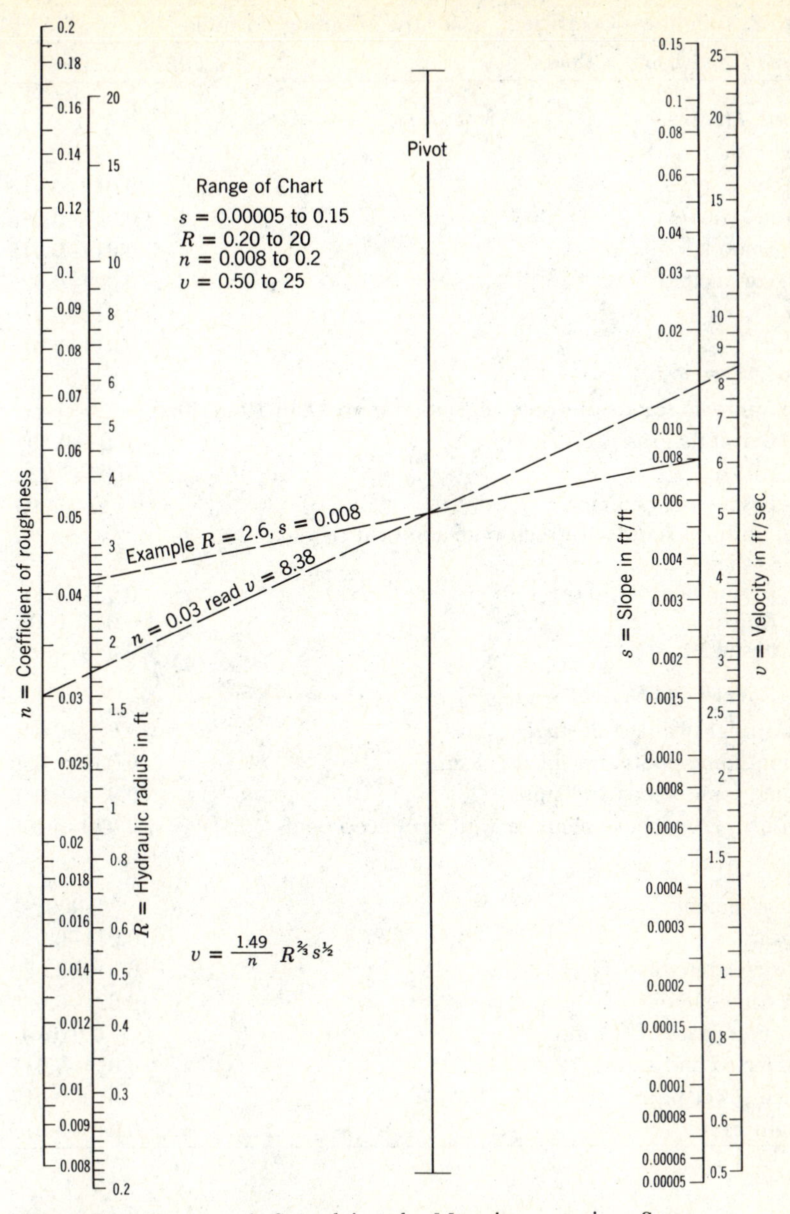

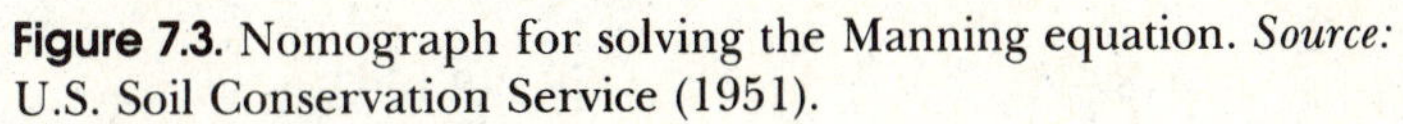

Figure 7.3. Nomograph for solving the Manning equation. *Source:* U.S. Soil Conservation Service (1951).

and

$$v\ (8\%) = C\ (0.08)^{1/2} = C \times 0.283$$

where C = a constant. Dividing the first equation by the second, C cancels and solving for

Table 7.2. Roughness Coefficient, *n*, for the Manning Formula

Type and Description of Channel	*n Values*
Channels, Lined	
Asphalt	0.015
Concrete	0.012–0.018
Concrete, rubble	0.017–0.030
Metal, smooth	0.011–0.015
Metal, corrugated	0.021–0.026
Plastic	0.012–0.014
Wood	0.011–0.015
Channels, Vegetated	
Dense, uniform stands of green vegetation more than 10 in. long	
Bermuda grass	0.04–0.20
Kudzu	0.07–0.23
Lespedeza, common	0.047–0.095
Dense, uniform stands of green vegetation cut to a length of less than 2.5 in.	
Bermuda grass, short	0.034–0.11
Kudzu	0.045–0.16
Lespedeza	0.023–0.05
Earth Channels and Natural Streams	
Clean, straight bank, full stage	0.025–0.040
Winding, some pools and shoals, clean	0.035–0.055
Winding, some weeds and stones	0.033–0.045
Sluggish river reaches, weedy or with very deep pools	0.050–0.150
Pipe	
Asbestos cement	0.009
Cast iron	0.011–0.015
Clay or concrete (4–12 in.)	0.010–0.020
Metal, corrugated	0.021–0.0255
Plastic, corrugated (2–4 in.)	0.016
Steel, riveted and spiral	0.013–0.017
Vitrified sewer pipe	0.010–0.017
Wrought iron	0.012–0.017

$$v\ (8\%) = 5 \times \frac{0.283}{0.20} = 7.1 \text{ fps} \quad (2.16 \text{ m/s})$$

The velocity thus varies as the half power of the slope. Since the slope is doubled, the velocity is proportional to $(2)^{1/2} = 1.41$, a 41 percent increase or $1.41 \times 5 = 7.1$ fps as above. In a similar way the roughness coefficient and/or the hydraulic radius to the 2/3 power can be changed and the new velocity can be estimated. Note that the roughness coefficient varies inversely as the velocity.

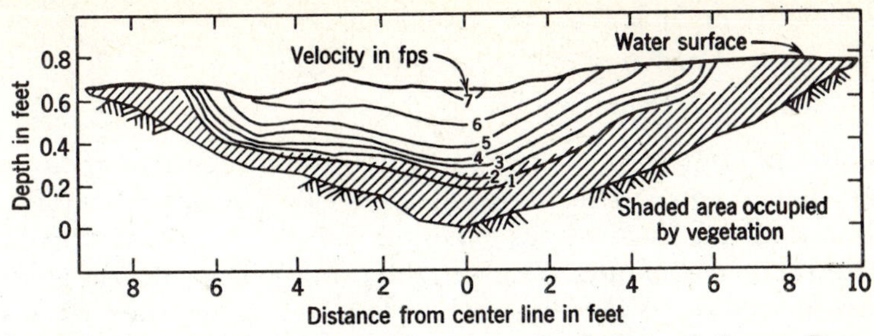

Figure 7.4. Velocity distribution in a grass-lined channel. *Source:* Ree (1949).

Table 7.3. Permissible Velocities for Vegetated Channels

Vegetative Cover	*Slope range (%)*	*Permissible Velocities in fps*	
		Easily Eroded Soils	*Erosion Resistant Soils*
Bermuda grass	0–5	6	8
	5–10	5	7
	over 10	4	6
Blue grama	0–5	5	7
Buffalo grass	5–10	4	6
Kentucky bluegrass	over 10	3	5
Smooth brome			
Tall fescue			
Grass mixture	0–5	4	5
	5–10	3	4
	over 10	Not recommended	
Annual crops for temporary protection	0–5	2.5	3.5
	over 5	Not recommended	
Alfalfa			
Crabgrass			
Kudzu			
Lespedeza sericea			
Weeping lovegrass			

Note: Use velocities over 5 fps only where good cover and proper maintenance can be obtained.
Source: Ree (1949).

7.5 Channel Flow Capacity

The discharge or capacity of a channel may be calculated from the equation,

$$q = av \tag{7.2}$$

where q = capacity in cubic feet per second, cfs
a = cross-sectional area of flow in square feet
v = velocity in feet per second

The channel must be proportioned to carry the design runoff at average velocities less than or equal to the permissible velocity. The design runoff should be obtained by applying the methods of runoff estimation, discussed in Chapter 5, to the watershed contributing to flow in the proposed vegetated waterway. A runoff return period of 10 years is suitable for most grassed waterways. Emergency spillways for large ponds and reservoirs are often designed for higher return periods.

A properly designed channel must satisfy both Eqs. 7.1 and 7.2. By assuming a shape for the channel, roughness coefficient, and slope, the variables are reduced to a and v, and a solution can be obtained. An example using this procedure is given in Example 7.1. In grassed waterways, research has shown that the roughness coefficient varies greatly with the velocity and the hydraulic radius R. The hydraulic radius is a/p in the Manning equation and for wide channels is about equal to the depth of flow. Fig. 7.5 shows the n–vR curves for retardance classes (to be described) varying from E when the grass is short to A when the grass is tall. These experimental curves were developed in Oklahoma for uniform growing vegetation. Because the n–vR relationship further complicates the design, nomographs have been prepared for parabolic and trapezoidal cross section waterways.

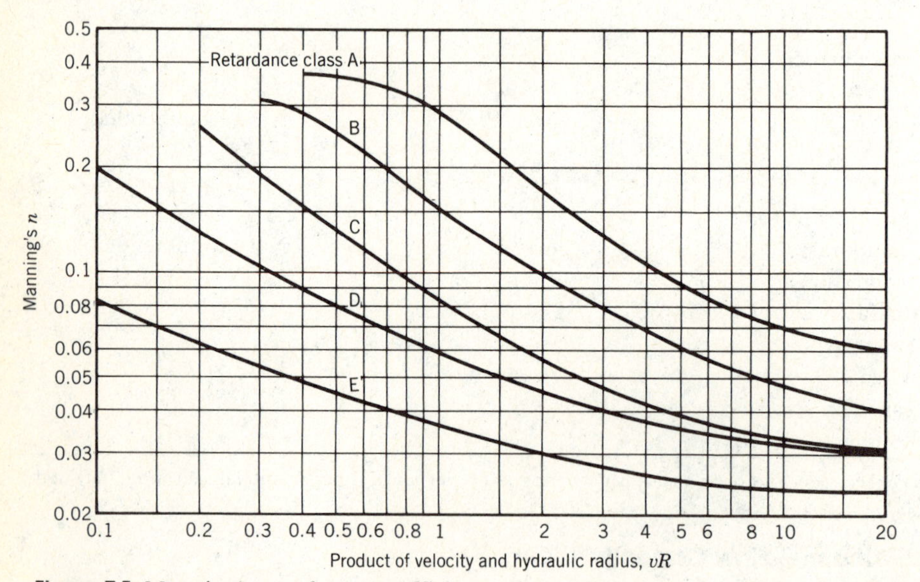

Figure 7.5. Manning's roughness coefficient n for various vegetal retardance classes. *Source:* U.S. Soil Conservation Service (1954).

7.6 Parabolic-Shaped Channel Design.

The recommended shape for a parabolic cross section is given in Fig. 7.6. This shape can be specified by the width W at 1-ft depth. The drawing shows W = 10 ft, which is about the minimum width that is practical. Side slopes for the bottom quarter, second quarter depth, and top half are shown for top widths from 10 to 50 ft. In staking for construction, $W/2$ is the width for the bottom quarter depth and $3W/4$ is the width for one half the depth. The top width for channels deeper than 1 ft may be estimated by using the side slope ratio for the top half depth. For example, in Fig. 7.6 the top width for a 1.3-ft depth channel is $10 + 2(3 \times 0.3) = 11.8$ ft rounded to 12 ft. The topography of the soil surface may dictate a more appropriate side slope. A freeboard of 0.3 ft should be added to the depth for all channels. It is a safety factor that allows for errors in construction or layout and for sediment accumulation or extra roughness. To obtain the minimum size channel, use the permissible velocity for the type of grass, channel slope, and soil erodibility as shown in Table 7.3. The minimum width W can be read directly from Fig. 7.7. Any greater width is satisfactory, since the velocity will be less than the permissible value. However, a wider waterway is usually not desirable, since more land is taken out of crop production.

The retardance class of the grass depends on the length and condition of the vegetation (Table 7.4). When the grass is short, it has a low retardance (low n

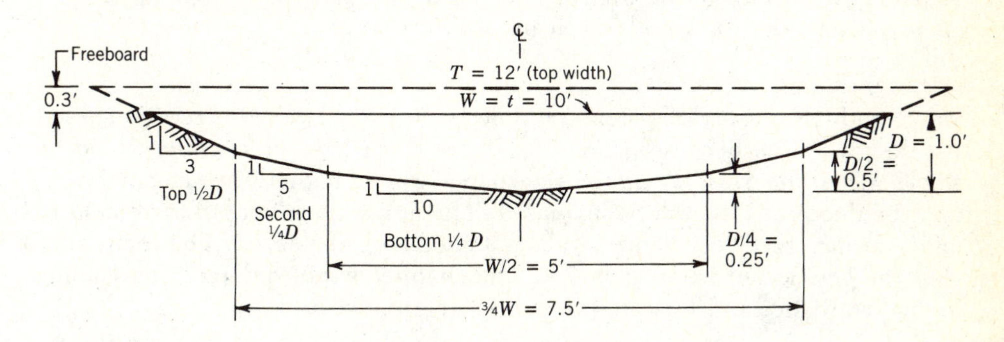

	Side Slope Ratio, Horizontal: Vertical		
Channel Width, W at 1-ft Depth	*Bottom Quarter Depth, 0–0.25 ft*	*Second Quarter Depth, 0.25–0.50 ft*	*Top Half Depth, 0.5–1 ft*
10 ft	10:1	5:1	3:1
15	15:1	7:1	4:1
20	20:1	10:1	5:1
30	30:1	15:1	7:1
40	40:1	20:1	10:1
50	50:1	25:1	12:1

Figure 7.6. Design parabolic cross-sectional shape for various channel widths for a 1-ft depth.

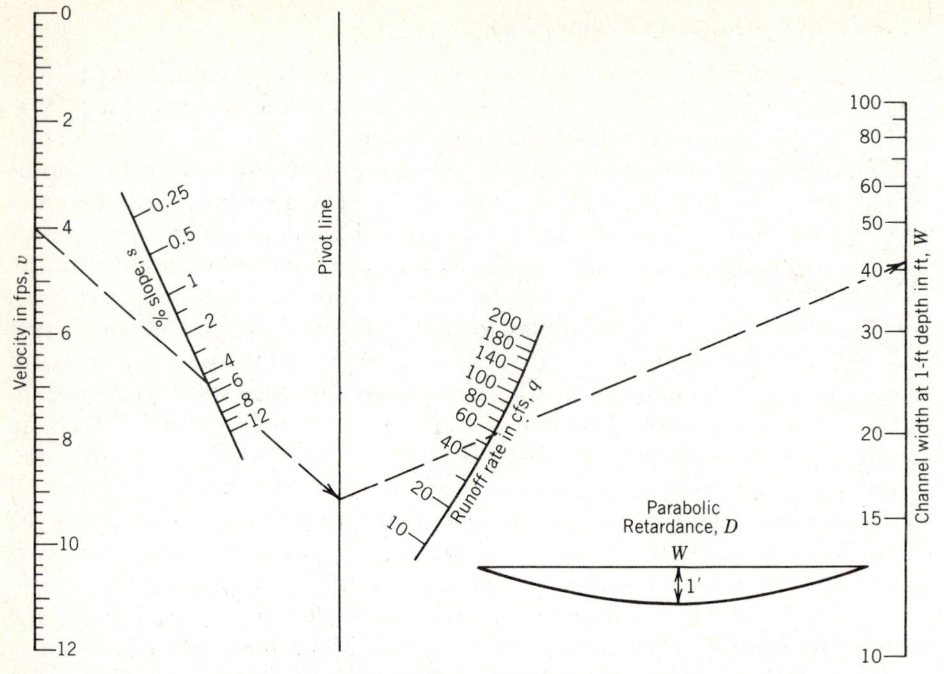

Figure 7.7. Nomograph for the average flow velocity in a parabolic grassed waterway with retardance class D. *Source:* Garton and Green (1981).

value) and the velocity of water would be high. Retardance *D* is recommended for design. If the permissible velocity is not exceeded, the channel would be stable. When the grass is tall, the retardance will be higher, the velocity lower, and the depth of flow will be increased. The approximate depth increment to allow for increased retardance can be obtained from Table 7.5. The retardance class can be selected from Table 7.4. The channel is thus designed for stability at low retardance and for capacity at high retardance.

Table 7.4. Guide to Selection of Vegetal Retardance Class

Average Length of Waterway Vegetation (in inches)	*Stand Conditions in Waterway*	
	Fair	*Good*
Less than 2	E	E
2–6	D	D
6–10	D	C
11–24	C	B
More than 30	B	A

Source: U.S. Soil Conservation Service (1954).

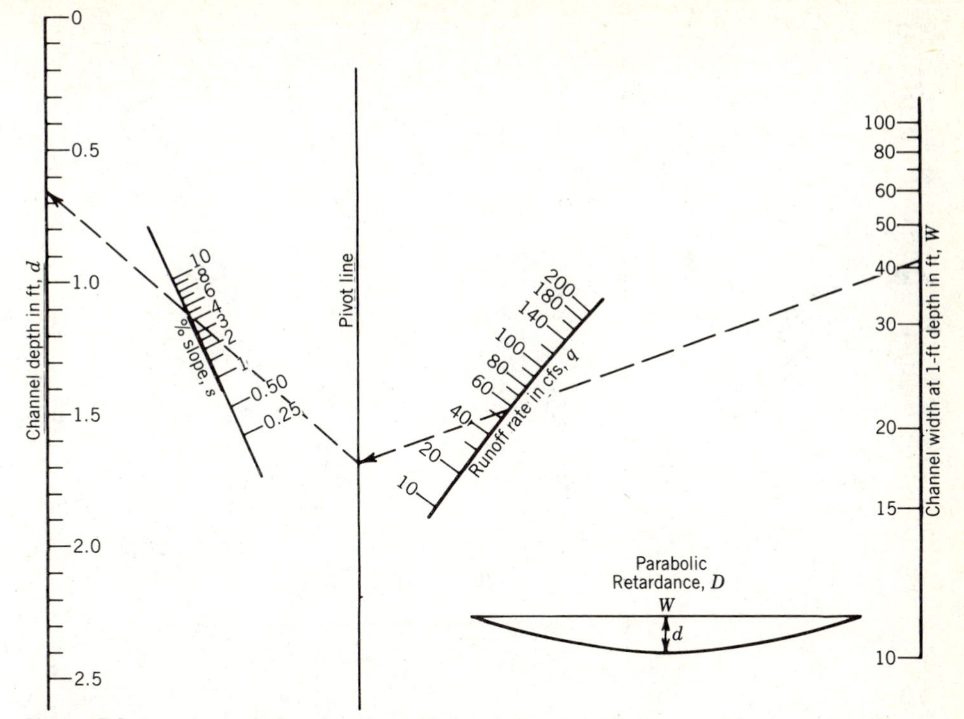

Figure 7.8. Nomograph for the channel depth in a parabolic grassed waterway with retardance class D. *Source:* Garton and Green (1981).

Table 7.5. Approximate Channel Depth Increment for Grassed Waterways with Increase in Retardance

Slope Range (%)	*Type of Channel Cross Section and Width*		
	Triangular or Trapezoidal, t or b = 6–19 ft	*Parabolic or Trapezoidal,* t or b = 20–89 ft	*Trapezoidal,* b = 90 ft or more
From Retardance Class D to Class C			
1 or more	0.1 ft	0.1 ft	0.1 ft
From Retardance Class C to Class B			
1–2	0.6	0.5	0.4
2–5	0.4	0.4	0.3
5 or more	0.3	0.3	0.2
From Retardance Class D to Class B			
1–2	0.7	0.6	0.5
2–5	0.5	0.5	0.4
5 or more	0.4	0.4	0.3

Source: Larson and Manbeck (1960) as computed by C. L. Larson.

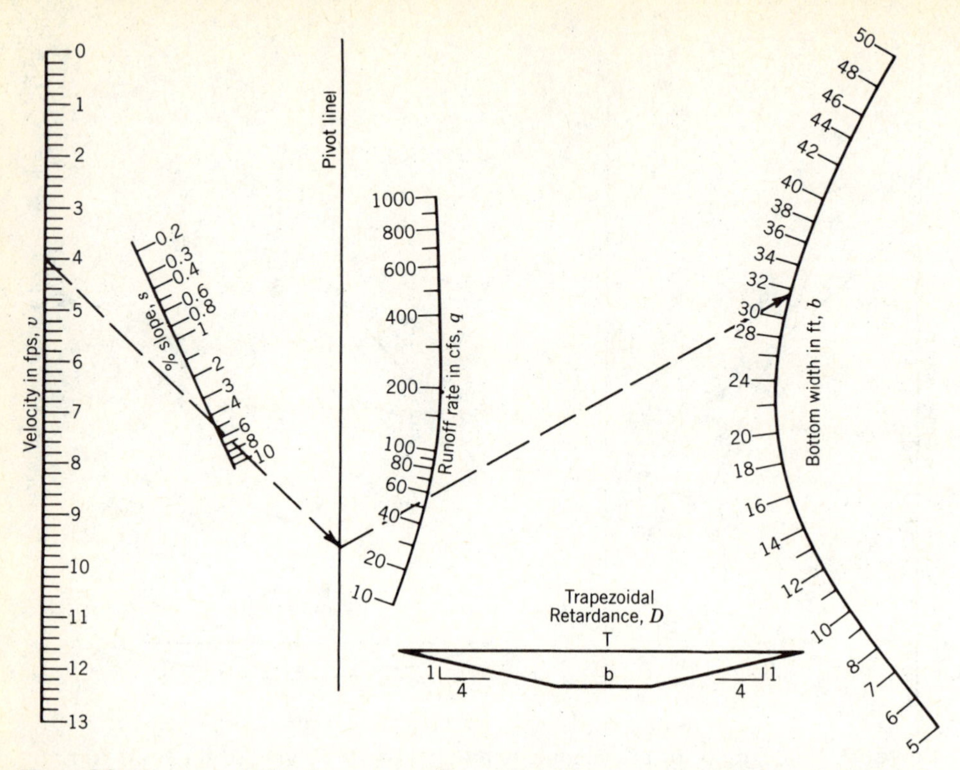

Figure 7.9. Nomograph for the average flow velocity in a trapezoidal grassed waterway with retardance class D. *Source:* Garton and Green (1981).

Design tables have been developed by the U.S. Soil Conservation Service (SCS) (1979) from which width and depth can be obtained for a wide range of conditions. Mathematical procedures and design curves have been developed by the U.S. Soil Conservation Service (1954) and Schwab et al. (1981). Nomographs developed by Garten and Green (1981) and shown in Figs. 7.7 and 7.8 simplify the design procedure.

Example 7.3. Determine the top width and depth for a parabolic-shaped grassed waterway in southeastern Iowa to handle a 10-year return period runoff rate of 55 cfs in a vegetated channel. The slope of the channel is 5 percent and the soil is easily eroded. Assume that the grass is in good condition, but will be allowed to grow to a 10-in. height before being cut to a 4-in. height. Design for stability and for adequate capacity when the grass is tall.

Solution

1. Select tall fescue grass from Table 7.1 for Iowa (area M shown in Chapter 1).
2. Select a permissible velocity of 4 fps from Table 7.3.

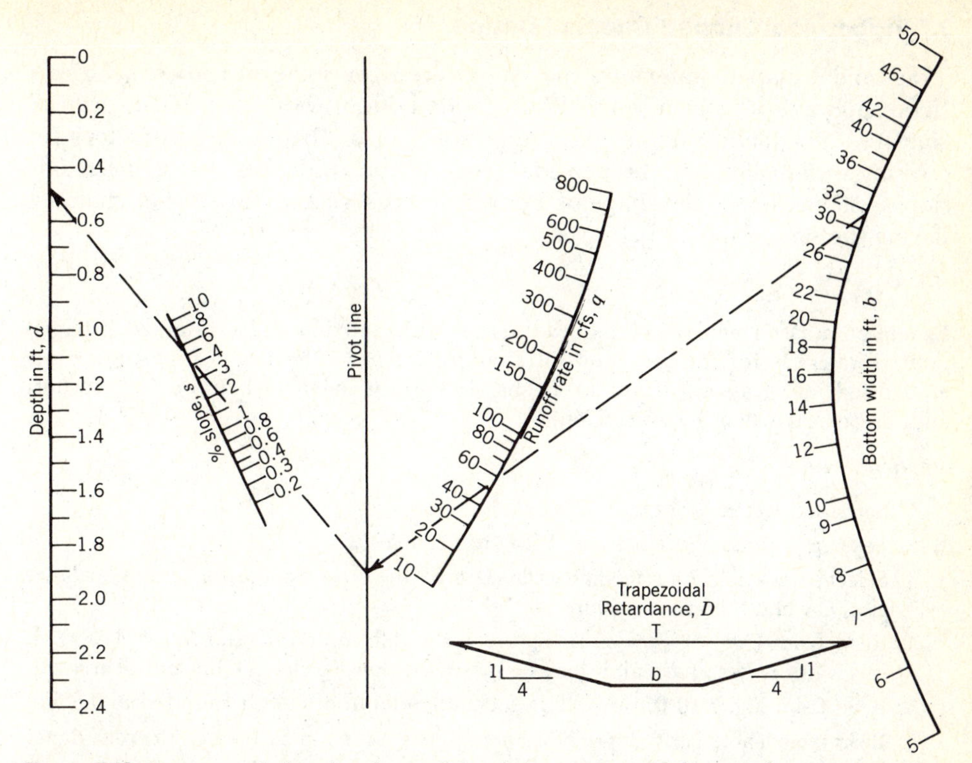

Figure 7.10. Nomograph for the channel depth in a trapezoidal grassed waterway with retardance class D. *Source:* Garton and Green (1981).

3. Select from Table 7.4 retardance class D when the grass is cut to 4 in. and retardance class C when the grass is 10 in. tall.
4. Read from Fig. 7.7 (v = 4 fps, s = 5 percent, and q = 55 cfs), W = 42 ft, channel width at 1-ft depth for retardance class D. Figure 7.7 is valid only for retardance class D. If the width is not practical, it may be increased (lower velocity), but not decreased.
5. Read from Fig. 7.8 (W = 42 ft, q = 55 cfs, and s = 5 percent) a channel depth of 0.65 rounded to 0.7 ft.
6. Read from Table 7.5, a depth increment of 0.1 ft ($W = t$ = 42 ft) when the retardance class increases from class D to C.
7. Adding 0.3-ft depth of freeboard, design depth of channel, $D = 0.7 + 0.1 + 0.3 = 1.1$ ft.
8. From Fig. 7.6 the side-slope ratio for the top half of a channel (W = 42 ft) is 10 : 1. The increased top width above a depth of 1 ft is $(1.1 - 1) \times 10 \times 2 = 2$ ft. Top width $T = 42 + 2 = 44$ ft.
9. The 44-ft channel will carry 55 cfs (d = 0.7 ft and retardance D) at 4 fps (a stable channel). When the grass is tall, the depth of flow will be 0.8 ft. The velocity will be less than 4 fps, but the channel will have adequate capacity.

7.7 Trapezoidal-Shaped Channel Design

Trapezoidal-shaped waterways may be desired for some purposes, such as a flood spillway for a farm pond. With a wide bottom width a large flow can be conveyed at a shallow depth. Side slopes of 4 : 1 or flatter are satisfactory for crossing with machinery. Trapezoidal cross sections with these constructed side slopes will be close to the shape of a parabolic cross section for smaller channels having the same top width.

Example 7.4. Determine the bottom width, top width, and depth of a trapezoidal-shaped grassed waterway for the conditions in Example 7.1 ($q = 55$ cfs, $s = 5$ percent, easily eroded soil, and grass 4 to 10 in. in height). Design for stability when the grass is short and for capacity when the grass is tall.

Solution

1. Select tall fescue grass from Table 7.1.
2. Select a permissible velocity of 4 fps from Table 7.3.
3. Select from Table 7.4 retardance class D when the grass is cut to 4 in. and retardance class C when the grass is 10 in. tall.
4. Read from Fig. 7.9, $b = 32$ ft. Figure 7.9 is valid only for retardance class D. If this width is not practical, it may be increased (lower velocity), but not decreased.
5. Read from Fig. 7.10 for $b = 32$ ft, a channel depth of 0.48 ft rounded to 0.5 ft.
6. Read from Table 7.5, a depth increment of 0.1 ft for $b = 32$ ft when the retardance class increases from class D to C.
7. Allowing 0.3 ft depth for freeboard, the design depth of the channel $D = 0.5 + 0.1 + 0.3 = 0.9$ ft (0.3 m).
8. Top width of channel for $b = 32$ ft and 4 : 1 side slopes, $T = 32 + 2(0.9 \times 4) = 39.2$ rounded to 39 ft (12 m).
9. The 39-ft channel will carry 55 cfs ($d = 0.5$ ft and retardance D) at 4 fps (a stable channel).

When the grass is tall, the depth of flow will be 0.6 ft. The velocity will be less than 4 fps, but the channel will have adequate capacity.

A channel may be called upon to carry runoff before vegetation is established. It is not practical to design for this extreme condition and it may be necessary to divert flow from the channel until vegetation is established. In some instances the possibility that high runoff will occur before vegetation is established must be accepted as an unavoidable risk.

7.8 Drainage

Waterways are often located where there is low flow over long periods of time. Tile outlets, springs, or seeps flowing into a waterway may cause a continually wet condition, which may prevent the development and maintenance of good vegetal cover. Diversion of such flow, either by subsurface or surface means, is essential to the success of the waterway.

Detailed discussion of subsurface drainage of wet draws will be found in Chapter 14, where the interception of seepage along the sides or upper end of a waterway is discussed. Surface water entering a waterway at some point may be intercepted by a catch basin and carried off by a pipe drain. In some situations it may be desirable to provide a concrete or asphalt trickle channel in the bottom of the waterway to carry prolonged low flow.

7.9 Establishment of Waterways by Seeding

After a properly designed waterway is constructed, adequate vegetation must be established. Lime and fertilizer should be applied in accordance with local recommendations for each soil type. A heavy application of manure worked into the seedbed will add nutrients and organic matter, making the soil more resistant to erosion.

Waterway seeding mixtures should include some quick-growing annual for temporary control as well as a mixture of hardy perennials for permanent protection. Seed should either be broadcast or drilled nonparallel to the direction of flow. Seeding rates twice those used for field seedings are recommended. Early fall is the best season for the establishment of grass seedings because of the normal assurance of gentle rains and cool weather. Waterways prepared in the summer may be given temporary seedings of sudan grass, corn, or oats which is later mowed down and disked lightly for fall grass seeding. Some types of vegetation may be established by planting crowns 5–10 ft apart. Bermuda grass may be started by planting stolons 2–4 ft apart as well as by direct seeding. Seedings should be cultipacked crosswise or zigzag to the channel to cover the seed and firm the seedbed. A light straw or strawy manure mulch added on top of the new seeding may mean the difference between success and failure. The straw protects against raindrops and, when anchored by the shoots of the faster growing grasses, gives temporary protection against moderate runoffs.

Many new materials have been developed to assist in the establishment of vegetation. Such materials include chemical soil stabilizers, plastic, fiber or other mesh or net covers, asphalt mulches, and plastic or other surface covers. These materials reduce the rate of drying and crusting, absorb the energy of raindrops or wind, and reduce the flow velocity.

7.10 Establishment of Waterways by Sodding

Direct sodding is the surest but most expensive means of establishing vegetation. Where there is no alternative but to turn heavy flow into the waterway immediately, a sodded channel may be desirable. In general, sodding is used only for high-velocity channels or for critical situations, such as short, steep "sod chutes." Sod should be laid on a fertile, well-prepared seedbed and may be anchored with a suitable material. Lay strips across the watercourse with joints staggered. Fill joints with loose soil and tamp or roll the sod. Sodding of narrow cross strips at frequent intervals will often give control until seedings are established. A sod strip down the center line of the channel will prevent gullying by low flows before seedings become established.

Some types of sod, such as Reed canary grass or Bermuda grass, may be established by "broadcast sodding." This method involves disking up the sod field and loading the cut sod into a manure spreader from which it is spread in the watercourse.

7.11 Waterway Construction

Prior to construction, flags or stakes should be set along the sides of the waterway to locate the top width. Where the center of the channel is not obvious, such as a new alignment, the center line may also be staked. Depth and grade of the channel can be established as described in Chapter 13 for surface drains. For parabolic-shaped waterways the bottom half width and the three-fourths width may be staked to secure the proper shape as shown in Fig. 7.6. For trapezoidal-shaped waterways, bottom width stakes may be set.

For waterways up to 1 to 2 ft in depth, farmer-owned equipment is satisfactory. The moldboard plow, disk plow, pull or push tractor-mounted blades, bulldozers, and small scrapers are most suitable. Final shaping of the waterway and seedbed preparation can be done with such machines and with tillage implements. Construction procedure with plows can be accomplished by following the round sequence described in Fig. 7.11. Observe that each successive series of 10 to 15 rounds starts outside the preceding series. The loose soil that piles up along the outside rounds can be spread with blade-type equipment.

For filling large gullies or for shaping emergency spillways on farm ponds, large earth-moving equipment, such as bulldozers, motor graders, and pan scrapers, are usually more satisfactory. The availability of equipment and the size of the job affect machine selection.

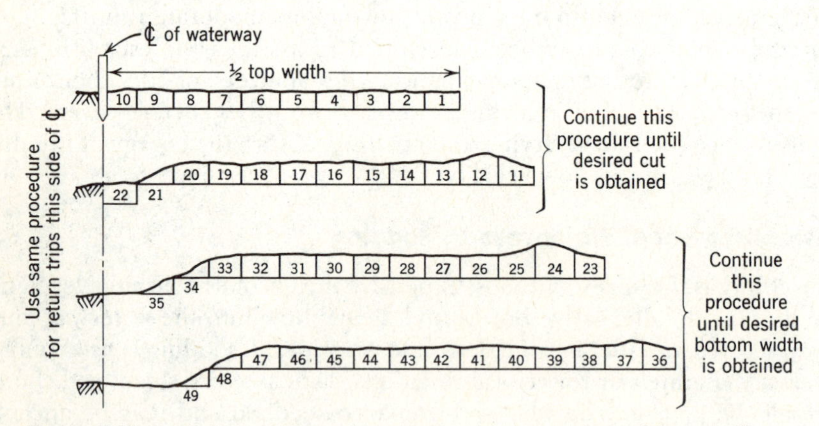

Figure 7.11. Round procedure for waterway construction with a small moldboard or a disk plow.

7.12 Waterway Maintenance

The condition of the vegetation is dependent not only on design and construction but also on subsequent management. Vegetation in waterways that are used as trails or lanes for livestock or vehicles will often be damaged to the point of waterway failure. Such use in wet weather is particularly damaging. Plow furrows ending against the vegetated strip should be staggered to prevent flow concentration down the edges of the watercourse. Implements should be raised when crossing waterways to avoid damaging the sod.

During the first few years after seeding, vegetation should be mowed several times during the growing season to stimulate spreading and growth and to control weeds. Dense cuttings and bunches of cut grass, especially from rotary mowers, should be removed to prevent the smothering of established vegetation. An annual application of manure and fertilizer will help maintain a dense sod. Waterways may be pastured in normal weather, but overgrazing is detrimental. Vegetation may be harvested as a hay or seed crop without interfering with its primary purpose of erosion control.

High runoff during the period vegetation is being established may result in small gullies. These should be filled in and the grass reestablished, or the channel may need to be reshaped and reseeded. Such risks of washout can be avoided by diverting the water to another channel during the period that grass is being established. In some situations this alternative is not possible.

Accumulations of sediment in the waterway will smother vegetation and restrict channel capacity. Sediment can best be controlled by good conservation practices on the watershed above. Rank or matted vegetal growth contributes to this problem. Flat slopes at the lower end of a waterway cause sediment fans to develop. Where this happens, the grassed waterway may be discontinued and a narrower drainage channel established to carry the runoff at a higher velocity. If crop damage is slight, the water may simply be allowed to spread out and the fan to develop. Extending vegetation well up the sides of the waterway and into terrace channels is recommended. Accumulated sediment should be removed to avoid damage to the vegetation and to prevent localized erosion.

REFERENCES

Garton, J. E., and J. E. P. Green (1981) "Improved Design Procedures for Vegetation-Lined Channels," *Okla. Agr. Exp. Sta. Research Report P 814.*

Larson, C. L., and D. M. Manbeck (1960) "Improved Procedures in Grassed Waterway Design," *Agr. Eng.* **41**:694–696.

Ree, W. O. (1949) "Hydraulic Characteristics of Vegetation for Vegetated Waterways," *Agr. Eng.* **30**:184–187, 189.

Schwab, G. O., R. K. Frevert, T. W. Edminster, and K. K. Barnes (1981) *Soil and Water Conservation Engineering*, 3rd ed., John Wiley & Sons, New York.

U.S. Soil Conservation Service (1979) *Engineering Field Manual for Conservation Practices* (Lithographed), Washington, D.C.

U.S. Soil Conservation Service (1951) "Hydraulics," *Natl. Eng. Handbook, Section 5* (Lithographed), Washington, D.C.

U.S. Soil Conservation Service (1954) *Handbook of Channel Design for Soil and Water Conservation*, SCS-TP-61 (Revised), Washington, D.C.

PROBLEMS

7.1 Determine the bottom width, top width, and total depth for a trapezoidal-shaped (4 : 1 side slopes) grass waterway to carry a flow of 50 cfs (1.42 m^3/s). The design velocity is 5 fps (1.52 m/s), the channel slope is 4 percent and good grass stand is cut to a 4-in. (10-cm) height.

7.2 Determine the top width and total depth for a parabolic-shaped grass waterway for the same conditions as in Problem 7.1.

7.3 Determine the top width and total depth for a trapezoidal-shaped grass waterway to be constructed in erosion resistant soil to carry a runoff of 40 cfs (1.13 m^3/s) where the channel slope is 10 percent. Design for the recommended grass in good condition, which is an 8-in. (20-cm) height for your present location.

7.4 Determine the top width and total depth for a parabolic-shaped grass waterway for the same conditions as in Problem 7.3.

7.5 From the appropriate nomograph determine the channel slope for a parabolic-shaped waterway to obtain a velocity of 7 fps (2.13 m/s). The channel width W is 20 ft (6.1 m) for a 1-ft (0.3 m) depth, the design runoff is 70 cfs (2 m^3/s), and the retardance for the grass is class D.

7.6 From the Manning equation compute the top width and depth of flow for a parabolic-shaped grass waterway using weeping lovegrass, a channel slope of 4 percent, a roughness coefficient of 0.03, and a flow of 18 cfs (0.51 m^3/s). The soil is silty clay and resistant to erosion. Assume the hydraulic radius $R = 2d/3$.

7.7 If the average velocity in a trapezoidal-shaped channel with 4 : 1 side slopes is 4 fps (1.22 m/s), the width of the water surface is 16 ft (4.88 m), and the depth of flow is 1.0 ft (0.3 m), what is the discharge in the channel?

7.8 If the channel in Problem 7.7 was parabolic, what would be the discharge?

7.9 The bottom width of a trapezoidal-shaped waterway for a pond flood spillway is to be 10 ft (3.1 m), the scraper width. If the design runoff is 20 cfs (0.57 m^3/s) and the channel slope is 6 percent, what is the average flow velocity? What is the depth of flow? Assume retardance class D for the grass.

7.10 The average velocity in a grass waterway is 6 fps (1.83 m/s) with a slope of 9 percent. Assuming no change in roughness and in the hydraulic radius R, what is the velocity if the slope is reduced to 4 percent? What is the velocity if the roughness coefficient is also reduced from 0.04 to 0.03?

CHAPTER 8

CONTOURING, STRIP CROPPING, AND TERRACING

As discussed in Chapter 6, contouring, strip cropping, and terracing are important conservation practices for controlling water erosion. These all require the adjustment of tillage and crop-row direction from uphill and downhill to contour operation. The small ridges and depressions increase the storage of water on the surface and provide for snow trapping, thus increasing infiltration and storage of water in the soil profile. Runoff rates are decreased, with a corresponding decrease in soil loss.

CONTOURING

8.1 Contour Guide Lines

In contouring, tillage operations are carried out as nearly as practical on the contour. A guide line is laid out for each plow land, and the backfurrows or deadfurrows are plowed on these lines. On small fields of uniform slope, one guide line may be sufficient; on larger, more irregular fields, several lines may be required to assure that all tillage rows remain within the usual limits of 1 to 2 ft of fall per 100 ft. Contouring operations must be laid out carefully if they are to be effective. Contour guide rows, terraces, and contour strip boundaries establish the pattern and accuracy of all subsequent contouring operations. In changing field boundaries for contour farming, fences should be relocated on the contour or moved so as to eliminate odd-shaped fields that would result in short, variable-length rows, called point rows.

On gently sloping land, contouring will reduce the velocity of overland flow. If ridge or lister cultivation is practiced, the storage capacity of the furrows is materially increased, permitting storage of large volumes of water. When contouring is used alone on steeper slopes or under conditions of high rainfall intensity and soil erodibility, there is an increased hazard of gullying because row breaks may release the stored water. Break-overs cause cumulative damage as the volume of water increases with each succeeding row. For slopes below about 10 percent, contouring reduces soil losses by 50 percent compared to up- and downhill operation, but it is much less effective on steeper slopes.

In plowing, the best plan is to start backfurrows on the guide lines and work around them until about one half the area is plowed. Next plow the lands between, leaving deadfurrows as shown in Fig. 8.1, Method 1. The advantage of this method is that it practically eliminates turning on plowed ground. If the field has been in an intertilled crop (corn or soybeans) and the point rows are visible, Method 2 in Fig. 8.1 may be more satisfactory. This procedure will establish the deadfurrows on the guide line where the backfurrows were located on the previous plowing.

Prepare the seedbed and plant small grain and row crops (Fig. 8.2) using the same general plan as in plowing. The method of harvesting the buffer crop areas to provide access routes should be given some consideration before starting to plant row crops. Except when there is a grass strip on either side, it is generally

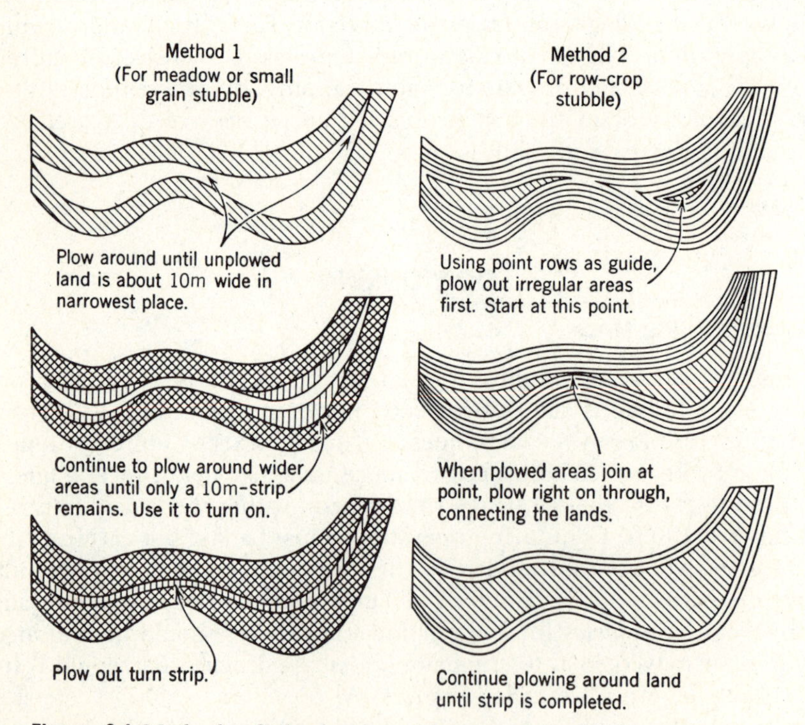

Figure 8.1. Methods of plowing out point-row areas. (*Redrawn from Hay, 1948.*)

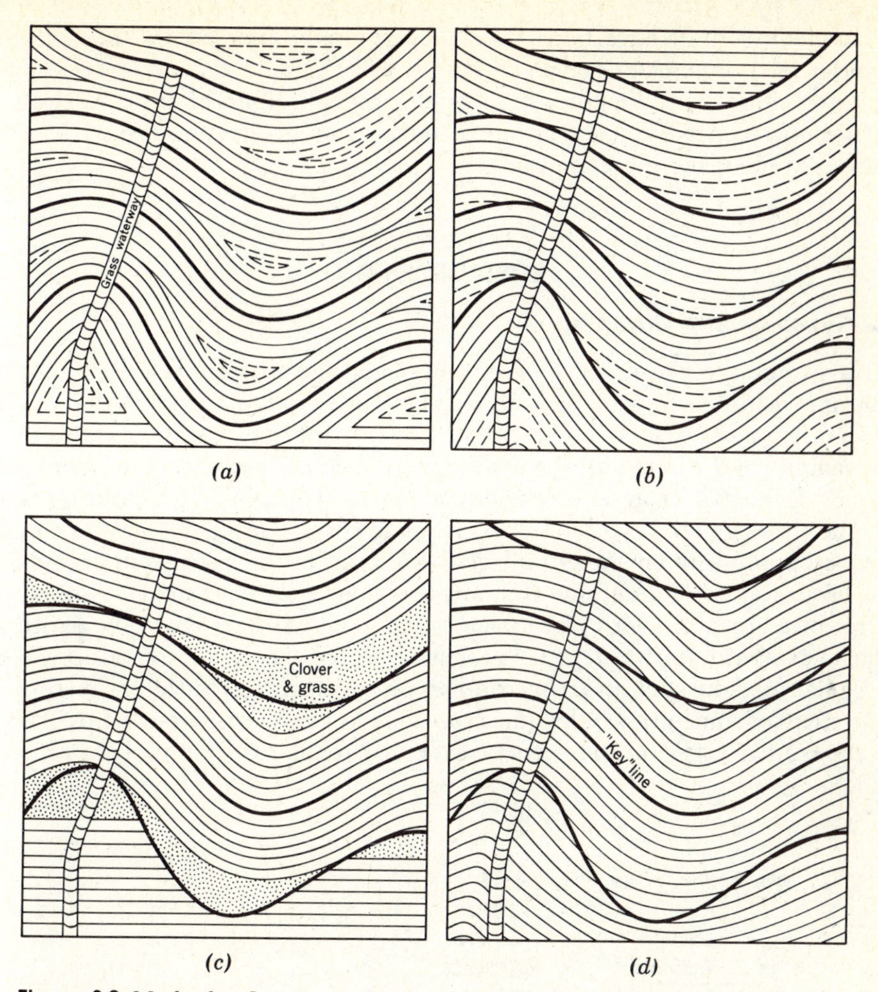

Figure 8.2. Methods of contour planting. (*a*) Point rows midway between lines. Entire field is in a cultivated crop. (*b*) Point rows end at next contour line below. Entire field is in a cultivated crop. (*c*) Point rows not farmed, but area is in small grain or hay crop. (*d*) "Key" line system, point rows only at corners. Field is in one crop. [Except for (*d*), these methods are well suited for terraces or strip cropping.] (*Redrawn from Hay, 1948.*)

preferable to have the point rows midway between the guide lines as in Fig. 8.2*a*. When harvesting row crops, the point rows would be harvested last.

Hand levels, when carefully adjusted, are satisfactory for establishing contour and strip cropping guide lines. With the Abney or hand level, greater accuracy can be obtained with a unipod stand and target shown in Fig. 8.3*a*. In areas with moderate to smooth topography, vehicle-mounted pendulum devices or slope meters are satisfactory. A slope meter (bubble tube) or pendulum device using a dampening fluid have been developed for mounting on tractors or other vehicles. A commercial slope meter is shown in Fig. 8.3*b*. It can also be fastened to a trenching machine for rough grade control. None of these devices are sufficiently accurate for the layout of terraces or diversions. Tripod levels or transits are best for these applications.

STRIP CROPPING

8.2 Types of Strip Cropping

Strip cropping is the practice of growing alternate strips of different crops in the same field. For controlling water erosion, the strips are always on the contour, but in dry regions, such as the Great Plains, strips are placed crosswise to the prevailing wind direction to control wind erosion. As discussed in Chapter 6, soil losses for strip cropping are about 50 percent less than for contouring alone.

The three general types of strip cropping shown in Fig. 8.4 are (*a*) contour strip cropping with layout and tillage held closely to the exact contour and with the crops following a definite rotational sequence, (*b*) field strip cropping with strips of a uniform width placed across the general slope; when used with adequate grassed waterways the strips may be placed where the topography is too irregular to make contour strip cropping practical, and (*c*) buffer strip cropping with strips of some grass or legume crop laid out between contour strips of crops in the regular rotation; they may be even or irregular in width; or they may be

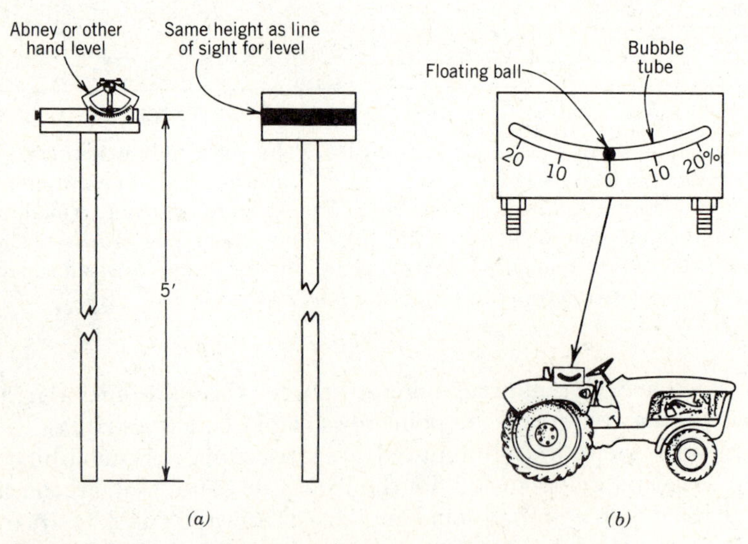

Figure 8.3. Contour layout equipment. (*a*) Level and target. (*b*) Tractor-mounted slope meter.

placed on critical slope areas of the field. Their main purpose is to give protection from erosion.

Maximum erosion protection is achieved when the point rows of a cultivated crop are placed in the middle of the strip width (Fig. 8.2*a*) with the first rows parallel to the edges of the strip. Point row areas may also be left in permanent meadow.

In some areas insect damage, due to long exposed crop borders, has proved to be a serious disadvantage of strip cropping. The establishment of rotations that give a minimum of protective harbor to insects, spray programs, and other approved insect-control measures reduce this problem. Crop damage from chinch bugs can often be avoided by growing corn with meadow or small grain with meadow in alternating strips in separate fields, thus eliminating the bordering corn and small grains.

Grass strips and large meadow areas can be grazed by using portable electric fencing. Grazing may also be facilitated by establishing interfield rotations that

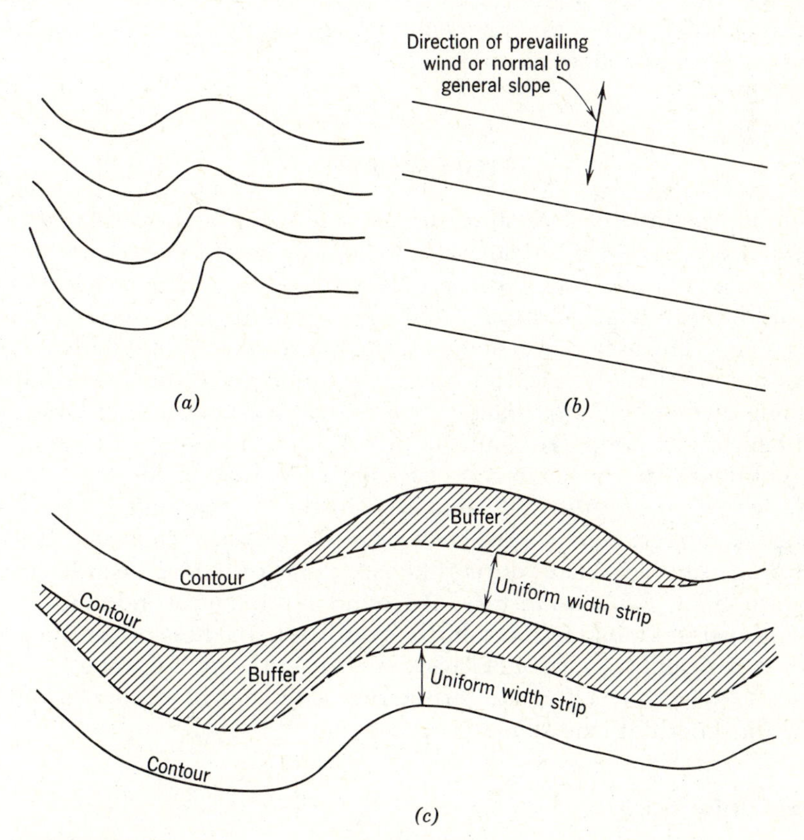

Figure 8.4. Types of strip cropping. *(a)* Contour. *(b)* Field. *(c)* (bottom) Buffer.

provide that each year one field will be entirely in meadow owing to the first and second years' meadows being adjoining.

8.3 Strip Cropping Layout

Strips may be established by any of the methods for laying out contour guide lines described in Fig. 8.2. Unlike contouring, adjacent strips are planted with different crops.

Strip widths vary with local conditions. When strip cropping is used with terraces, the width usually conforms to the terrace interval. In dry farming or other critical wind-erosion areas, strip widths vary according to soil-drifting conditions. Since strips should be a width that is convenient to farm, they should approach some multiple of implement width. For an eight-row planter with 30-in. rows, the strip width could be any multiple of 20 ft [(30/12) × 8]. For 42-in. rows the width should be any multiple of 28 ft. In the East and South the minimum strip width is usually considered to be 56 ft, in the Middle West the widths range from 50 to 130 ft, and in the Southeast the widths vary from 42 to 150 ft for slopes from 3 to 18 percent. These spacings should be selected to keep soil losses within allowable limits.

TERRACING

Terracing is a method of erosion control accomplished by constructing broad channels or benches across the slope. It has been estimated that over 90 million acres of cropland in the United States could be more effectively protected from runoff and erosion damage through the use of well-designed and maintained terrace systems. The use of diversions, a form of terrace, to protect bottomlands, buildings, and special use areas from damaging runoff and erosion from hillsides is becoming increasingly important. The first terraces consisted of large steps or level benches as compared with the broadbase terraces now common. For several thousand years, bench terraces have been widely adopted over the world, particularly in Europe, Australia, and Asia. In the United States, ditches that functioned as terraces were constructed across the slopes of cultivated fields by farmers in the southern states during the latter part of the eighteenth century.

As shown in Fig. 8.5, most terraces are found in the central and south central states with a considerable number in the Southeast. In those areas where soil moisture is deficient, terraces hold back runoff and conserve water. In more humid areas the primary function of terraces is to control erosion, mainly by reducing the length of land slopes.

8.4 Types of Terraces

The two major types of terraces are the bench terrace, which reduces land slope, and the broadbase terrace, which removes or retains water on sloping land (see

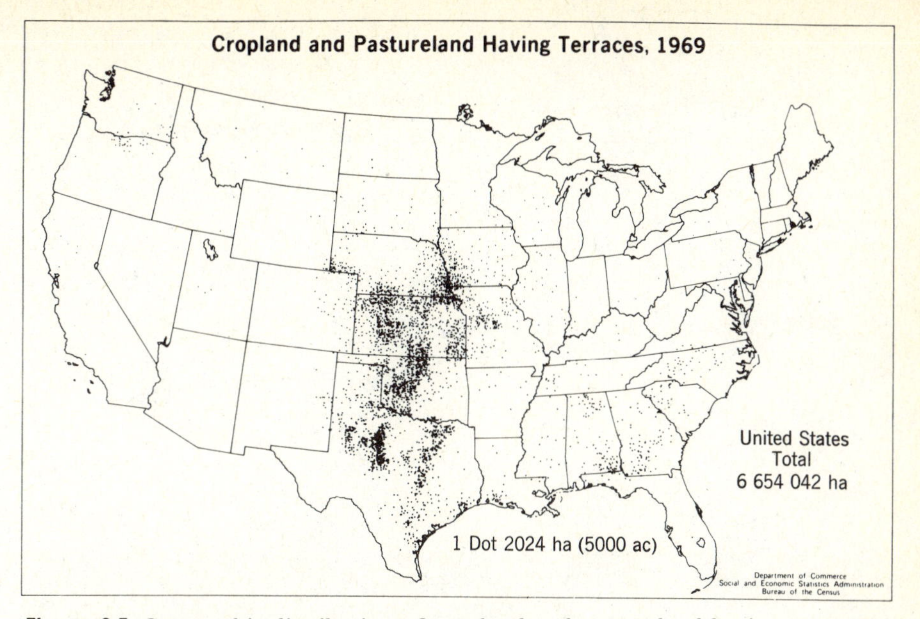

Figure 8.5. Geographic distribution of cropland and pastureland having terraces in the United States. (*Courtesy U.S. Bureau of the Census.*)

Fig. 8.6). The original bench terraces were adapted to slopes of 25 to 30 percent, were costly to construct, and were not always well adapted to modern cultivation equipment. The modern conservation bench terrace is adapted to slopes up to 6 to 8 percent and may be changed to meet cultivation machinery needs. The conservation bench terrace is designed for semiarid regions, where maximum moisture conservation is required. The runoff from the upper side of the terraced interval is held in the lower portion, where it spreads out and infiltrates into the soil.

The broadbase terrace shown in Fig. 8.6*a* may be constructed with no grade in the channel (level) or with a slope (graded) so that water will flow to one end or the other. The level terrace is designed primarily for moisture conservation and is suitable only where rainfall is limited and/or where the soil has a sufficient rate of infiltration so that the runoff will not overtop the terrace ridge. The channel may be made open at the ends, or (where maximum runoff is to be retained) may be closed at the ends. Because water is spread over a relatively small area with level terraces on slopes over 2 percent less water will be stored than is possible with the conservation bench terrace.

The graded terrace is designed primarily to minimize erosion by reducing the length of slope. The side slopes of the channel and the ridge should be kept as flat as possible to facilitate modern farm-machinery operations. By varying the percent of slope in the channel of graded terraces, the horizontal spacing may be kept constant. Such systems are called parallel terraces. Point rows between terraces are thus eliminated.

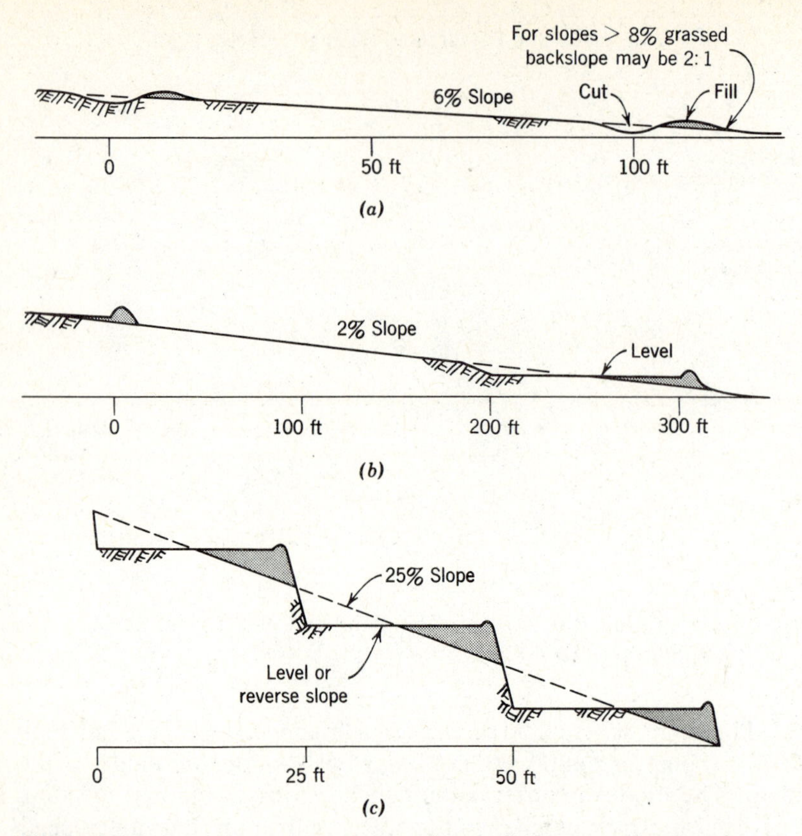

Figure 8.6. Types of terraces. (*a*) Broadbase. (*b*) Conservation bench. (*c*) Bench.

8.5 Terrace Spacing

The spacing for broadbase terraces is usually expressed as the vertical distance (difference in elevation) between consecutive terraces, primarily because a level instrument is required for layout and location. Graded terrace spacing is computed from the empirical equation,

$$\text{V. I.} = XS + Y \tag{8.1}$$

where $V.\,I.$ = vertical interval between corresponding points on consecutive terraces or from the top of the slope to the bottom of the first terrace in feet

S = land slope above the terrace in percent

X = constant for geographical location as given in Fig. 8.7

Y = constant for soil erodibility, cropping systems, and crop management as follows:

$Y = 1$ for low intake rates and little cover

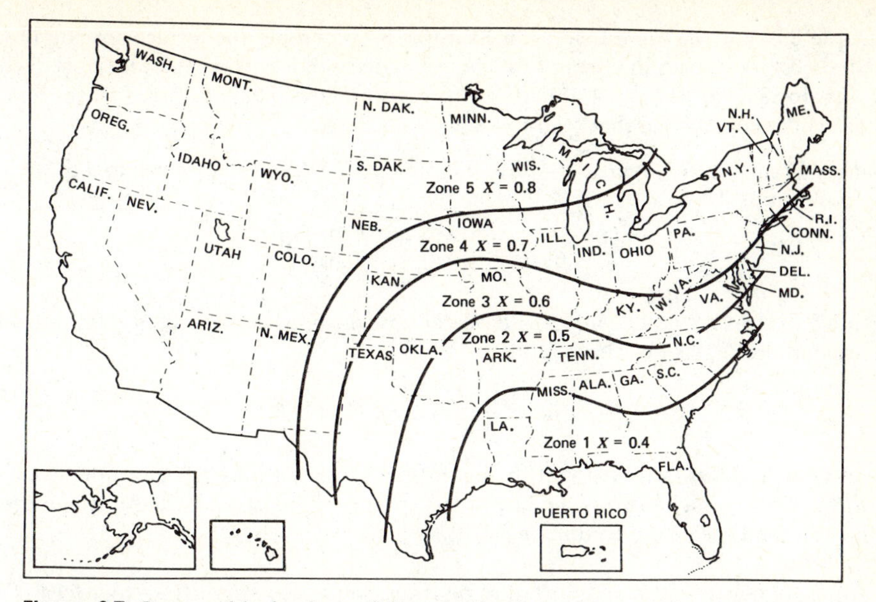

Figure 8.7. Geographical values of X and Y in the terrace spacing equation. (*Modified from U.S. Soil Conservation Service, 1979.*)

$$Y = 2.5 \text{ for average intake rates and medium cover}$$
$$Y = 4 \text{ for high intake rates and good cover}$$

Spacings thus computed may be varied as much as 25 percent to allow for soil, climatic, and tillage conditions. Terraces are seldom recommended on slopes over 20 percent, and in many regions slopes from 10 to 12 percent are considered the maximum.

Where soil loss data are available, spacings should be based on slope lengths using contouring and the appropriate cropping-management factor which will result in soil losses within the permissible loss as outlined in Chapter 6. Estimation of terracing spacing is illustrated in the following examples.

Example 8.1. Compute the terrace spacing from Eq. 8.1 for a field near Memphis, Tennessee, with an average slope above the proposed terrace of 8 percent and soil with a low intake rate and medium cover.

Solution. From Fig. 8.7 read for Zone 2, $X = 0.5$ and interpolate for soil erodibility, and so forth, assuming $Y = 2$. Substituting in Eq. 8.1,

$$\text{V. I.} = 0.5 \times 8 + 2 = 6.0 \text{ ft} \quad (1.8 \text{ m})$$

Substituting in Eq. 8.2, the slope distance between the terraces is

$$\text{H. I.} = 6.0 \times \frac{100}{8} = 75 \text{ ft} \quad (22.9 \text{ m})$$

Example 8.2. For the same field as in Example 8.1, compute the terrace spacing based on the USLE described in Chapter 6. The estimated soil loss from the unterraced field is 7.4 t/a for $K = 0.12$ t/a, $y = 400$ ft, $s = 8$ percent, $C = 0.2$, and $P_c = 0.5$ (for loss to terrace channel). Assume that tolerable soil loss is 3 t/a.

Solution. From Fig. 6.5, read $LS = 2.0$. The maximum LS to reduce soil loss to 3 t/a is (by direct ratio)

$$LS = 2.0 \times \frac{3.0}{7.4} = 0.81$$

From Fig. 6.5 for $LS = 0.81$ and $s = 8$ percent, read slope length of 68 ft (20.7 m). The corresponding V. I. is (by direct ratio)

$$\text{V. I.} = 68 \times \frac{8}{100} = 5.44 \text{ ft} \quad (1.66 \text{ m})$$

(Because the soil loss factors are difficult to estimate, this method may give unreasonable spacings.) For parallel terraces the horizontal spacing should be adjusted to some even-numbered multiple of the equipment width.

For level terraces the spacing may be the same as for broadbase terraces. The cross-sectional area of the conservation bench or level terrace must also be sufficient to hold the design runoff volume for the land area between the terraces.

Where terraces are to be made parallel to accommodate multiple-row farm equipment, the horizontal spacing should correspond to the appropriate vertical interval. The horizontal interval, which is nearly the same as the slope distance on flat slopes, can be computed from the equation,

$$H.I. = V.I. \times 100/s \tag{8.2}$$

where $H.I.$ = horizontal interval in feet

s = average slope in percent

8.6 Terrace Channel Grades and Lengths

The gradient in the channel should be sufficient to provide good drainage and to remove the runoff at nonerosive velocities. Minimum and maximum grades are given in Table 8.1. For graded terraces the zero minimum should not extend for long lengths. Level terraces have zero grade for their entire length. Maximum grades decrease as the terrace length increases, so as to prevent high velocities and erosion in the channel. Conversely, for short terraces or for the upper end of long terraces, grades can be greater because the runoff to be carried is lower. These higher grades allow for more flexibility to make adjacent terraces parallel.

The maximum length of the graded terrace should be about 1000 to 1800 ft. If suitable outlets are available at either end, a long terrace may be laid out so that it drains toward both ends. The maximum length applies only to that portion of the terrace that drains toward one of the outlets.

Table 8.1. Maximum and Minimum Terrace Grades

Terrace Length or Length from Upper End of Long Terraces	*Maximum Slope in Percent*
99 ft or less	2.0
100 to 199 ft	1.2
200 to 499 ft	0.5
500 to 1199 ft	0.35
1200 ft or more	0.3
	Minimum Slope in Percent
Soils with slow internal drainage	0.2
Soils with good internal drainage	0.0

Source: Beasley (1963).

There is no maximum length for level terraces, particularly where blocks or dams are placed in the channel every 400 or 500 ft. These dams prevent total loss of water from the entire terrace and reduce gully damage should a break occur. The ends of the level terrace may be left partially or completely open to prevent overtopping in case of excessive runoff.

8.7 Terrace Cross Section

The terrace cross section should have adequate channel capacity, good farmability, and be economical to construct. Cross sections for design purposes can be simplified as triangular shaped as shown in Fig. 8.8*a*. The height from the bottom of the channel to the top of the ridge is the depth of flow, d, plus the freeboard of 0.25 ft. The cross section will become curved and have the shape shown in Fig. 8.8*b* after 10 years of farming.

In designing the cross section, select the frontslope width (W_f) equal to the machinery width of row crop equipment. The depth of flow is determined from the runoff rate for a 10-year return period storm. For level or storage type terraces the depth will depend on the runoff volume. When the cutslope (W_c), the frontslope (W_f), and the backslope (W_b), are made equal, the cut or fill from geometry is

$$c = f = h + \frac{sW}{2} \tag{8.3}$$

where c = cut in feet
f = fill in feet
h = depth of channel = depth of flow + 0.25 ft
s = original land slope in feet per foot
W = width of cutslope, frontslope, or backslope in feet

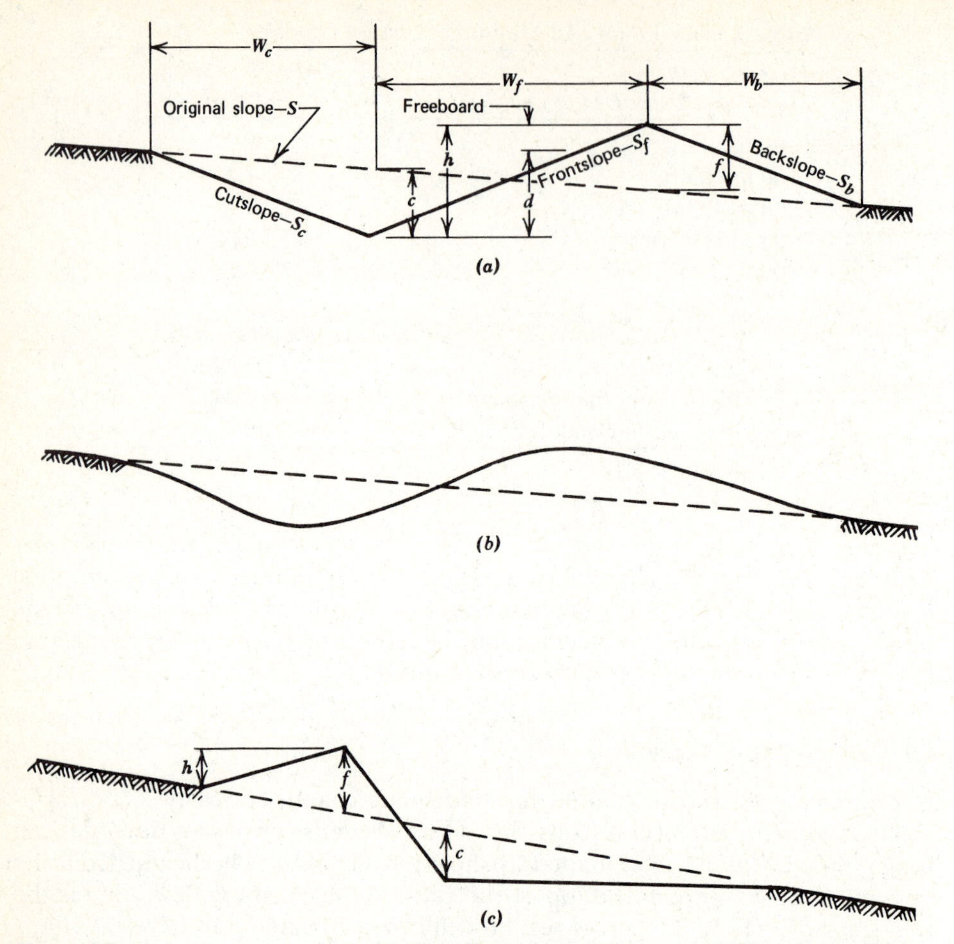

Figure 8.8. Terrace cross sections. (*a*) Broadbase design elements. (*b*) Broadbase after 10 years of farming. (*c*) Grassed backslope.

Table 8.2 gives the solution of Eq. 8.3 for a slope width of 14 ft (four 42-in. row widths) and a freeboard of 0.25 ft. For a balanced cross section, Table 8.2 gives either cut or fill from which the side slopes may be computed. The volume of cut or fill is needed where earthwork is necessary to make terraces parallel.

Example 8.3. Determine the depth of cut and earth volume for 100 ft of terrace length for a flow depth of 0.8 ft and a terrace slope width to accommodate four-row equipment (42-in. row width). For a land slope of 6 percent, compute the frontslope (horizontal to vertical) of the terrace.

Solution. From Table 8.2 or Eq. 8.3 for $h = 0.8 + 0.25 = 1.05$, $s = 6$ percent, and $W = 4 \times (42/12) = 14$ ft,

$$c = \frac{1.05 + 0.06 \times 14}{2} = 0.9 \text{ ft} \quad (0.27 \text{ m})$$

Table 8.2. Depth of Cut in Terrace Channel in Feet[a]

Land Slope in Percent	*d = Depth of Flow in Feet*							
	0.7	*0.8*	*0.9*	*1.0*	*1.1*	*1.2*	*1.3*	*1.4*
1	0.5	0.6	0.6	0.7	0.7	0.8	0.8	0.9 ft
2	0.6	0.7	0.7	0.8	0.8	0.9	0.9	1.0
3	0.7	0.7	0.8	0.8	0.9	0.9	1.0	1.0
4	0.8	0.8	0.9	0.9	1.0	1.0	1.1	1.1
6	0.9	0.9	1.0	1.0	1.1	1.1	1.2	1.2
8	1.0	1.1	1.1	1.2	1.2	1.3	1.3	1.4
10	1.2	1.2	1.3	1.3	1.4	1.4	1.5	1.5
12	1.3	1.4	1.4	1.5	1.5	1.6	1.6	1.7

[a]Assumes $W = 14$ ft and a freeboard of 0.25 ft.

From the triangular cross section (two triangular areas), the cross-sectional area of flow,

$$a = \left(W + \frac{W}{2}\right)\left(\frac{c}{2}\right) = \left(14 + \frac{14}{2}\right)\left(\frac{0.9}{2}\right)$$

$$= 9.45 \text{ ft}^2 \ (0.88\text{m}^2)$$

$$\text{Volume per 100 ft} = 9.45 \times 100$$

$$= 945 \text{ ft}^3 \quad (26.8 \text{ m}^3)$$

$$\text{Frontslope} = \frac{W}{(d + 0.25)}$$

$$= \frac{14}{1.05} = 13.3$$

or 13.3 : 1

The cross-sectional area for a level or storage type terrace should normally be larger than that for a graded terrace. The minimum cross-sectional area should not be less than about 12 sq ft, or it should be sufficient to prevent overtopping from a 10-year return period storm. A minimum capacity of 2 in. of runoff from the interterraced area to the next terrace above should be provided.

8.8 Parallel Terraces

Most modern terrace systems are installed parallel so as to eliminate point rows and to facilitate farming with large equipment. In fields with uniform slopes, terraces can be made parallel by land grading before terrace construction and/or by adjusting the grade in the terrace channel, especially at the upper end (see Table 8.1). For fields that have considerable variation in land slopes, terraces with underground pipe outlets have been developed. With pipe outlets, grassed waterways can be eliminated. Figure 8.9 shows a field with rather irregular slopes

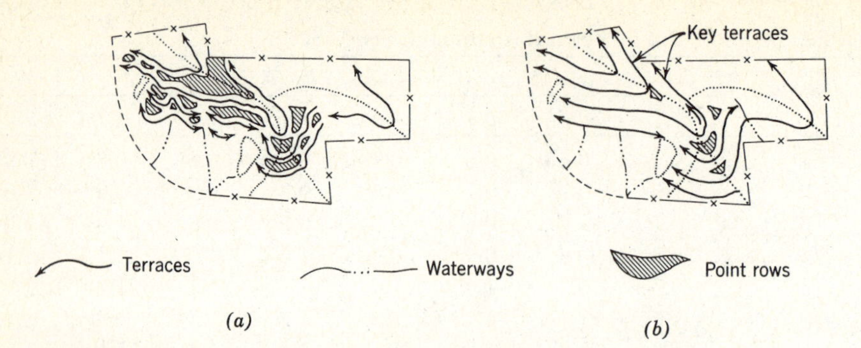

Figure 8.9. Terrace layout. (*a*) Graded terraces with many point rows. (*b*) Paralled terraces in the same field with few point rows. (*Redrawn from Jacobson, 1961.*)

in which the parallel terraces eliminated most of the point rows that were necessary with conventional terraces.

An underground pipe outlet terrace is shown in Fig. 8.10*a*. By straightening the terrace at natural channels with an earth fill, it is easier to make adjacent terraces parallel. The pipe outlet shown in Fig. 8.10*b* has an orifice plate to restrict the outflow. The size of the orifice is designed to control a 10-year, 24-hr storm, so that the water temporarily stored may be removed within 48 hr. Orifice flow is discussed in Chapter 10. The pipe outlet should not be smaller than 4 in. in diameter and should have a capacity equal to or greater than that for the orifice plate. Pipe capacity is discussed in Chapter 14. To provide sufficient storage in the terrace, the top ridge may be constructed at the same elevation for most of its length even though the bottom of the channel may have a slope to the pipe inlet. During the time water is detained in the terrace, most of the sediment will settle out near the inlet. This provides pollution control downstream and filling in of the channel so that the terrace will become more farmable with time. With a high ridge the terrace backslope may be made steep, provided it is seeded to grass as shown in Fig. 8.10*a*. The pipe inlet is usually a perforated or slotted vertical pipe riser. Flow will continue even though sediment has been deposited around it.

8.9 Planning the Terrace System

The terrace system should be coordinated with the overall water disposal plan for the farm, including the entire watershed and areas where terraces may be constructed at a later date.

Selection of Outlets. Graded terraces are normally outletted into natural or constructed grassed waterways, as discussed in Chapter 7. Since level terraces do not require outlets, their location and layout are greatly simplified. Some prefer constructed waterways along the edge of a field rather than in the natural

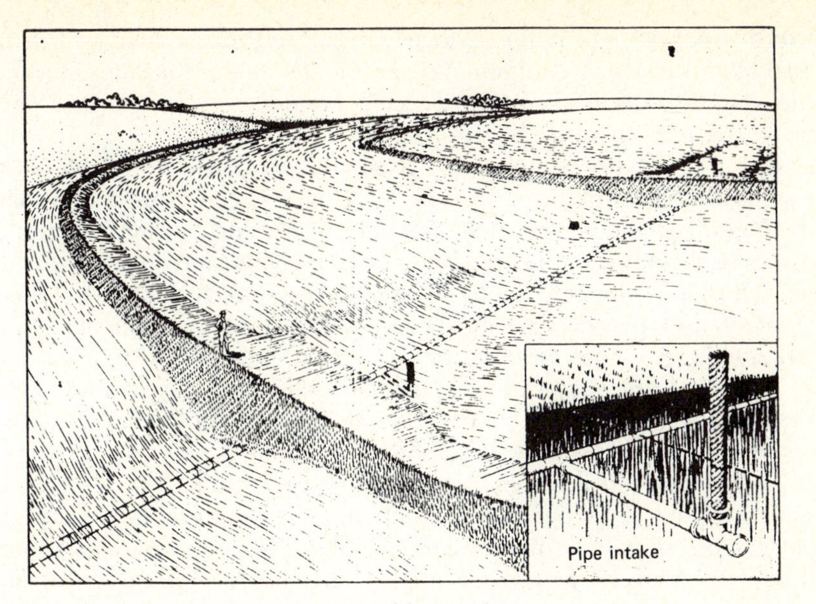

(a)

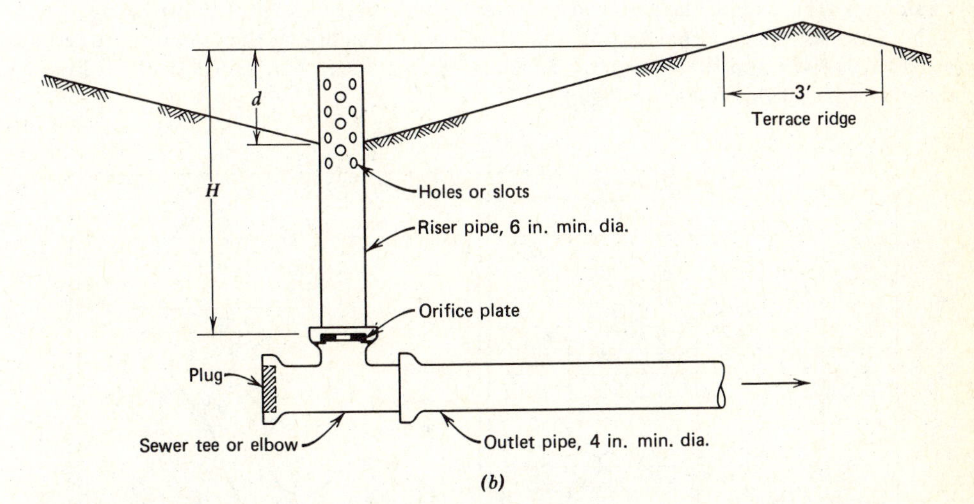

(b)

Figure 8.10. Underground pipe outlet terrace. (*a*) Plan showing grassed backslope. (*b*) Details of controlled-flow pipe intake. (*Redrawn from U.S. Soil Conservation Service, 1979.*)

channel. A location at the edge of the field will provide a turn area for equipment and the elimination of the grassed waterway in the field. Stabilized natural draws, permanent grassland, road ditches, and pipe drains are other types of outlets. The proper disposal of water from terraces is necessary, since they concentrate the flow and increase the danger of gullies developing.

Terrace Location. Normally, terraces are located in the field with a tripod level and other surveying equipment. Example field notes for planning, staking, and checking terraces are shown in Figs. 8.11 and 8.12. Topographic maps are especially useful for parallel terrace location.

The staking procedure for graded terraces that drain into an established grass waterway is illustrated in Fig. 8.13. The difference in elevation of the stake line and the bottom of the constructed channel will depend on the land slope, the shape of cross section, and the location of the staked line in relation to the terrace channel. Normally, the center line of the ridge is staked as shown. On slopes more than about 6 percent, the staked line may be lower than the bottom of the channel. In Fig. 8.13 the vertical scale has been exaggerated to show y more clearly. As shown, the rod reading at $0 + 50$ assuming $y = 0.5$ and 0.2 ft slope in 50 ft is $(6.8 - 0.5 - 0.2) = 6.1$ ft. Note that the remaining rod readings decrease by 0.2 ft for each 50-ft station to establish the desired grade of 0.4 percent. After the terrace lines are staked, some realignment is necessary to eliminate sharp curves for greater convenience in farming. Whenever practical, graded terraces should be made parallel, in which case the above procedure would apply only to the *key* terrace, or to a single terrace on the same slope.

Example 8.4. Determine the level rod readings for staking terrace no. 1 shown in Fig. 8.11 with a grade of 0.3 percent. Assume that the first rod reading at the outlet in the grass waterway is 8.7 ft and the bottom of the constructed terrace channel will be 0.5 ft lower than the staked ridge line (y in Fig. 8.13).

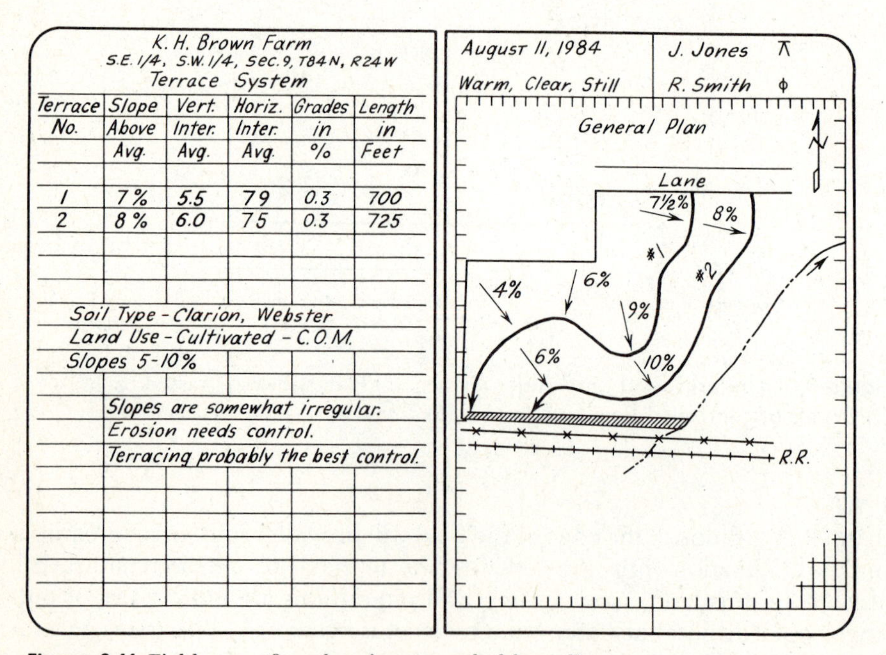

K. H. Brown Farm
S.E. 1/4, S.W. 1/4, Sec. 9, T84N, R24W
Terrace System

Terrace No.	Slope Above Avg.	Vert. Inter. Avg.	Horiz. Inter. Avg.	Grades in %	Length in Feet
1	7%	5.5	79	0.3	700
2	8%	6.0	75	0.3	725

Soil Type - Clarion, Webster
Land Use - Cultivated - C.O.M.
Slopes 5-10%
Slopes are somewhat irregular.
Erosion needs control.
Terracing probably the best control.

August 11, 1984 | J. Jones
Warm, Clear, Still | R. Smith

Figure 8.11. Field notes for planning a graded broadbase terrace system.

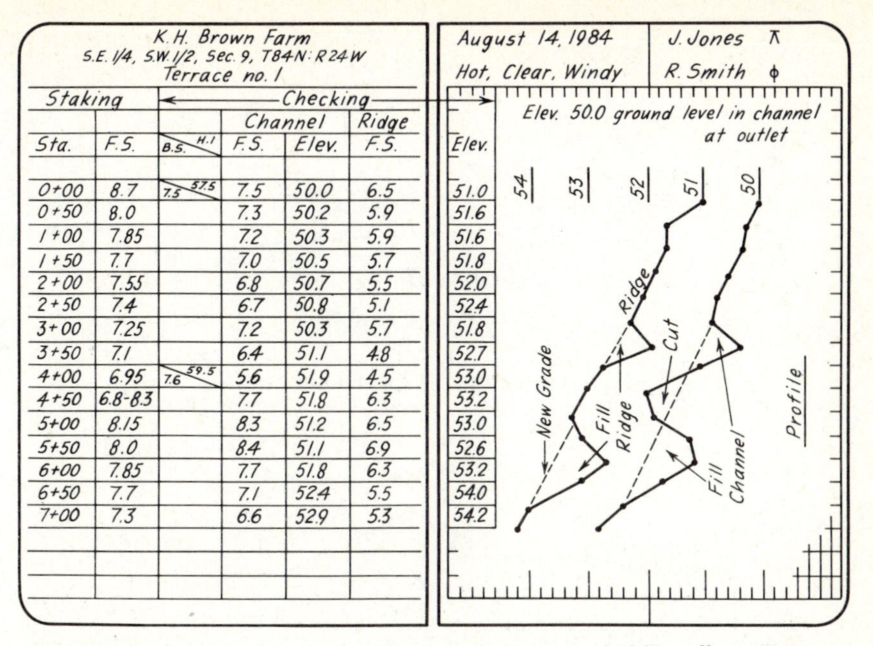

K. H. Brown Farm
S.E. 1/4, S.W. 1/2, Sec. 9, T84N: R24W
Terrace no. 1

August 14, 1984 | J. Jones
Hot, Clear, Windy | R. Smith

Staking		Checking				
			Channel		Ridge	
Sta.	F.S.	B.S. / H.I.	F.S.	Elev.	F.S.	Elev.
0+00	8.7	7.5 / 57.5	7.5	50.0	6.5	51.0
0+50	8.0		7.3	50.2	5.9	51.6
1+00	7.85		7.2	50.3	5.9	51.6
1+50	7.7		7.0	50.5	5.7	51.8
2+00	7.55		6.8	50.7	5.5	52.0
2+50	7.4		6.7	50.8	5.1	52.4
3+00	7.25		7.2	50.3	5.7	51.8
3+50	7.1		6.4	51.1	4.8	52.7
4+00	6.95	7.6 / 59.5	5.6	51.9	4.5	53.0
4+50	6.8-8.3		7.7	51.8	6.3	53.2
5+00	8.15		8.3	51.2	6.5	53.0
5+50	8.0		8.4	51.1	6.9	52.6
6+00	7.85		7.7	51.8	6.3	53.2
6+50	7.7		7.1	52.4	5.5	54.0
7+00	7.3		6.6	52.9	5.3	54.2

Figure 8.12. Field notes for staking and checking a graded broadbase terrace.

Solution. The desired rod readings are shown in column 2 in Fig. 8.12. The grade change in 50 ft is $0.003 \times 50 = 0.15$ ft. The reading at 0 + 50 is $(8.7 - 0.5 - 0.15 = 8.05)$ rounded to 8.0 ft. The reading at each successive station is then decreased by 0.15 ft. Note that 4 + 50 is a turning point: the reading 8.3 being from the new setup. The last station 7 + 00 is raised 0.4 ft to assure that water will not drain out of the upper end.

Unless there are obvious reasons for doing otherwise, the top terrace is laid out first, starting from the outlet end. It is important that the top terrace be located in the proper place, so that it will not overtop and cause failure of other terraces below.

Some general rules for the location of the top terrace are: (1) the drainage area above the top terrace should not ordinarily exceed three acres, (2) if the top of the hill comes to a point, the interval may be increased to one and one-half times the regular vertical interval, (3) on long ridges, where the terrace approximately parallels the ridge, the regular vertical interval should be used, and (4) if short abrupt changes in slope occur, the terrace should be placed just above the break.

Obstructions or topographic features below the top terrace, such as a boulder or tree, may necessitate locating a *key* terrace at that point first. Terraces above the *key* terrace, which are located by determining the vertical interval as before, may require an adjustment in spacing in order to place the top terrace at the proper location. Terraces below the key terrace are located by using the normal vertical interval.

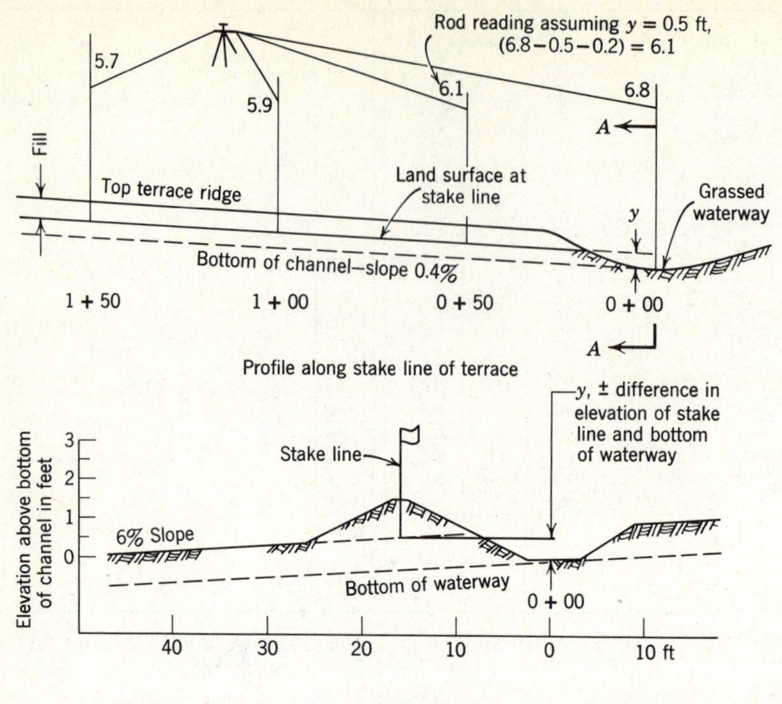

Figure 8.13. Staking procedure from the outlet of a graded terrace.

Parallel terraces may be located by adjusting the stakes set for uniform graded terraces, by planning on a topographic map, or by other procedures. A realignment of stakes as shown in Fig. 8.14 should be limited to not more than 1 ft below the bottom of the normal channel or a ridge height not to exceed 1.5 ft. For parallel pipe outlet terraces these heights may be increased. For the terrace shown in Fig. 8.14, a pipe outlet at point *B* would be a possible outlet rather than the grassed waterway.

8.10 Terrace Construction

Most terraces are built by contractors with motor graders, pans, and bulldozers. Procedures and techniques have been developed that are particularly suited to their equipment and local soil conditions. Moldboard and disk plows are suitable for small projects. Terraces are constructed by moving the soil laterally (only for uniform cross section terraces) or by moving soil both laterally and longitudinally (when cut and fill are required, as with parallel terraces). With a lateral movement of the soil, the ridge-fill material may be obtained from the uphill, downhill, or both sides of the ridge. Terrraces requiring longitudinal movement of soil are the most difficult to build as well as to lay out.

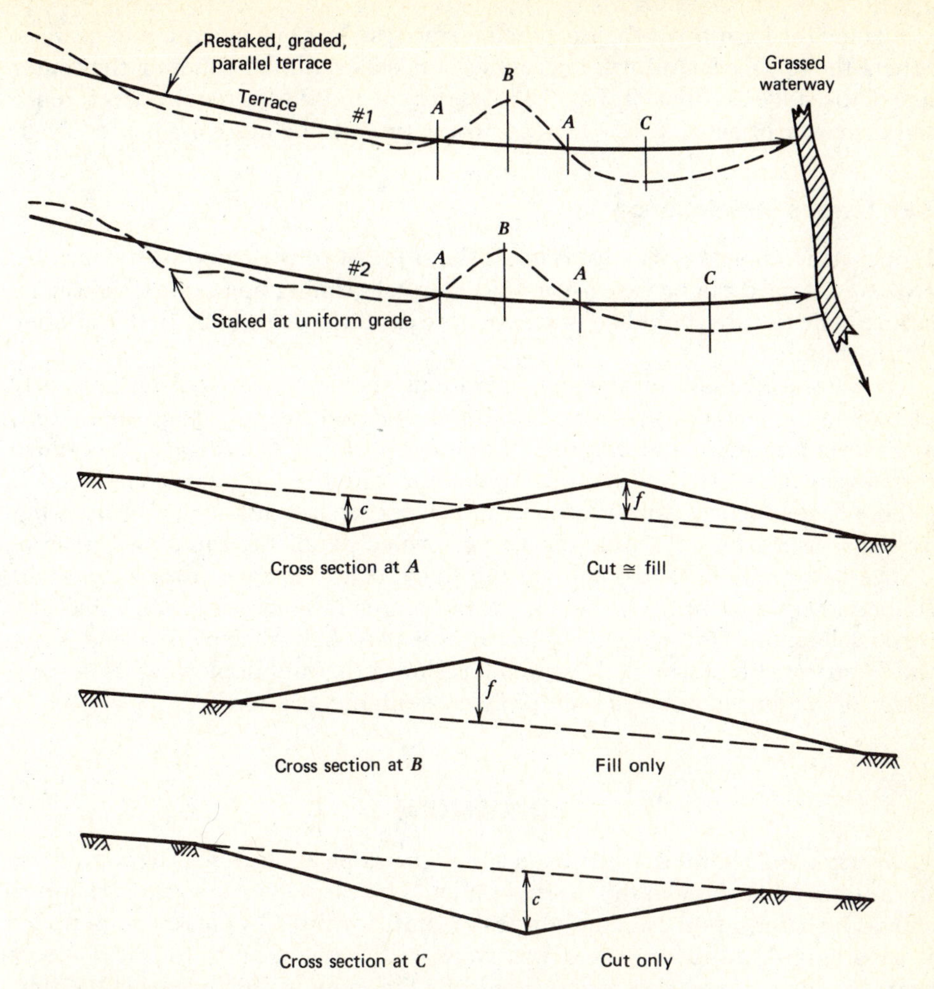

Figure 8.14. Layout of parallel terraces with varying cuts and fills.

The rate of construction decreases as the land slope increases. The terrace ridge must be constructed with a 10 to 25 percent excess-fill height to allow for settlement. Factors that affect the amount of settlement include the soil and moisture conditions, the type of construction equipment, the construction procedure, and the amount of vegetation and crop residue. Crop residue should be removed before construction, as it will make the ridge less compact and more subject to washout and settlement. Land grading and smoothing between the terraces after construction will reduce the accumulation of water in low places along the rows and local sedimentation in the terrace channel.

Where cuts are deep and unfavorable subsoil is exposed, topsoil should be saved and respread over the subsoil area. Such practice is more important where the topsoil is shallow.

Staking the location of the proposed terrace can be done in a number of ways. Where the terrace is uniform in cross section, stakes can be set along the center line of the ridge as shown in Fig. 8.13. Where cut and fill are required, reference stakes may be offset from the terrace, where they can be more easily preserved.

8.11 Terrace Maintenance

Proper maintenance is as important as the original construction of the terrace. However, it need not be expensive, since normal farming operations will usually suffice. The terrace should be watched more carefully during the first year after construction.

In a terraced field all farming operations should be carried out as nearly parallel to the terrace as possible. The most evident effect of tillage operations, after several years, is the increase in the base width of the terrace. Procedure for plowing out point rows is similar to that for contoured areas given in Fig. 8.1. Other tillage practices, such as stubble-mulch operations, disking, and harrowing as well as listing and planting, can be performed parallel to the plow furrows.

The best method of maintaining the shape of the terrace cross section and counteracting erosion from the interterraced area is by plowing with a two-way (reversible) plow. Unfortunately, the two-way plow is more expensive and is not widely accepted by farmers. Research has shown that uphill plowing is the only tillage operation that moves soil uphill to counteract erosion.

DIVERSIONS

A diversion is a channel constructed across the slope and given a slight gradient to cause water to flow to the desired outlet. The capacity of diversion channels should be based upon estimates of peak runoff for the 10-year return period if it is to empty into a vegetated waterway. If the diversion is to outlet into a permanent structure, the design should be the same as for the structure. The design procedures for diversions are the same as for vegetated waterways discussed in Chapter 7.

Most effective control of gullies is by complete elimination of runoff into the gully or the gullied area. This may often be accomplished by diverting runoff from above the gully and causing it to flow in a controlled manner to some suitably protected outlet. A diversion, referred to in some areas as a diversion terrace, is particularly effective for this purpose. They also effectively protect bottomland from hillside runoff and divert water from uncontrolled areas away from buildings, strip cropped fields, and other special purpose areas. A typical application is shown in Fig. 8.15.

Cross section design may vary to fit soil, land slope, and maintenance needs. Side slopes of 4 to 1 with bottom widths that permit mowing are frequently used. Since sediment deposition is often a problem in diversions, the designed velocity of water flow should be kept as high as the channel protection will

Figure 8.15. Diversion terrace to protect cultivated bottomland from damaging hillside runoff. *Source:* Phillips (1963).

permit. Permissible velocities for vegetated channels are given in Chapter 7. In the event that the channel cross section has been designed to permit cultivation, the velocity of flow must be based on bare soil conditions, that is, a maximum of 1.5 to 2.0 fps. Construction and maintenance of diversions is similar to that described for grass waterways.

REFERENCES

American Society of Agricultural Engineers (ASAE), Terrace and Related Slope Modification Committee (1983) "Design, Layout, Construction and Maintenance of Terrace Systems," *ASAE Yearbook*, R268.1.

Beasley, R. P. (1963) "A New Method of Terracing," *Missouri Agr. Expt. Sta. Bull. 699* (Revised July).

Hay, R. C. (1948) "How to Farm on the Contour," *Univ. Ill. Agr. Ext. Service, Cir. 575.*

Jacobson, P. (1961) "A Field Method for Staking Cut and Fill Terraces," *Agr. Eng.* **42**:684–687.

Phillips, R. L. (1963) Committee Report on Surface Drainage Systems, *Trans. Am. Soc. Agr. Eng.* **6**(4):313–317, 319.

Schwab, G. O., R. K. Frevert, T. W. Edminster, K. K. Barnes (1981) *Soil and Water Conservation Engineering*, 3rd ed., John Wiley & Sons, New York.

U.S. Soil Conservation Service (1979) *Engineering Field Manual for Conservation Practices* (Lithographed), Washington, D.C.

PROBLEMS

8.1 Compute the slope distance between two consecutive contour lines on a slope of 2 percent if the slope distance for the same two contours was 100 ft (30.5 m) on a slope of 6 percent.

8.2 Determine the strip width for strip cropping a field at your present location to accommodate four-row equipment if the land slope is 6 percent and the row width is 42 in. (1.1-m). Width should correspond to that for complete rounds of the equipment.

8.3 Determine the vertical interval and the horizontal interval for graded terraces at your present location if the prevailing land slope above the terrace is 6 percent and the soil has an average intake rate with medium cover.

8.4 If the maximum soil loss to the terrace channel is 4 t/a, determine the vertical interval for terraces using the soil loss equation given in Chapter 6. Location is northern Missouri, $K = 0.4$ t/a, slope is 6 percent, and $C = 0.15$. *Hint*: Use Fig. 6.5 for LS.

8.5 Solve Problem 8.4 for your present location with a soil factor K from Chapter 6. Assume silt loam soil with 0.5 percent organic matter.

8.6 Determine the difference in elevation between the upper end and the outlet of a 1000-ft (305-m) terrace if the maximum grade was selected for each section of the terrace length.

8.7 Compute the volume of water stored per foot of length in a standard cross section of a level terrace on a slope of 10 percent. If the terrace spacing is 100 ft (30.5 m), how many inches of runoff would it hold? Assume water depth is 1.0 ft (0.3 m) and channel side slopes are 8 : 1 (horizontal to vertical).

8.8 Compute the discharge rate in cfs from a terrace if the water flow depth is 1.0 ft (0.3 m) and the average velocity of flow is 2 fps (0.6 m/s). The cutslope ratio is 5.2 : 1 and the frontslope is 11.2 : 1.

8.9 A terrace 300 ft (90 m) long with a uniform grade of 0.4 percent is to outlet into an established waterway. If the first rod reading in the outlet is 6.7, what readings should be obtained in locating each of the remaining 50-ft (15-m) stakes? Assume that the bottom of the constructed channel will be 0.5 ft (0.15 m) lower than the stake line.

8.10 If the terrace in Problem 8.9 is to outlet into a waterway that has not been constructed, what should be the rod readings? Assume $y = 0$.

8.11 The top terrace is to be constructed on a hill that comes to a point. If the average slope is 5 percent, compute the vertical interval for low intake soil and little cover at your present location. Assume that the drainage area is less than 3 acres above the terrace.

8.12 If the rod reading at the outlet of the first terrace is 5.2 ft (1.6 m) and the vertical interval for the second terrace is 4.8 ft (1.5 m), what is the correct rod reading at the outlet of the second terrace if taken from the same instrument setup?

8.13 Determine the depth of cut, depth of channel, and cut volume per foot of terrace length for a depth of flow of 0.8 ft (0.24 m) where the land slope is 3 percent, the freeboard is 0.25 ft (8 cm), and the terrace slope widths (W) for the cross section are all equal to 14 ft (4.3 m) for equipment operation.

8.14 Compute the cutslope and frontslope ratios (horizontal to vertical) for the terrace in Problem 8.13.

CHAPTER 9

SOIL EROSION BY WIND AND CONTROL PRACTICES

In the arid and semiarid regions of the United States, large areas are affected by wind erosion. The Great Plains region, an area especially subject to soil movement by wind, represents about 20 percent of the total land area in the United States. Wind erosion not only removes soil but also damages crops, fences, buildings, and highways. The geographic distribution of the wind-erosion hazard in the western United States is shown in Fig. 9.1. In many areas these hazards only apply after the soil has been cultivated and left unprotected by vegetation. About 39 million acres of cropland and 19 million acres of rangeland in the 10 Great Plains states were eroding at 5 t/a or more in 1977 (SCS, 1978). An example of the effects of severe erosion on cropland is shown in Fig. 9.2.

Contrary to popular opinion, many humid regions are also damaged by wind. The areas most subject to damage are the sandy soils along streams, lakes, and coastal plains and the organic soils. Peats and mucks comprise about 25 million acres located in 34 states.

Very little attention was given to wind erosion until the severe dust storms of the 1930s. In the Great Plains these storms produced tremendous clouds of dust that were blown as far as the East coast. The removal of natural vegetation, long and continued drought, the deterioration of the soil structure, high winds, and the presence of bare soil in eroded areas all were contributing causes.

EROSION BY WIND

Wind erosion, unlike water erosion, cannot be divided into types, as wind erosion varies only by degree.

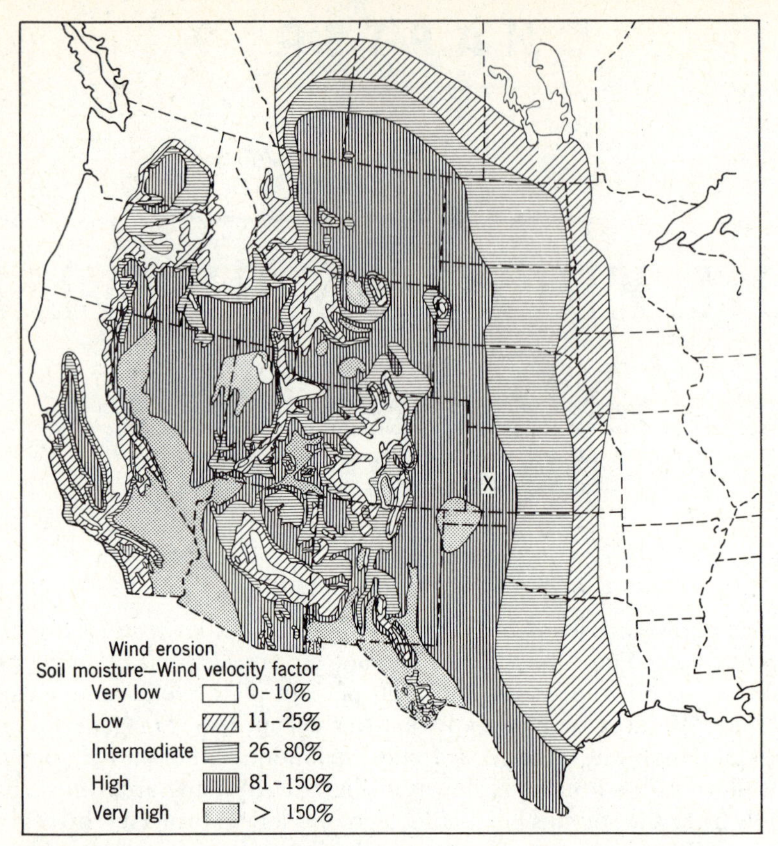

Figure 9.1. Annual wind erosion climatic factor as a percentage of that in the vicinity of Garden City, Kansas, marked by *X*. In the eastern United States and Canada the factor is less than 18 percent. *Source:* Chepil, Siddoway, and Armbrust (1962).

9.1 Types of Soil Movement

Soil movement caused by wind may occur by suspension of particles in the air, by skipping and bouncing along the surface (saltation), and by rolling or sliding in almost continuous contact with the surface (surface creep). These three types of movement usually occur simultaneously. Studies have shown that about 50 percent of the soil eroded by wind is moved in saltation, 20 percent in surface creep, and 10 percent in suspension. Particles moved by suspension are generally less than 0.1 mm in diameter, by saltation 0.1 to 0.5 mm, and by surface creep 0.5 to 1.0 mm.

9.2 Characteristics of Wind and Soil Movement

1. On smooth, bare soil, the wind velocity increases from zero at the surface to 6 mph at a height of 1 ft if the velocity is 10 mph at a height of 10 ft.

Figure 9.2. Severe wind erosion in New Mexico. (*Courtesy U.S. Soil Conservation Service.*)

The velocity profiles above a snow surface and above vegetation are shown in Fig. 9.3.

2. Winds are quite variable in velocity and direction, producing gusts and eddies that lift and transport soil particles.
3. Soil movement is initiated as a result of the turbulence and velocity of wind. The velocity required to start movement increases as the weight and size of the particles increase. For many soils, this velocity is about 18 mph at a height of 30 ft above the ground. The velocity required to sustain movement, once started, is less than that required to start blowing.
4. The major portion of soil movement occurs near the soil surface at heights not greater than 3 ft. Laboratory studies by Chepil (1945) have shown that the amount carried near the surface may vary from 62 to 97 percent of the total.
5. Saltating soil particles tend to rise almost vertically upon impact and then descend to the surface at an angle of 6 to 12 degrees from the horizontal. The distance moved is about 10 to 15 times the height of rise.
6. Soil particles that are suspended in the air may be moved hundreds or even thousands of miles before they are deposited. Large quantities of dust can be carried in the air, for example, dust loads from 53 to 1290 tons per cubic mile were measured in the Great Plains in the 1950s.

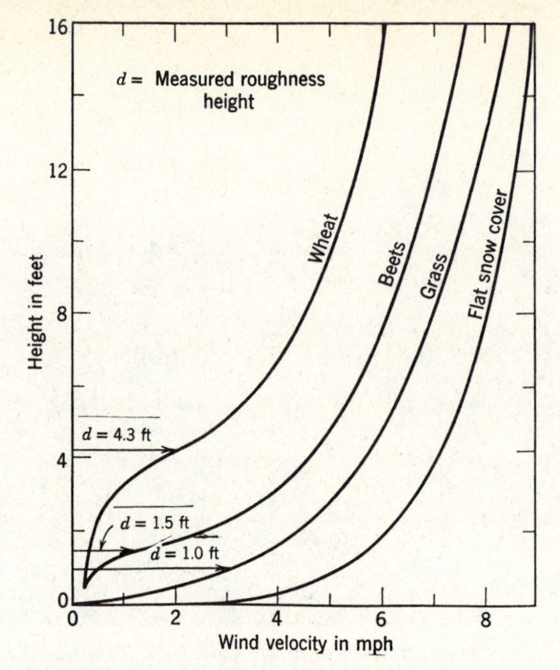

Figure 9.3. Wind-velocity distribution over different types of plant cover and soil surfaces. (*Redrawn from Geiger, 1957. Original from W. Paeschke.*)

7. Soil movement increases as the length of the field is increased, provided the field does not have barriers to obstruct wind movement. Fine particles drift and accumulate on the leeward side of the field or pile up in dunes. Eroding particles accumulate on the surface with distance downwind causing progressively greater concentrations of impacting particles that abrade the soil surface and vegetation, if present.

9.3 Estimating Soil Losses

The relationships between the amount of wind erosion and the various field and climatic factors (a wind erosion equation) have been developed from more than 40 years of wind-tunnel and field studies, primarily in the Great Plains area. The relationships among the many factors are complicated, but charts, tables, maps, and nomographs have been prepared by Skidmore and Woodruff (1968) for 212 sites located in 39 states from which average annual wind erosion estimates may be determined. They are too lengthy and involved to be presented here. A computer solution of the equation has been developed by Skidmore et al. (1970), and a slide rule calculator is available commercially. Soil loss estimates of average erosion for periods shorter than one year were developed by Bondy et al. (1980). The following factors are considered in the wind erosion equation for making soil loss estimates:

1. *Soil erodibility*. It is determined by the percentage of dry soil fractions greater than 0.84 mm in diameter, which is inversely related to soil loss. Land slope is also considered.
2. *Climate*. Relative climatic percentages have been developed, which are based on wind velocity and surface soil moisture (see Fig. 9.1). Annual and monthly values have been computed for the entire United States. For most areas April has the highest monthly values.
3. *Soil roughness*. It is the natural or artificial roughness of ridges or small undulations, but not that for clods or vegetation. Ridges about 2 to 5 in. in height are the most effective in reducing soil loss.
4. *Field length*. This factor is based on a weighted travel distance across the eroding surface and includes the magnitude, the prevailing wind direction, and the ratio of the wind erosion forces parallel to the prevailing wind direction to those perpendicular to the prevailing wind direction.
5. *Vegetation*. This factor depends on the kind of vegetation, its orientation and height, and the amount. The height, density, and surface area of the vegetation are effective in reducing wind velocity and thus erosion.

In addition to evaluating the average annual soil loss, the charts and tables previously mentioned are useful for predicting the effect of ridge roughness, vegetation, or strip width, singly or in combinations. In this way the effectiveness of various alternatives can be evaluated prior to use. An example of the relative effects of the last two factors is illustrated in Table 9.1 for Garden City, Kansas.

9.4 Nutrient and Organic Matter Losses

Because the wind removes the finer and lighter organic material more easily than the heavier particles, wind erosion may cause serious reduction in productivity. Therefore, the effect of wind on the loss of plant nutrients is likely to be many times greater during the first few years after cultivation than later on when the fines have been removed. Because plant nutrients and organic matter are normally transported from the eroded area, wind erosion may be serious even when soil loss is small. A comprehensive model has been developed by Williams et al. (1984) to predict the effects of erosion (both wind and water) on soil productivity.

There have been many reports concerning the extent and distance of soil movement by wind. Following a dust storm in 1937, sediment originating in the Texas–Oklahoma Panhandle area was collected on the snow in Iowa and compared to the soil in dunes near Dalhart, Texas. The dust, which was carried a distance of 500 miles, contained 10 times as much organic matter, 9 times as much nitrogen, 19 times as much phosphoric acid, and about $1\frac{1}{2}$ times as much potash as the dune soil. Studies in Oklahoma showed that each time a soil was

Table 9.1. Average Annual Soil Loss from Wind Erosion in Tons per Acre[a]

Unsheltered Distance (in feet)	*Flat Small Grain Residue Equivalent (Lbs/Ac)*					
	0	*500*	*1000*	*1500*	*2000*	*2500*
10000	86	61	30	9	3	0.3
8000	86	61	30	9	3	0.3
6000	86	61	30	9	3	0.3
4000	86	61	30	9	3	0.3
2000	83	58	28	8	3	0.3
1000	76	53	25	7	2	
800	74	52	24	7	2	
600	69	48	22	6	2	
400	62	42	19	5	1	
200	51	34	15	3	1	
100	40	26	10	2	0.4	
80	37	23	9	2		
60	31	16	7	1		
40	24	15	5	1		
20	16	9	3			

[a]Computed from the wind erosion equation for Garden City, Kansas; (C = 100), for sandy loam soil (I = 86), and for soil ridge roughness factor, K = 0.7.

Source: U.S. Soil Conservation Service, provided by L. Lyles.

shifted more plant nutrients were removed and finally the dunes consisted entirely of sand, regardless of the original texture.

9.5 Principles of Control

The two major types of wind erosion control consist of (1) those measures that reduce surface wind velocities, and (2) those that affect soil characteristics, such as the conservation of moisture and tillage. Many measures, such as permanent grass and contouring, provide both types of control. For example, vegetation not only retards surface winds but also improves soil structure.

In brief, the general practices to be followed if wind erosion is to be kept within reasonable limits are: (1) keep the soil covered with growing vegetation or crop residues as much of the time as possible, (2) reestablish permanent vegetation on unproductive cultivated soils, (3) till bare soil after rains to prevent blowing until vegetation provides protection, (4) limit cultivation to that necessary to obtain control, (5) avoid working the soil when dry, (6) select tillage implements that leave a rough surface, (7) cultivate the soil before blowing is expected, (8) utilize such emergency measures as may be necessary during the period of blowing, and (9) avoid the overpasturing of rangeland and cereal crops.

CONTROLLING SURFACE WIND VELOCITY

The three principal methods of reducing surface wind velocities are vegetative measures, tillage practices, and mechanical methods. Vegetation is the most effective method because roots and organic matter help hold the soil against the force of the wind as well as reduce the wind velocity. The effect of vegetation on wind velocity near the ground is shown in Fig. 9.3. The velocity increases very rapidly above the roughness height. This height is that above which the velocity varies as the logarithm of the height.

Woody plants, such as shrubs and trees, may be planted to reduce wind velocities over large areas. Wind erosion hazards are often most severe when fields are bare of growing crops. Cultural methods of control, though temporary, are effective under these conditions. When practical, cultural practices should be performed before blowing starts, for wind erosion is easier to prevent than to arrest. In general, long periods of drought will obliterate the effects of cultural treatments. Mechanical barriers, such as slat fences and matting, are most practical for small areas, such as sand dunes, or for the protection of high-value crops.

9.6 Cultivated Crops

In general, close-growing crops are more effective for erosion control than are intertilled crops. The effectiveness of crops is dependent upon the stage of growth, the density of cover, the row direction, the width of rows, the kind of crop, and the climatic conditions. Pasture or meadow tends to accumulate soil if there is a good growth of vegetation, the soil coming from neighboring cultivated fields and being deposited by sedimentation. Good management practices such as rotation grazing are important.

Intertilled crops such as corn, cotton, and vegetables offer some protection. The best practice is to seed the crop normal to prevailing winds. In the Great Plains region the rows are usually in an east-west direction. A good crop rotation that will maintain soil structure and conserve moisture should be followed. Crops adapted to soil and climatic conditions and providing as much protection against blowing as is practical are recommended. For instance, in the Great Plains region, cane and sudan grass are often grown, as they are quite resistant to drought and are effective in preventing wind erosion. Stubble mulch farming and cover crops between intertilled crops in more humid regions aid in controlling blowing until the plants become established. In some dry regions, emergency crops may be planted on summer fallow land before seasons of high intensity winds. In muck soils where vegetable crops are grown, "buffer strips," consisting of rows of small grain, are sometimes planted. The stabilization of sand dunes with vegetation may be accomplished by establishing grasses and then reforesting. Spray-on adhesives, such as asphalts, latexes, and resins, are being applied on sand dunes and in some vegetable-growing areas.

9.7 Shrubs and Trees

The U.S. Soil Conservation Service estimated that eight million acres (1970) were in need of protection by shelterbelts and windbreaks. Although a windbreak is defined as any type of barrier for protection from winds, it is more commonly associated with mechanical or vegetative barriers for buildings, gardens, orchards, and feed lots. A shelterbelt is a longer barrier than a windbreak, usually consisting of shrubs and trees, and is intended for the conservation of soil and moisture and for the protection of field crops. Over 200,000 miles of windbreaks and shelterbelts have been planted in the United States since 1850. Not only are windbreaks and shelterbelts valuable for wind erosion control, but they also save fuel, increase livestock gains, reduce evaporation, prevent firing of crops from hot winds, catch snow during the winter months, provide better fruiting in orchards, and make the spraying of fruit trees for insect control more effective as well as provide farm woodlots and wildlife refuges.

The relative wind velocity at a height of 1.3 ft above the soil surface near a windbreak is shown in Fig. 9.4. The data were obtained for a slat fence barrier 19 times longer than its height and having an average porosity of 50 percent, but more open in the lower than in the upper half. Similar results could be expected from a tree shelterbelt with an equivalent density. In Fig. 9.4 the wind

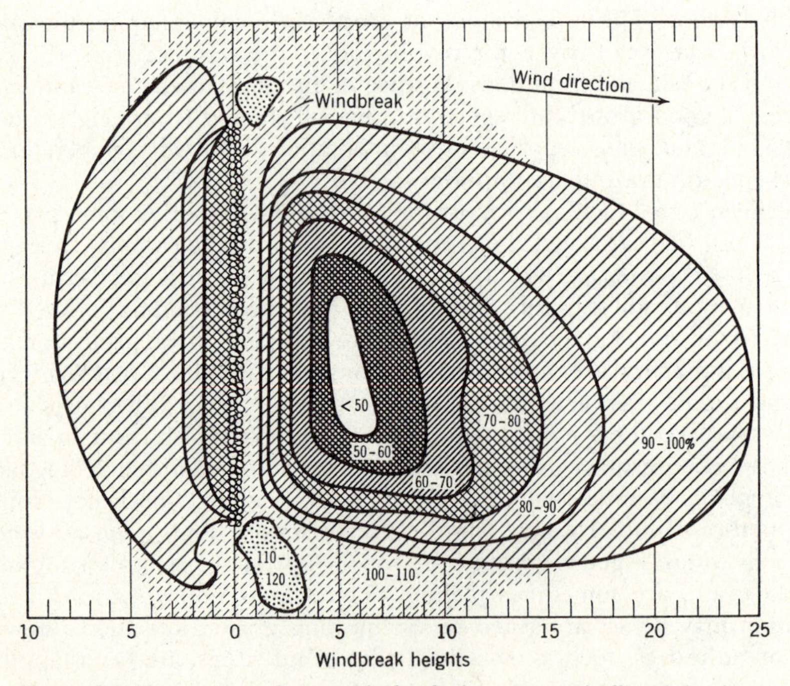

Figure 9.4. Percentage of open wind velocity near a windbreak having an average porosity of 50 percent. (*Redrawn from Bates, 1944.*)

velocity is affected for a distance of about eight times the windbreak height on the windward side and twenty-four times on the leeward side. Wind velocities are reduced on the windward side from five- to ten-tree heights from the protecting shelterbelt; on the leeward side the reduction varies from ten to twenty times its height.

Shelterbelts should be moderately dense from ground level to treetops if they are to be effective in filtering the wind and lifting it from the surface. Because the wind velocity at the ends of the belt as given in Fig. 9.4 is as much as 20 percent greater than velocities in the open, it is evident that long shelterbelts are more effective than short ones. An opening or break in an otherwise continuous belt results in a similar increase in velocity and will reduce the area protected. Roads through shelterbelts should therefore be avoided, and, when essential, they should cross wide belts at an angle or they should be curved. The direction of shelterbelts should be as nearly as possible at right angles to the prevailing direction of winds.

The distance of full protection from a windbreak may be computed from the equation,

$$W = (365\ h/v) \cos A \qquad (9.1)$$

where W = distance of full protection in feet
h = height of the barrier in feet or any units, the same as W
v = actual wind velocity at 50-ft height in mph, but less than 40 mph
A = the angle of deviation of prevailing wind direction from the perpendicular to the windbreak

The above equation was developed from wind-tunnel tests, and is applicable only for a smooth bare surface after erosion has been initiated but before wetting by rainfall and subsequent surface crusting. The width of strips may be estimated from Eq. 9.1 by using the crop height in the adjoining strip.

The types of trees and row arrangement for a shelterbelt are shown in Fig. 9.5. The number of rows will usually vary from one to ten. Single-row belts are preferred in many areas because fewer trees and less land are required and the shelterbelt is easier to cultivate and maintain. Studies in North Dakota indicated that shelterbelts with a density of much less than 50 percent were effective in trapping snow for distances up to 300 ft with a 15-ft single row of trees. Local recommendations should be followed for varieties, spacings, and other practices.

Shelterbelts are designed principally to control erosion, but crop production may be increased even though some land is occupied by the trees. Under most situations in the Great Plains, Stoeckeler (1965) reported that where shelterbelts occupy 1 to 5 percent of the gross land area, optimum crop gains are most likely to be obtained. Among the crops most responsive to shelterbelt protection are garden crops, flower bulbs, and citrus and fruit trees. Those with the least response are the drought-hardy small grains and corn.

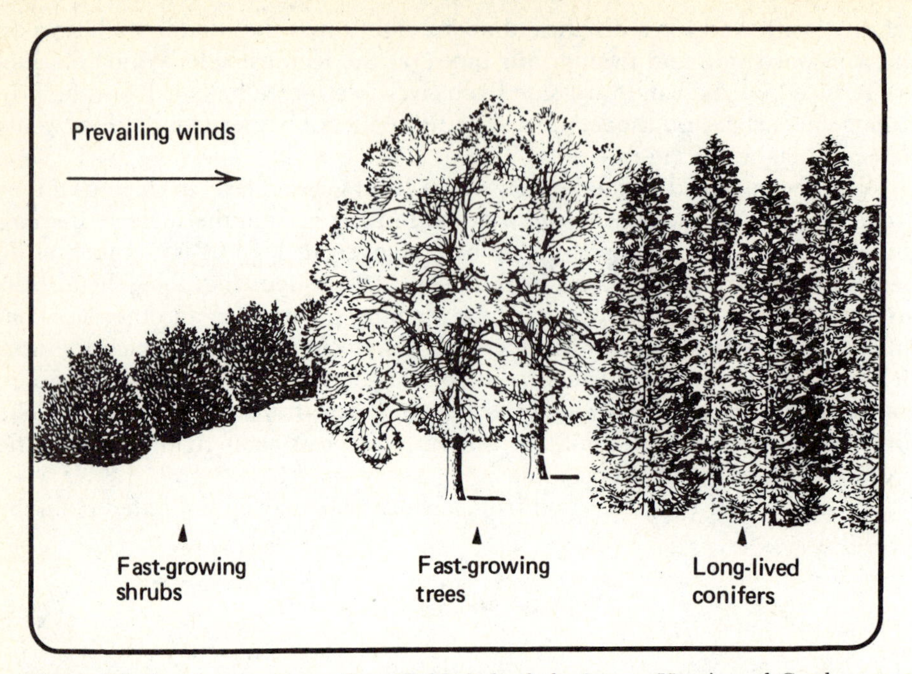

Figure 9.5. Row arrangement for a field shelterbelt. *Source:* Harris and Carder (1969).

9.8 Field and Contour Strip Cropping

Field and contour strip cropping consists of growing alternate strips of clean-cultivated and close-growing crops in the same field. Field strip cropping is laid out parallel to a field boundary or other guide line, whereas ordinary strip cropping operations are on the contour. In some of the plains states strips of fallow and grain crops are alternated. The chief advantages of strip cropping are (1) physical protection against blowing by vegetation trapping soil particles, (2) soil erosion limited for a distance equal to the width of strip, (3) greater conservation of moisture, particularly from snowfall, and (4) the possibility of earlier harvest. The chief disadvantages are (1) machine problems in farming narrow strips, and (2) greater number of edges to protect in case of insect infestation.

The strips should be of sufficient width to be convenient to farm, yet not so wide as to permit excessive erosion. The width of strips depends on the intensity of wind, row direction, crops, and erodibility of the soil. Average strip widths are shown in Table 9.2.

9.9 Primary and Secondary Tillage

The objectives of primary and secondary tillage for wind erosion control are to produce a rough, cloddy surface, to maintain surface residue, or to control weeds

Table 9.2. Average Width of Strips for Wind-Erosion Control[a]

	Width of Strips (ft) for Wind Angles from the Strip[b]		
Soil Texture	*90°*	*70°–110°*	*45°–135°*
Sand	20	18	14
Loamy sand	25	22	18
Granulated clay	80	75	54
Sandy loam	100	92	70
Silty clay	150	140	110
Loam	250	235	170
Silt loam	280	260	190
Clay loam	350	325	250

[a]From Chepil and Woodruff, 1963.

[b]For negligible surface roughness, average soil cloddiness, 1-ft high stubble, and soil movement of 1.2 tons per 100 ft.

or both. To obtain maximum roughness, bare land normally should be cultivated as soon after a rain as possible. Large clods as well as a high percentage of large aggregates are desirable.

Tillage may be quite effective as an emergency control measure. Soil blowing usually starts in a small area where the soil is less stable or is more exposed than in other parts of the field. Where the entire field starts to blow, the surface should be put in a rough and cloddy condition as soon as practicable. Such emergency tillage can be accomplished by making widely spaced trips through the field. After the field has been so tilled, the areas between the tilled strips may then be cultivated.

Crop residues exposed on the surface are an effective means of control, especially when combined with a rough soil surface. This practice may be called stubble mulch tillage, reduced tillage, or conservation tillage.

Crop residues act in two ways: they reduce wind velocity and trap eroding soil. Short stubble is less effective than long stubble. A mixture of straw and stubble on the surface provides good protection against erosion. The higher the wind velocity the greater the quantity of crop residue required.

9.10 Mechanical Methods

Artificial barriers, such as windbreaks, are of little importance for field crops, but they are frequently employed for the protection of farmsteads and small areas. Some of the mechanical methods of control include slat or brush fences, board walls, and vertical burlap or paper strips, as well as surface protection, such as brush matting, rock, and gravel. These barriers may be classed as semi-permeable or impermeable. Semipermeable windbreaks are usually more effective than impermeable structures because of diffusion and eddying effects on the leeward side of the barrier. A slat snow fence is a good example of this action. Slat fences, picket fences, and vertical burlap or paper strips are sometimes used for the protection of vegetable crops in organic soils. Brush matting, spray-on adhesives, debris, rock, and gravel may be suitable in stabilizing sand dune areas.

Terraces have some effect on wind erosion, mostly because they conserve moisture. In the Texas Panhandle, terraces lost less soil than the interterraced area and in some instances gained soil. Most of the soil that was lost from the interterraced area was collected in the terraces. Where vegetation was growing on the ridge and in the channel, terraces were even more effective.

CONTROLLING SOIL FACTORS

The principal soil factors influencing wind erosion control include the conservation of moisture to improve vegetative growth and conditioning of the surface soil to improve aggregation.

9.11 Conserving Soil Moisture

The conservation of moisture, particularly in arid and semiarid regions, is of utmost importance for wind erosion control as well as for crop production. The means of conserving moisture fall into three categories: increasing infiltration, reducing evaporation, and preventing unnecessary plant growth. In practice these can be accomplished by such measures as level terracing, contouring, conservation tillage, mulching, and selecting suitable crops.

The greater the amount of mulch on the surface the greater the quantity of moisture conserved. The effect of crop residue and tillage practices on the conservation of moisture is shown in Fig. 9.6. These data show the percentage of rainfall conserved from April to September. Where the straw was disked, or plowed in, some of the residue was left on the surface, and, where the soil was plowed and disked, no straw was present. Although there was no runoff for the basin-listed area, the quantity of moisture conserved was about half of that for the treatment with 2 tons of straw on the surface, probably owing to greater evaporation. For the same reason, disking and plowing were less effective than any of the mulch treatments.

The time of seedbed preparation in some regions has considerable effect on the conservation of moisture. This effect is shown in Fig. 9.7 for winter wheat at Hays, Kansas. Late plowing (September–October) resulted in a decrease in available moisture as compared with early plowing (July) or summer fallow. Wheat yields for late and early plowing were 37 and 65 percent respectively of that for summer fallow.

Organic soils do not blow appreciably if the soil is moist. Where subsoil is wet, rolling the soil with a heavy roller will increase capillary movement and moisten

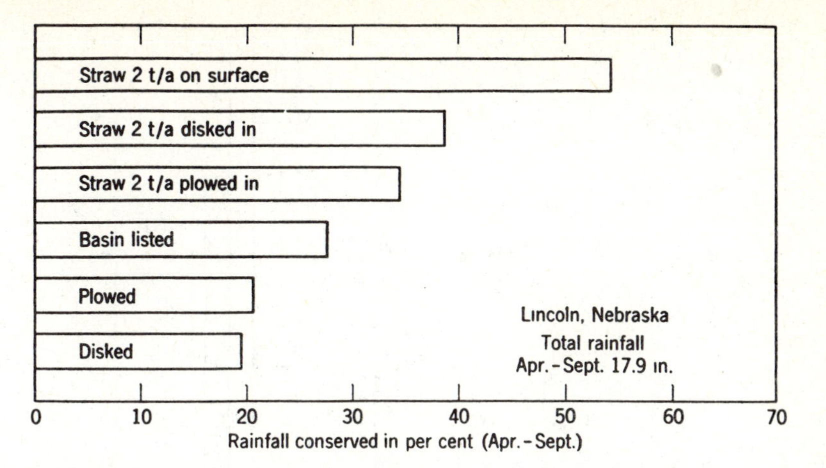

Figure 9.6. Effect of straw and tillage practices on storage of soil moisture. (*From data by Duley and Russel, 1939.*)

the surface layer. Controlled drainage where the water level is maintained at a specified depth may also reduce blowing. In irrigated areas the surface is often wet down by overhead sprinkling at critical periods.

Terracing is a good moisture conservation practice where level terraces are suitable and where the slopes are gentle enough so that the water can be spread over a relatively large area. Such practices as contouring, strip cropping, and mulching are effective in increasing the total infiltration and thereby the total soil moisture available to crops. Field strip cropping generally does not conserve as much moisture as contour strip cropping, but it is somewhat more effective in reducing surface wind velocities.

9.12 Conditioning Topsoil

Because wind erosion is influenced to a large extent by the size and apparent density of aggregates, an effective method of conditioning the soil against wind erosion is by adopting practices that produce nonerodible aggregates or large clods.

Tillage may or may not be beneficial to soil structure, depending on the moisture content of the soil, type of tillage, and number of operations. For optimum resistance to wind erosion in semiarid regions it is desirable to perform primary tillage as soon as practical after a rain. The number of operations should be a minimum because tillage has a tendency to reduce soil aggregate size and to pulverize the soil. Secondary tillage for seedbed preparation should be delayed as long as practical.

Good crop and soil management practices are necessary to maintain good soil structure. In the central and southern Great Plains a typical recommended rotation consists of an intertilled crop (grain sorghum), fallow, and winter wheat.

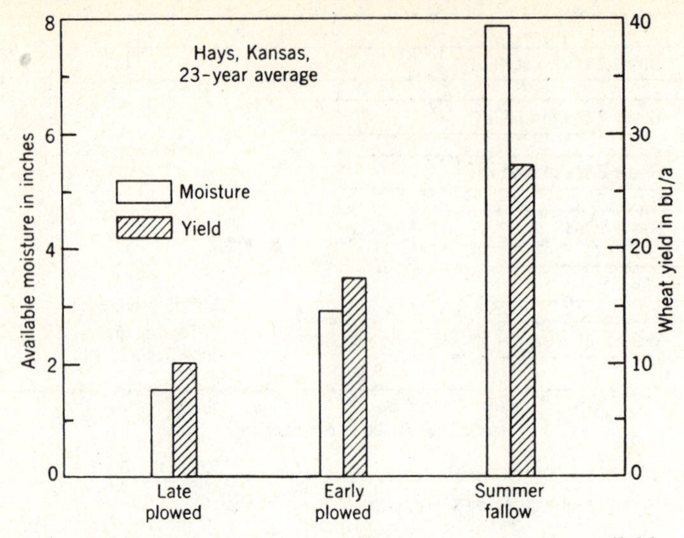

Figure 9.7. Effect of time of seedbed preparation on available moisture at seeding time and on wheat yields. (*From data by Hallsted and Mathews, 1936.*)

The organic matter in the soil should be held to a high level and lime and fertilizers applied where necessary.

Soil structure is affected by the climatic influences of the season, rainfall, and temperature. Freezing and thawing generally have a beneficial effect in improving soil structure where sufficient moisture is present. However, in dry regions the soil may be more susceptible to erosion because of the rapid breakdown of the clods into smaller aggregates.

REFERENCES

Bates, C. G. (1944) "The Windbreak as a Farm Asset," *U.S. Dept. Agr. Farmers' Bull. 1405* (Revised).

Bondy, E., L. Lyles, and W. A. Hayes (1980) "Computing Soil Erosion by Periods Using Wind Energy Distribution," *J. Soil and Water Conservation* **35**(4):173–176.

Chepil, W. S. (1945) "Dynamics of Wind Erosion: I. Nature of Movement of Soil by Wind," *Soil Sci.* **60**:305–320.

Chepil, W. S. (1960) "How to Determine Required Width of Field Strips to Control Wind Erosion," *J. Soil and Water Conservation* **15**(2):72–75.

Chepil, W. S., F. H. Siddoway, and D. V. Armburst (1962) "Climatic Factor for Estimating Wind Erodibility of Farm Fields," *J. Soil and Water Conservation* **17**(4):162–165.

Chepil, W. S., and N. P. Woodruff (1963) "The Physics of Wind Erosion and Its Control," *Advances in Agronomy* **15**:211–302. Academic Press, New York.

Duley, F. L., and Russel, J. C. (1939) "The Use of Crop Residues for Soil and Moisture Conservation," *Am Soc. Agron. J.* **31**:703–709.

Ferber, A. E. (1969) "Windbreaks for Conservation," U.S. Soil Conservation Service, U.S. Department of Agriculture, *Agr. Information Bull. 339.*

Geiger, R. (1957) *The Climate Near the Ground*, Translation of the 2nd German edition by M. N. Steward and others, Harvard Univ. Press, Cambridge.

George, E. J. (1966) "Shelterbelts for the Northern Great Plains," *U.S. Dept. Agr. Farmers' Bull. 2109* (Revised).

Hallsted, A. L., and O. R. Mathews (1936) "Soil Moisture and Winter Wheat with Suggestions on Abandonment," *Kans. Agr. Exp. Sta. Bull. 273.*

Harris, R. E., and A. C. Carder (1969) "Shelterbelts for the Peace River Region," *Canada Dept. Agr. Publ. 1384.*

Read, R. A. (1964) *Tree Windbreaks for the Central Great Plains*, U.S. Dept. Agr. Handb. 250, U.S. Government Printing Office, Washington, D.C.

Skidmore, E. L., and N. P. Woodruff (1968) "Wind Erosion Forces in the United States and Their Use in Predicting Soil Loss," U.S. Department of Agriculture, Agricultural Research Service, Agr. Handb. 346, April, U.S. Government Printing Office, Washington, D.C.

Skidmore, E. L., P. S. Fisher, and N. P. Woodruff (1970) "Wind Erosion Equation: Computer Solution and Application," *Soil Sci. Soc. Am. Proc. 34*: 931–935.

Stoeckeler, J. H. (1965) "The Design of Shelterbelts in Relation to Crop Yield Improvement," *World Crops*, pp. 3–8. Grampian Press Ltd., England.

U.S. Soil Conservation Service (1969) "Soil and Water Conservation Needs Inventory." Preliminary report, base year 1967. *Soil Conservation* **35**:99–109. (See also *USDA Stat. Bull. 461*, 1971.)

U.S. Soil Conservation Service (1978) *1977 National Resource Inventories*, Washington, D.C.

Williams, J. R., C. A. Jones, and P. T. Dyki (1984) "A Modeling Approach to Determining the Relationship between Erosion and Soil Productivity," *Trans. Am. Soc. Agr. Eng.* **27**:129–144.

Woodruff, N. P., L. Lyles, F. H. Siddoway, and D. W. Fryrear (1977) "How to Control Wind Erosion," U.S. Department of Agriculture, *Agr. Inf. Bull. 354*, U.S. Government Printing Office, Washington, D.C.

Woodruff, N. P., and F. H. Siddoway (1965) "A Wind Erosion Equation," *Soil Sci. Soc. Am. Proc.* **29**:602–608.

PROBLEMS

9.1 Determine the spacing between shelterbelts that are 40 ft (12.2 m) high if the five-year return period wind velocity at 50-ft (15.2-m) height is 30 mph (48 km/h) and the wind direction deviates 15° from the perpendicular to the shelterbelt line.

9.2 Estimate the strip-crop width from the windbreak spacing equation if the prevailing 40-mph (64 km/h) wind at a height of 50 ft (15.2 m) is perpendicular to the strip and the height of the crop in the adjoining strip is 5 ft (1.5 m) tall. Compare this spacing with that in Table 9.1 for loamy sand. How do these widths compare? If they are not similar, why are they not?

9.3 Estimate the maximum width of the protected area from a 4-ft (1.2 m) high slat fence with an average density of 50 percent. Compute both the windward and the leeward distances from the fence using the 90 percent wind velocity line shown in Fig. 9.4.

9.4 Determine the full protection strip-crop width if the crop in the adjoining strip is wheat 3 ft (0.9 m) tall and the wind velocity at a 100-ft (30.5-m) height (taken at nearest airport) is 30 mph (48 km/h) perpendicular to the field strip. Assume the wind velocity varies as the logarithm of the height.

9.5 For the conditions given in Table 9.1, how much flat small grain residue equivalent is needed to reduce the annual soil loss to 5 t/a (11.2 Mg/ha) for a 1000-ft (305-m) unsheltered field length?

CHAPTER 10

CONSERVATION WATER-CONTROL STRUCTURES

Water-flow channels designed to convey irrigation water, drainage flow, or flood runoff often require stabilization by structures made of concrete, metal, treated wood, or some other durable material. Such stabilization is required where the flow gradient without such structures is large enough to cause high velocities and erosion of the channel. Conservation structures are designed to reduce these velocities to nonerosive levels. Much of the fall in the channel is taken up by structures that dissipate the energy of the falling water. Ideally, the channel sections between structures are designed at a grade to maintain nonsilting and nonscouring velocities.

10.1 Gully Erosion Control

Any concentration of surface runoff is a potential source of gullying. Such channeling of runoff may be initiated by stock trails, implement tracks, dead-furrows, tillage furrows, or other small depressions on sloping land. Although gully erosion is often not as serious as sheet erosion, gullying is more spectacular and thus more noticeable. Gullying is the easiest and most economical to control during early stages of development.

The major factors to consider in describing gullies and in planning for their control include the size of the gully, its drainage area, and the slope of the channel. An arbitrary description of gullies is given in Table 10.1.

Gullies may be stabilized with vegetation, such as grassed waterways (Chapter 7) trees, and shrubs, by reduction of runoff with terraces, strip cropping, contour rows, and diversions (Chapter 8), and by controlling water velocity with structures that reduce the channel gradient. A combination of measures is usually required. Where conservation practices alone are inadequate or cannot be applied because

Table 10.1. Description of Gullies and Drainage Areas

Description	*Gully Depth (ft)*	*Drainage Area (acres)*
Small	3 or less	5 or less
Medium	3–15	5–50
Large	15 or more	50 or more

of lack of cooperation among adjacent landholders, the direct stabilization of a limited length of the gully may be necessary.

Runoff Reduction. The first step in gully control is the reduction of the runoff, through the use of good crop-management practices, contouring, terraces, and diversions. Diversions may eliminate most of the runoff, provided that it can flow to another suitably protected outlet. Diversions should be established with grass in the channel. This grass will permit higher velocities than farmed terrace channels.

Diversions should be located far enough above the gully overfall so that sloughing of the gully head will not threaten the diversion. Availability of outlets must also be considered in the location of the diversion. In some situations water may be discharged to spread over pasture or woodland. Other situations may require the establishment of a vegetated outlet to serve the diversion.

The most common cause of failure of diversions is sedimentation in the channel. This can best be prevented by a proper combination of conservation practices on the contributary watershed. Small sediment deposits should be removed with light equipment, such as a shovel or scraper. Extensive sedimentation should be removed to restore the channel to its original capacity. Other causes of failure are those given as being common to terraces. Rodent holes and other small breaks should be repaired as detected. Vegetation should be mowed annually to prevent the growth of woody plants which clog the channel and reduce its capacity.

Vegetative Control. Where vegetation is to be established, the gully banks should be shaped to a stable slope and the bottom of the channel should be shaped to the desired cross section. Where trees and shrubs are to be established in the gullied area, rough sloping of the banks to about 1 : 1 should be sufficient. In cultivated fields where gullies are to be reclaimed as grassed waterways, sloping of banks must be done to an extent that will permit crossing the drainageway with farm machinery and operating in the channel for seeding, fertilizing, and harvesting the grass crop. Side slopes of 4 : 1 or flatter are usually desired in such situations.

Natural revegetation may be stimulated by diverting flow and fencing livestock out of the gullied area. Steep gully banks may be roughly sloped to provide improved conditions for natural seeding and to prevent caving. A mulch to conserve moisture and to protect young plants may help to accelerate the natural revegetation process. When a suitable environment is established, a gradual succession of plant species will eventually protect the gullied area with grasses, vines, shrubs, or trees native to the region. The opportunity to provide protective cover by natural revegetation is frequently overlooked and unnecessary expenditures are made for structures and plantings.

In some situations trees and shrubs may be more suitable for gully control than grasses and legumes. This is particularly true if the gully is not to be shaped to permit operations of farm implements. Trees planted on gullied areas may be utilized for fence posts or rough timber. Only adapted species should be planted. Shrubs are desirable for establishing the gullied area as a wildlife refuge. Variety in plantings will reduce the danger of destruction of vegetation by disease. Trees, shrubs, and vines for gully control may often best be selected from species native to the locality and found growing on sites similar to the area to be reclaimed.

Structures. Where the gradient in the channel must be reduced and other means of gully control are not possible, structures should be considered, but only as the last resort. Structures may be required at overfalls, at abrupt changes in grade, at junctions with branch gullies, or at other critical points. Typical locations for a drop spillway, an earth dam and drop inlet, and a chute spillway are illustrated in Fig. 10.1. The earth dam may provide for permanent water storage, for flood-water storage, or for a sediment basin. Such a dam would have the same requirements as do farm ponds (discussed in Chapter 12).

Temporary structures made of rock, logs, brush, woven wire, and other nondurable materials are generally not recommended. Studies have shown that vegetal protection can usually be established without temporary structures. During the 1930s, many such structures were built, but they have declined in popularity because of high labor costs and their unsatisfactory performance.

10.2 Requirements for Permanent Structures

All channel stabilization structures should meet the following functional requirements: (1) be constructed with durable materials of adequate structural strength, (2) have sufficient hydraulic capacity to handle the design discharge, (3) dissipate within the confines of the structure the kinetic energy of the discharge in a manner and to a degree that will protect both the structure and the downstream channel from damage, and (4) be designed to prevent water seepage under and along the sides of the structure so as to avoid a "washout." Small-scale models are used extensively to verify the hydraulic characteristics of structures. In addition to good design, proper construction techniques are necessary to meet many of the above requirements.

Standard design plans have been developed by agencies, such as the Soil Conservation Service and the Bureau of Reclamation. Several of the more common types of structures will be described in the remainder of this chapter. In

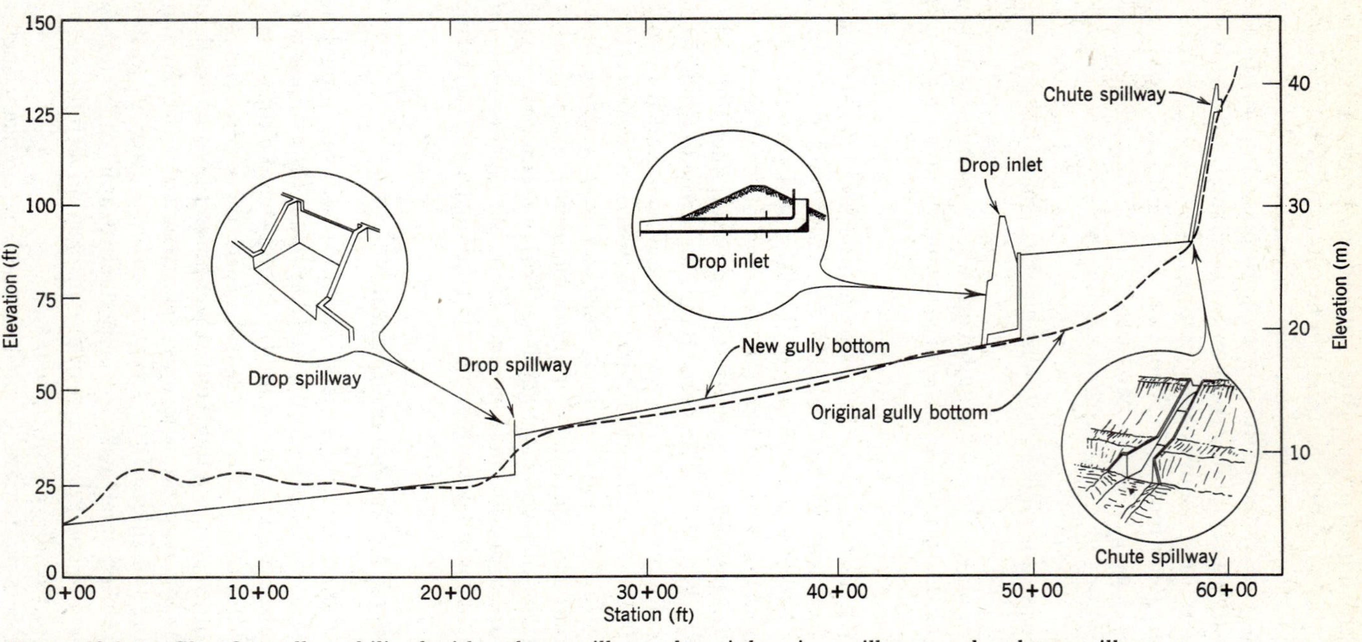

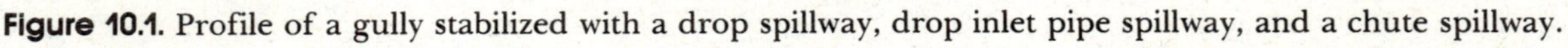

Figure 10.1. Profile of a gully stabilized with a drop spillway, drop inlet pipe spillway, and a chute spillway.

this text the actual design of the structure itself will not be covered, as this should be made by a competent engineer.

10.3 Drop Spillways

A straight drop spillway with a rectangular notch is illustrated in Fig. 10.2. The weir that controls the flow is a rigid structure with a stable cross section smaller than the cross sectional area of the channel above. The structure shown is known as a broadcrested weir, since the bottom of the channel above is level with the bottom of the weir. By measuring the depth of flow, the flow rate of the channel may be computed.

Water flows through the weir notch and falls to the base of the structure, where a stilling basin or apron dissipates the energy of the water before it flows into the channel below. The length of the structure in Fig. 10.2 could be reduced by recessing the apron to form a stilling basin. These structures may be constructed of monolithic reinforced concrete, coated steel, or concrete blocks. Where the stream channel is narrow, the straight weir could be replaced with a box inlet. The length of the weir would be the length of the three sides of the box.

The hydraulic capacity of these structures must be adequate to carry the peak runoff from the watershed or the design flow of the channel above. The width and depth of flow in the weir notch control the discharge, which is tabulated in Table 10.2.

The dimensions of the structure are functions of the depth of the notch and the drop height, as shown in Fig. 10.2. For channel stabilization, drop spillways are usually limited to drops of 10 ft, flumes or pipe spillways being suitable for greater drops. A drop spillway with a small drop may be used as a toe wall at the bottom of a steep sod chute spillway.

Example 10.1. Determine the dimensions of a drop spillway to carry a 50-year return period storm of 200 cfs. The bottom width of the stream channel is 22 ft, and the drop in the structure to stabilize the channel must be 5 ft.

Solution. Assume a length of weir opening L of 18 ft. From Table 10.2, an 18-ft weir length with a flow depth of 2.5 ft will discharge 227 cfs, which is adequate. The dimension H in Fig. 10.2 is 5 ft. Other dimensions of the structure from Fig. 10.2 are as follows:

$$\text{Total depth of notch, } d = h + 0.5 = 2.5 + 0.5 = 3.0 \text{ ft}$$
$$\text{Length of apron} = 0.75(H + d) + H = 0.75(5 + 3) + 5 = 11 \text{ ft}$$
$$\text{Length of head wall extension} = H + d + 3 = 5 + 3 + 3 = 11 \text{ ft}$$
$$\text{Length of wing wall} = 2.25h = 2.25(2.5) = 5.6 \text{ ft}$$

10.4 Chutes and Flumes

Chutes and flumes convey water down steep grades. They are lined with various materials, such as concrete, metal, and wood, to prevent scour. Water flows down the chute at a high velocity and its energy is dissipated at its lower end before

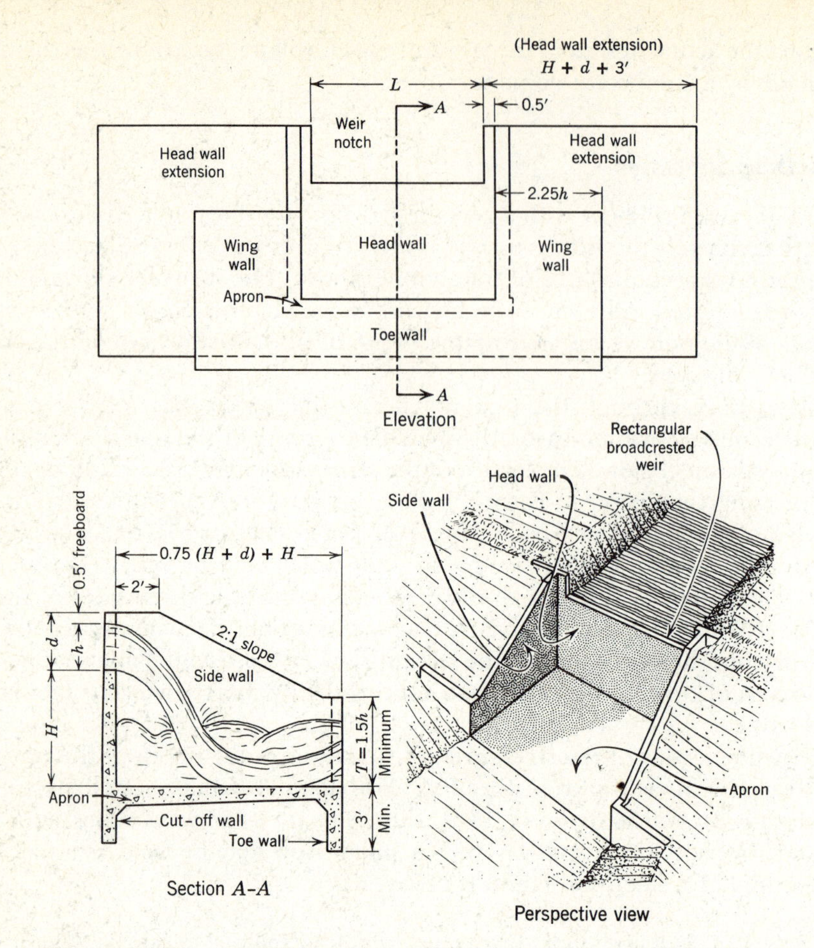

Figure 10.2. Straight drop spillway with a rectangular broadcrested weir.

Table 10.2. Flow Capacity of Broadcrested Weirs in Cubic Feet per Second (cfs)[a]

Head, h	*Length of Weir, L (ft)*							
(ft)	*6*	*8*	*10*	*12*	*14*	*16*	*18*	*20*
1.0	19	26	32	38	45	51	58	64
1.5	35	47	59	71	82	94	106	118
2.0	54	72	91	109	127	145	163	181
2.5	76	101	126	152	177	202	227	253
3.0	100	133	166	200	233	266	299	333

[a] $q = 3.2Lh^{3/2}$. (The 3/2 power is the square root of the cube.)

Note: For box inlets, L is the length of three sides of the box.

flowing into the channel below. Examples of these structures are shown in Figs. 10.3 and 10.4. An elevated section of an irrigation canal is called a flume or an aqueduct.

The hydraulic capacity of a chute, such as shown in Fig. 10.3, is normally controlled by the inlet section. Inlets may be straight, as shown in Fig. 10.2 for the drop spillway, a box-inlet, or a curved weir, in which cases the discharge can be determined from Table 10.2. Chutes are applicable for drops up to 16 or 20 ft, which is greater than for drop spillways. They usually require less concrete than do drop-inlet structures of the same capacity and drop. However, the danger of undermining of chutes by rodents is greater, and in poorly drained locations seepage may be a greater threat to the foundation.

Several types of energy dissipators have been developed for the outlet section, such as the SAF (Saint Anthony Falls) stilling basin shown in Fig. 10.3. Where

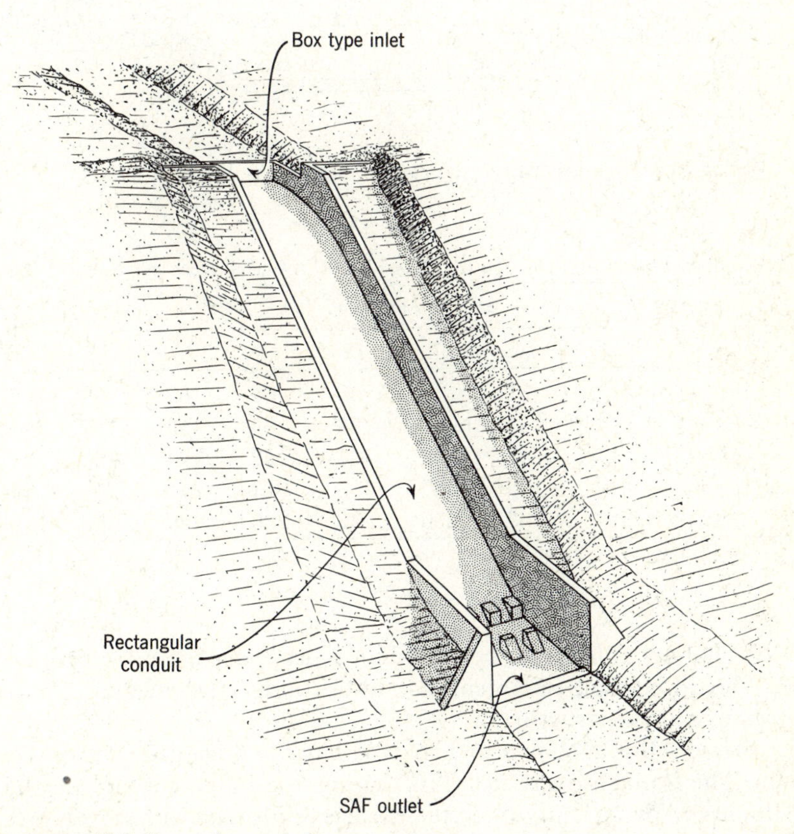

Figure 10.3. Box inlet and chute structure.

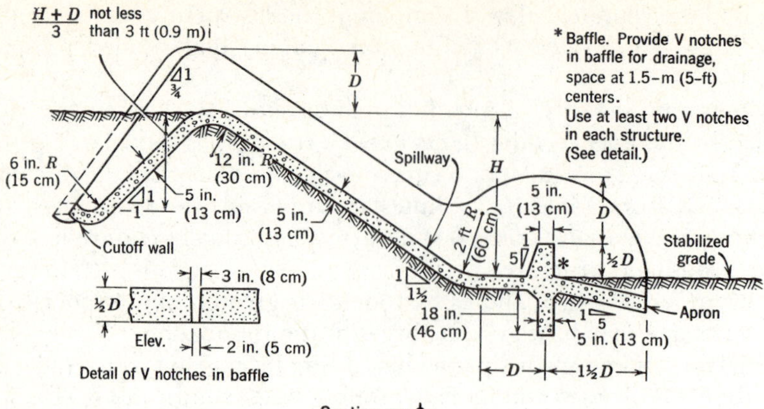

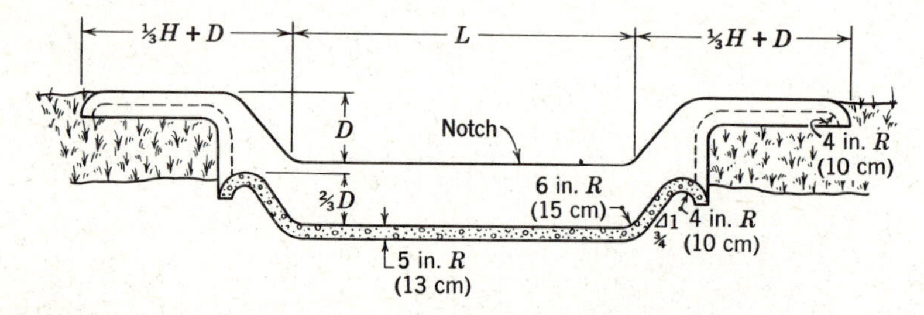

Figure 10.4. Formless flume structure. *Source:* Wooley et al. (1941).

Table 10.3. Flow Capacity of Formless Flume in Cubic Feet per Second (cfs)[a]

Depth of Notch, D	Length of Notch, L (ft)							
	6	8	10	12	14	16	18	20
1.0	23	31	39	46	54	62	69	77
1.5	42	57	71	85	99	113	127	142
2.0	65	87	109	131	152	174	196	218
2.5	91	122	152	183	213	244	274	304
3.0	120	160	200	240	280	320	360	400

[a] $q = 3.85LD^{3/2}$.

the outlet channel is likely to be unstable, a propped outlet similar to that in Fig. 10.5 should be used. Sedimentation in the outlet channel does not decrease the capacity of chute structures.

The formless flume shown in Fig. 10.4 has the advantage of low-cost construction. The flume is constructed by shaping the soil to conform to the shape of the flume and by pouring concrete, which is reinforced with woven wire mesh. No forms are needed; thus, the construction is simple and inexpensive. Such

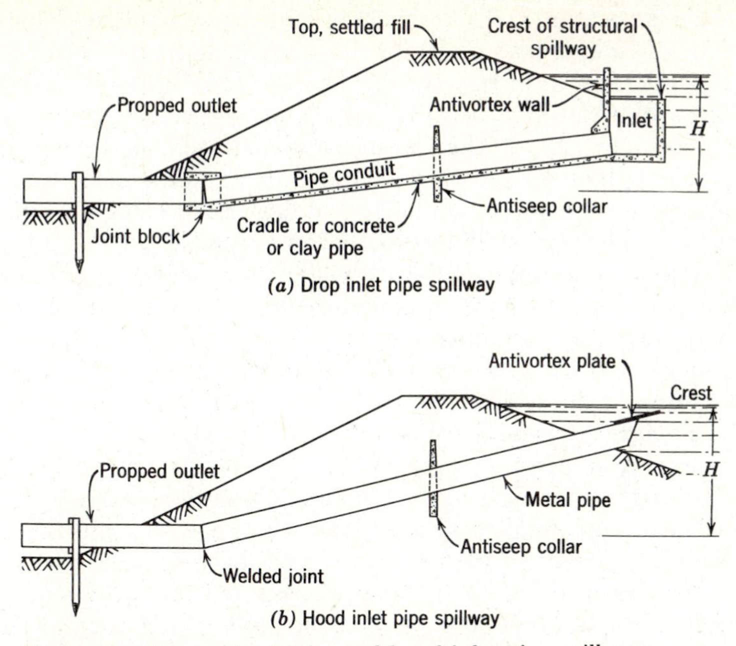

Figure 10.5. Square drop inlet and hood inlet pipe spillways.

structures are recommended where the fall does not exceed 7 ft, where water is not impounded upstream (danger of undermining by seepage), and where freezing does not occur at great depth. The depth D of the notch in Fig. 10.4 should be increased by 0.25 to 0.5 ft to allow for freeboard so that the design flow does not overtop the ends of the flume. Because the ends of the notch are sloped outward at the top, the discharge is somewhat greater (Table 10.3) than that for the rectangular weir notch.

10.5 Pipe Spillways

These structures may vary from a simple culvert under a road to a pipe through an earth embankment with various types of inlets as illustrated in Fig. 10.5. Pipe-spillway conduits may be round, square, rectangular, or arch-shaped in cross section. Pipes with the drop-inlet structure as shown in Fig. 10.5*a* are suitable for gully control or for farm ponds. For small-sized pipes, typical of most farm ponds, the hooded inlet shown in Fig. 10.5*b* is common because of simpler construction and lower cost.

As shown in Table 10.4, pipe spillways are relatively low-capacity structures compared to drop spillways and chutes, because the flow varies as the one-half power of the head rather than as the three-halves power. With a box-inlet structure, the weir controls flow when the flow depth is small. When the pipe fills with water, it backs up in the box, submerging the weir, and the pipe controls the flow. The discharge-head curve for the box-inlet structure is similar to that for the hood inlet shown in Fig. 10.6*d*. The head causing flow through a pipe is the difference in elevation between the center of the pipe at the outlet and the water surface (Fig. 10.5). The entrance loss coefficient (K_e), the pipe roughness, and the length of the pipe are other factors that influence the discharge. For a square corner entrance on the pipe and for a flush wall even with the end of the pipe, $K_c = 0.6$. The roughness coefficient K_c is related to n in the Manning velocity equation. As shown in Table 10.4, the roughness of clay and concrete pipe is sufficiently lower than that of corrugated metal pipe to permit a reduction of one size to carry the same discharge.

Pipe spillways and culverts with well-designed entrances will normally flow full if the vortex or the whirling action of the water can be eliminated at the entrance. This vortex causes air to enter, thereby reducing the discharge. Pockets of air and water are known as slug flow. With drop-inlet structures, the antivortex wall is the side of the box next to the dam and above the crest (Fig. 10.5*a*). On corrugated metal pipe risers, a splitter plate above and through the center of the riser serves to prevent a vortex. Three types of antivortex devices for the hood inlet are shown in Fig. 10.6. These devices will cause the pipe to prime and flow full on slopes up to 30 percent in the pipe. As indicated in Fig. 10.6*d*, the pipe will not begin to flow full until the depth of the water above the crest of the pipe is 1.4 times the diameter of the pipe.

Example 10.2. Determine the discharge of a 12-in. diameter corrugated metal pipe spillway 50 ft in length with a hood inlet having an entrance loss coefficient of 1.0. The pipe through the farm pond dam is straight, with a uniform slope of 14 percent. The depth of the water above the pipe crest of the inlet (also called invert) is 2 ft.

Table 10.4. Flow Capacity of a Pipe Spillway 100 ft in Length in Cubic Feet per Second (cfs)[a]

Corrugated Metal Pipe		*Head H (ft) (Water Surface to Center of Outlet Pipe)*				*Clay and Concrete Pipe*
Pipe Diameter (in.)	K_c	*10*	*20*	*30*	*40*	*Diameter (in.)*
6	0.292	1	1	1	1	Minimum 6
8	0.199	2	2	3	3	6
10	0.148	3	4	6	6	8
12	0.116	5	7	9	11	10
15	0.086	10	14	17	20	12
18	0.067	15	22	27	31	15

[a]Calculated to the nearest cfs from the equation, $q = 8a[H/(1 + K_e + K_cL)]^{1/2}$, where a = cross-sectional area in square ft, H = head in ft, $K_e = 0.6$, and K_c corresponds to a roughness coefficient of $n = 0.025$ for corrugated metal pipe.

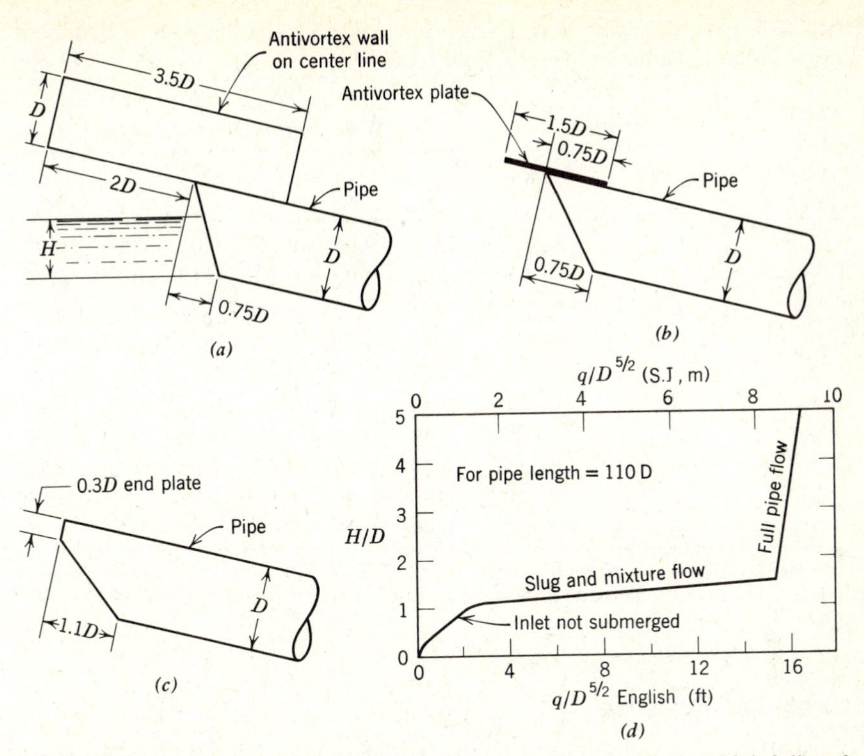

Figure 10.6. Three types of hood inlets for pipe spillways. *Source:* Blaisdell and Donnelly (1958) and Beasley et al. (1960).

Solution. The head measured from the water surface to the center of the outlet pipe is

$$H = 2 + (50 \times 0.14) - 0.5 = 8.5 \text{ ft}$$

From the pipe-flow equation and $K_c = 0.116$ from Table 10.4, the discharge is

$$q = 8 \times 0.785 \sqrt{\frac{8.5}{1 + 1 + 0.116 \times 50}} = 6.6 \text{ cfs}$$

Orifices in a pipe spillway may restrict the flow, such as in a pipe outlet of a terrace system (see Chapter 8). The circular orifice equation and the discharge for a range of sizes and heads are shown in Table 10.5. In a similar way this type of structure is suitable for a debris or sediment storage basin. An orifice placed in the discharge pipe from an irrigation well is a common method of water measurement. Such orifices are available commercially and are provided with a manometer calibrated to obtain the flow rate (see Chapter 11). For this application the equation in Table 10.5 is not appropriate because it applies only for a free discharge to the pipe below without back pressure.

Table 10.5. Flow Capacity of an Orifice for a Pipe Outlet Terrace or Sediment Storage Basin in Cubic Feet per Second (cfs)[a]

Riser Pipe Diameter (in.)	*Orifice Diameter (in.)*	*Orifice Area (ft^2)*	*Head, h (ft) (Water Surface to Orifice Plate)* 1	2	3	4	5	6
6	1.0	0.006	0.03	0.04	0.05	0.05	0.06	0.07
6	1.5	0.012	0.06	0.08	0.10	0.12	0.13	0.15
6	2.0	0.022	0.11	0.15	0.18	0.21	0.24	0.26
6	2.5	0.034	0.16	0.23	0.28	0.33	0.37	0.40
6	3.0	0.049	0.24	0.33	0.41	0.47	0.53	0.58
6	3.5	0.067	0.32	0.45	0.56	0.64	0.72	0.79
8	4.0	0.087	0.42	0.59	0.73	0.84	0.94	1.03
8	4.5	0.110	0.53	0.75	0.92	1.06	1.18	1.30
8	5.0	0.136	0.65	0.93	1.13	1.31	1.46	1.60
10	6.0	0.196	0.94	1.33	1.63	1.88	2.10	2.30

[a]Calculated from the circular orifice equation, $q = 0.0263\,d^2h^{1/2}$, where d = orifice diameter in inches, h = head in feet, and q = flow rate in cfs. (Orifice coefficient assumed as 0.6.)

10.6 Drainage Structures

Controlled drainage structures, such as that shown in Fig. 10.7, are suitable for maintaining a high water table in peat or muck soil to reduce subsidence, for subirrigating soils that have an impervious layer below and high horizontal permeability, for flooding cropland for the production of such crops as rice and cranberries, and for regulating the water level for management of fish-spawning areas and water-fowl habitat. Such structures with an adjustable headgate are suitable for controlling water in open channels. An example is shown in Fig. 10.8. Crest boards may be used in place of the headgate.

10.7 Irrigation Structures

Many types of control structures, including all of those previously described, are used to convey and to control irrigation water. Structures for distributing water on the farm will usually have some of the features shown in Figs. 10.8 and 10.9. Water is also conveyed through surface canals and low-head pipelines placed underground. Special structures for these purposes are too numerous to include. Some of the smaller field irrigation distribution devices will be discussed in Chapter 16.

The controlled drainage structure shown in Fig. 10.7 is suitable for subirrigating crops through existing tile drainage systems. Water may be pumped from a drainage ditch into the tile lines during the dry season. Systems of this type have been used successfully in the Saginaw Bay area in Michigan for a number

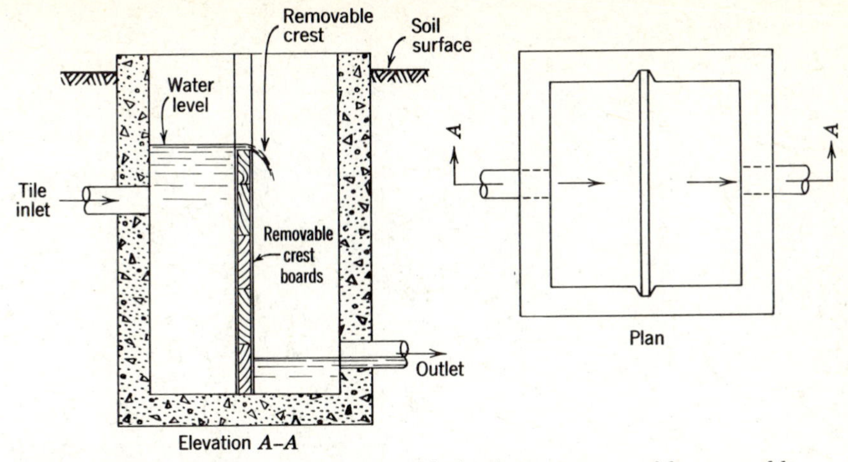

Figure 10.7. Controlled drainage or subirrigation structure with removable crest boards (stop logs).

of years. Where the land is rolling, several of these structures in series may be required.

10.8 Flood Control and Water Supply Structures

Dry dams for flood control and small reservoirs (Chapter 12) for water supply usually have chutes (Fig. 10.3) or pipe spillways (Fig. 10.5) for conveying flood water through or over the dam. Flood control dams with a conservation pool for water supply are known as multipurpose structures. The conservation pool

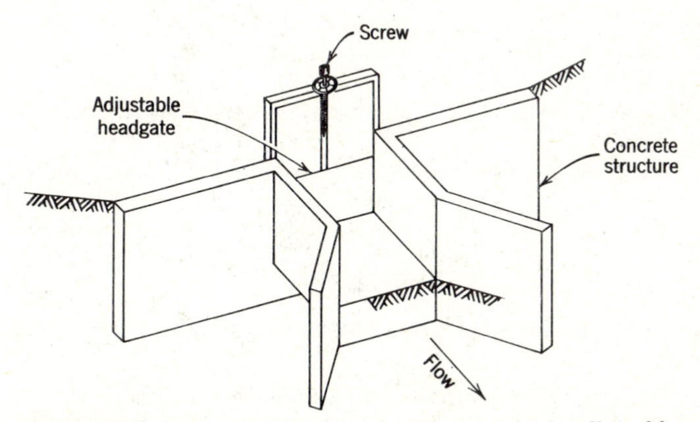

Figure 10.8. Irrigation flow-control structure with adjustable headgate. (*Redrawn from U.S. Soil Conservation Service, 1979*).

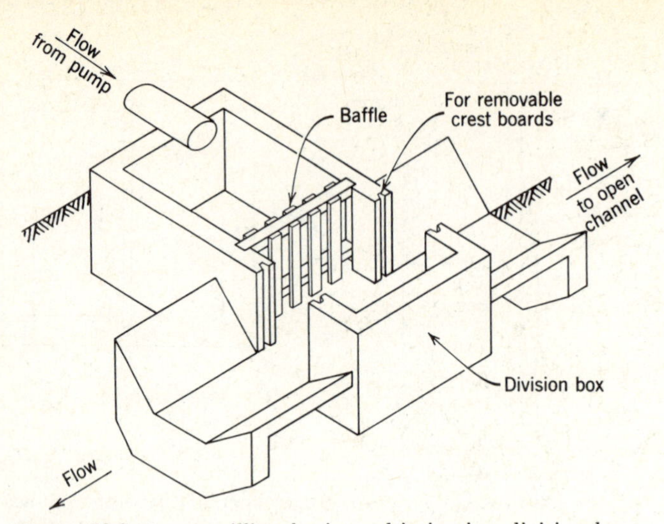

Figure 10.9. Pump stilling basin and irrigation division-box structure. (*Redrawn from U.S. Soil Conservation Service, 1979.*)

may also be designed for sediment storage, but it does not reduce flood flows. Such sediment basins trap a high percentage of the stream's sediment and thus reduce downstream pollution.

REFERENCES

Beasley, R. P. et al. (1960) "Canopy Inlet for Closed Conduits," *Agr. Eng.* **41**:226–228.

Blaisdell, F. W., and C. A. Donnelly (1958) "Hydraulics of Closed Conduit Spillways," Part X, *Univ. Minn. St. Anthony Falls Hydraulics Lab. Tech. Paper No. 20-B*.

Schwab, G. O. et al. (1981) *Soil and Water Conservation Engineering*, 3rd ed., John Wiley & Sons, New York.

U.S. Bureau Reclamation (1977) *Design of Small Dams*, 2nd ed., U.S. Government Printing Office, Washington, D.C.

U.S. Soil Conservation Service (1979) *Engineering Field Manual for Conservation Practices*, Washington, D.C.

Wooley, J. C. et al. (1941) "The Missouri Soil Saving Dam," *Mo. Agr. Exp. Sta. Bull. 434*.

PROBLEMS

10.1 What is the discharge of a 7-ft (2.1-m) straight-drop spillway when the depth of flow is 1 ft (0.3 m)?

10.2 What depth of flow is required over a straight-drop spillway 6 ft (1.8 m) in length to carry a design runoff of 76 cfs (2.15 m^3/s) from a 20-acre (8.1-ha) watershed? What depth is required if the weir length is 10 ft (3.0 m)?

10.3 For the 6-ft (1.8-m) weir length in Problem 10.2, what should be the length of the headwall extension, apron, and width and depth of the wingwall for a drop of 8 ft (2.4 m)?

10.4 What depth of notch is required for a formless flume to carry 76 cfs (2.15 m^3/s) if the notch length is 10 ft (3.0 m)? Allow 0.4 ft (0.12 m) for freeboard.

10.5 Determine the discharage capacity of a 4-ft (1.2-m) square box inlet (inflow into three sides only) and chute if the depth of flow is 2.5 ft (0.76 m), assuming that the box inlet controls the flow.

10.6 Determine the discharge of a box-inlet pipe spillway if the box is 2 by 2 ft (0.6 × 0.6 m), the pipe is 12 in. (30 cm) in diameter, the head from the water surface to the center of the outlet pipe is 40 ft (12.2 m), the length of the pipe is 100 ft (30 m), and K_c = 0.116 (English units).

10.7 In Problem 10.6, what depth of water over the crest of the box [L = 6 ft (1.8 m)] is required for the pipe to begin to flow full (weir flow)?

10.8 Determine the discharge of a corrugated metal hood inlet pipe spillway 10 in. (254 mm) in diameter if K_e = 1.0, the pipe length is 40 ft (12.2 m), and the head from the water surface to the center of the outlet pipe is 25 ft (7.6 m).

10.9 In Problem 10.8, determine the discharge for concrete pipe for which K_c = 0.046 (English units).

10.10 Determine the total head causing flow through an 8-in. (203-mm) diameter hood inlet pipe spillway if the depth of water over the crest at the inlet is 1.5 ft (0.46 m), 50 ft (15.2 m) of pipe having a slope of 20 percent, and the last 20 ft (6.1 m) of pipe having a slope of 1 percent.

10.11 Determine the flow rate through a 2.0-in. (51-mm) diameter orifice from a pipe outlet terrace when the maximum head above the center of the orifice is 4.0 ft (1.22 m). What quantity of water will flow from the terrace in one day?

10.12 Determine the flow rate from an irrigation pump that discharges into a tank with a 3-in. diameter (76-mm) orifice on its side, from which the water flows freely into a concrete-lined distribution ditch. The height of the water above the center of the orifice is 2.5 ft (0.76 m).

CHAPTER 11

WATER SUPPLY AND ITS DEVELOPMENT

Precipitation is presently our only practical source of a continuing fresh water supply for all agricultural, industrial, and domestic use. Increasing attention being given to the large-scale desalinization of brackish or salty water may eventually result in reasonable supplies of water for high-value uses in some locations, but precipitation will remain the dominant source of water. Nearly 5 million ac-ft of water falls on the 48 contiguous United States annually. Developed water supplies use only 4 percent of this amount. Developed water supplies are only 13 percent of the residual precipitation, after allowing for evaporation and transpiration from natural plants and unirrigated crops. Thus, there is actually ample water for our needs. However, it is often not available at the time and place of need. The development of water resources involves storage and conveyance of water from the time and place of natural occurrence to the time and place of beneficial use.

11.1 Sources of Water

Water in one or more of its three physical states, solid, liquid, or gaseous, is present in greater or lesser quantities in or on virtually all the earth, its atmosphere, and all things living or dead. Water, which is important from the standpoint of water-resource development, falls into the categories of atmospheric moisture, surface water, and ground water. Atmospheric moisture is relatively unimportant as a source of water. It is normally about 1 in. annually in the form of dew which condenses on soil or plant surfaces. Dew reduces evapotranspiration by the same amount and thus conserves moisture.

Surface Water. Surface waters exist in natural basins and stream channels. Where minimum flows in streams or rivers are large in relation to water demands of adjacent lands, towns, and cities, the development of surface waters is accomplished by direct withdrawal from the flow. On many streams and rivers, however, flow fluctuates widely from season to season and from year to year. Further, peak demands from many major rivers occur at seasons of minimum flow and in fact require that as much of the annual flow as possible be conserved and diverted for beneficial use. This situation requires the construction of reservoirs to hold the flow during seasons or years of high runoff for later release to beneficial use. Reservoirs range in size from many million acre-feet for large multiple-purpose reservoirs in the West to small ponds with but a few acre-feet of storage.

Ground Water. Subsurface water available for development is normally referred to as ground water. Ground water predominantly results from precipitation that has reached the zone of saturation in the earth through infiltration and percolation. Ground water is developed for use through wells, springs, or dugout reservoirs. In many areas where ground water is an important source of water supply, it is being withdrawn much faster than it is being replenished from infiltration and percolation of precipitation. Correction of this imbalance is a major challenge facing the conservationist.

11.2 Water Quality

The quality of water depends on the amount of suspended sediment and the chemical and biological constituents in the water. Quality requirements for water depend on the intended use of the water.

The effect of sediment in irrigation water is influenced by the nature of the soil of the irrigated area. Where fine sediment is deposited on sandy soil, the textural composition and fertility may be improved. However, if the sediment has been derived from eroded areas, it may reduce fertility or decrease soil permeability. Sedimentation in canals or ditches increases maintenance costs. Normally, ground water or water from reservoirs does not contain enough sediment to cause trouble in irrigation water, but for domestic supplies sediment is a problem.

The chemical properties of water affect its suitability for many uses. This discussion is limited to the chemical properties important in irrigation waters. A most important reference in relation to water quality in irrigation is USDA Handbook No. 60 (1954). The most important characteristics of irrigation water are (1) total concentration of soluble salts, (2) proportion of sodium to other cations, (3) concentration of potentially toxic elements, and (4) bicarbonate concentration as related to the concentration of calcium plus magnesium. Recent studies have shown that the classification system for irrigation water as given in this USDA Handbook has considerable limitations. For this reason, this system has not been included in the text.

In determining the suitability of water for irrigation, the crop, soil properties, irrigation management, cultural practices, and climatic conditions must be considered. The dependence of average root-zone salinity on electrical conductivity of irrigation water (salinity) and leaching fraction for conventional irrigation management is shown in Fig. 11.1. Root-zone salinity is obtained by taking water

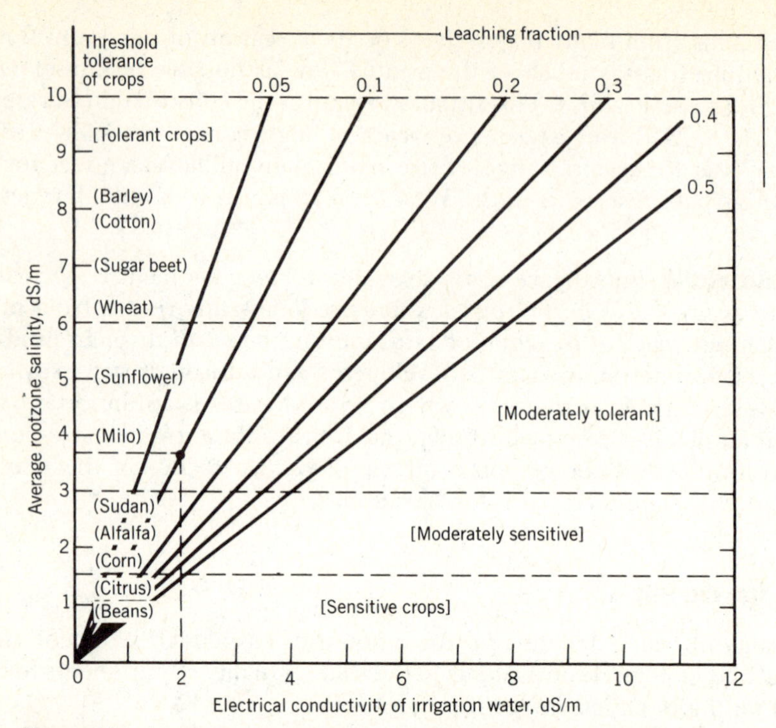

Figure 11.1. Average root-zone salinity (saturation extract basis), electrical conductivity of irrigation water, and leaching fraction for conventional irrigation management. *Source:* Rhoades (1983).

from a saturated soil paste according to standard procedures and then determining the electrical conductivity of the water. Electrical conductivity is expressed in decisiemens per meter, dS/m (formerly millimhos per centimeter). A mho or siemens is the unit of conductance equal to the reciprocal of the resistance in ohms. Electrical conductivity is related to total soluble salts and is used primarily because it is easier to measure than soluble salts. Irrigation water salinity is measured in the same way as soil water salinity. The leaching fraction is the ratio of the depth of leaching water (drainage) to the depth of irrigation water. It is the same as the ratio of the electrical conductivity of the irrigation water to the electrical conductivity of the leaching water. Figure 11.1 is useful for evaluating the salinity hazard. For example, irrigation water with a conductivity of 2 dS/m and 0.1 leaching fraction would result in an average root zone salinity of about 3.7 dS/m, suitable for milo or other more tolerant crops. Extensive lists of the salt tolerance of crops have been published by Ayers and Westcot (1976).

The proportion of sodium to other cations or the sodium hazard of water is indicated by the sodium-adsorption ratio, or *SAR*, calculated from

$$SAR = \frac{Na^+}{\sqrt{(Ca^{++} + Mg^{++})/2}} \tag{11.1}$$

where Na^+, Ca^{++}, and Mg^{++} represent the concentrations in milliequivalents per liter of the respective ions. The suitability of water for irrigation on the basis of conductivity and the sodium-adsorption ratio is shown in Fig. 11.2. The SAR of the soil and the total soluble salts in the irrigation water influence the soil permeability. Sodium acts as a dispersing agent allowing the fine soil particles to be dispersed rather than remain aggregated as larger soil granules. A high SAR could be caused by a deficiency of calcium or a high sodium level. The salinity of irrigation water influences the threshold value of the SAR of the soil that would cause a permeability hazard. For example in Fig. 11.2, irrigation water with a conductivity of 2 dS/m would cause a permeability hazard if the soil SAR is greater than about 20. These water quality procedures should be considered only as guidelines to be adjusted as experience and local conditions dictate.

Boron, though essential to the normal growth of plants, is toxic under some conditions in concentrations as low as $\frac{1}{3}$ part per million. High concentrations of bicarbonate ions may result in precipitation of calcium and magnesium bicarbonates from the soil solution, increasing the relative proportions of sodium and thus the sodium hazard. The USDA Handbook No. 60 (1954) should be referred to if there are potential problems with toxic elements or bicarbonate-ion concentration.

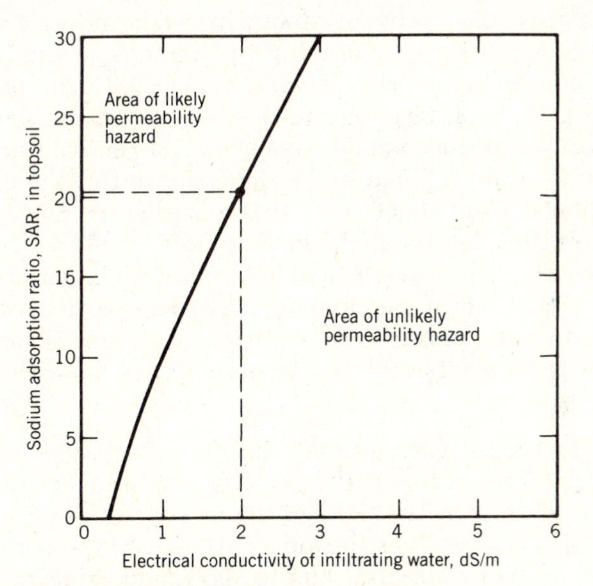

Figure 11.2. Threshold values of sodium-adsorption ratio of topsoil and electrical conductivity of infiltrating water for maintenance of soil permeability. *Source:* Rhoades (1983).

11.3 Nonpoint Sources of Water Pollution

Water pollution may be classified as either from point or nonpoint sources. Point sources include hazardous waste sites, municipal and industrial treatment plant discharges, and others where wastes discharge at a point directly into streams. Nonpoint sources are those where the polluted water does not enter the stream at one point, such as runoff from cropland, livestock pastures, and forests.

The pollution of our water, soil, and air has largely resulted from a high standard of living, from agricultural and industrial growth, and from a rapid increase in population. People's attitude toward pollution over the years has also changed because of better detection methods and a greater public awareness. Many chemicals get into the food chain, some of which are hazardous even in small quantities. However, foods themselves contain many thousands of chemicals, some of which may also be harmful.

Water, soil, and air can absorb or dispose of pollutants in varying amounts and rates. Pollution is often called, the problem of misplaced waste, which could be partially solved by recycling. In the 1970s, state and federal agencies were formed to investigate and to determine pollution policies.

Nonpoint sources of pollution from agriculture can be attributed largely to fertilizers, pesticides, soil erosion, and animal wastes. Rural pollution problems have become greater because of the increased erosion, the greater use of chemicals, and the larger numbers of livestock feed lots.

Fertilizers. The substantial increase in fertilizer usage in the United States since 1950 has been accompanied by an upward trend in nitrate concentrations in drainage waters. High nitrogen fertilizer applications generally cause greater nitrate losses, but the relationship is not always clear. Many other factors are involved, such as rainfall, farming practices, organic matter, and the time of application. Losses of nitrogen in tile drainage water generally have been higher than from surface runoff. Some studies have shown that the annual nitrogen content in rainfall is about equal to that in drainage water. Such nitrogen would be beneficial to plants, but is a pollutant in runoff.

Phosphorous, another major element in fertilizers, is important to water quality. Because it is attached to soil particles, phosphorus losses are closely related to soil erosion. In lakes, phosphorous contributes to algae blooms. Losses from surface runoff are usually greater than from tile drains, but both are much lower than nitrogen losses.

Pesticides. Pesticides, which include fungicides, herbicides, and insecticides, have been increasing in use for many years, especially with the greater acceptance of conservation tillage systems. Many of the early pesticides, such as DDT and those containing heavy metals are no longer approved. New pesticides are being developed that are highly specific, safer to apply, and biodegradable by organisms, in the environment.

Many pesticides can be found in surface waters, usually in low concentrations. The greatest danger is to humans, animals, fish, and aquatic life rather than damage to plants by irrigation. Losses of pesticides in drainage waters are usually less than 10 percent of the amount applied. Pesticides are more often attached to sediment rather than in soluble form. Because of government regulations, pesticides are likely to be kept within tolerable limits, but they do pose a potential problem.

Sediment. Sediment is by far the major single water pollutant. Sediment loss by erosion depletes the land from which it comes and impairs water quality. In addition to transporting soil, water with sediment may carry plant nutrients, pesticides, toxic metals, organic materials, and biological organisms. Fine sediment will adsorb or desorb elements on its surfaces. Water flowing in an erodible channel will attempt to transport sediment up to its energy ability. It may erode its bed or banks to obtain this material. Erosion is a natural process that can never be eliminated, but man's activities have accelerated the process.

Animal Wastes. The growing numbers of livestock, especially those confined in large feed lots, have accelerated the disposal of animal manure. Where these wastes are spread either as slurry or as solid material on land, the land is considered a nonpoint source of pollution. Liquid or solid storage is needed in the northern states to avoid spreading where the soil and weather conditions would produce polluted runoff.

The soil has a high capacity to accept animal wastes because of its ability to hold phosphorous and to transform nitrogen. Phosphorous in animal wastes becomes insoluble shortly after application whereas nitrogen in nitrate and nitrite forms is water soluble and will move with soil water, often contaminating ground water. The contamination of ground water by land disposal of animal wastes has becme a major problem in some areas. Runoff from barnyards, manure pits, and feed lots was identified as a source of the high concentrations of nitrogen compounds in shallow wells in many midwestern states. Nitrogen in irrigation water is considered a benefit to crop production rather than a pollutant, but heavy applications of nitrogen may contribute to ground water having nitrate concentrations above the 10 mg/l public health standard.

WATER SUPPLY DEVELOPMENT

11.4 Surface Water

Reservoir development is basic to surface-water utilization. In some situations surface waters may be recharged into ground water reservoirs, but more commonly surface reservoirs are constructed.

The three types of surface reservoirs are (1) dugout reservoirs fed by ground water, (2) on-stream reservoirs fed by continuous or intermittent flow of surface runoff, streams or springs, and (3) off-stream reservoirs. The details of the construction of small reservoirs are given in Chapter 12.

Dugout Reservoirs. Dugout reservoirs are limited to areas having slopes of less than 4 percent and a prevailing reliable water table within 3 to 4 ft of the ground surface. Design is based on the storage capacity required, depth to the water table, and the stability of the side-slope materials.

On-Stream Reservoirs. The on-stream type of reservoir depends on the inflow of surface runoff for replenishment. The designed storage capacity must be based both on use requirements and on the probability of a reliable supply of runoff. Where heavy usage is expected, the design capacity of the reservoir must be adequate to supply several years' needs in order to assure time for recharge in the event of a sequence of one or more years of low runoff. Spring- or creek-fed reservoirs consist of either a scooped-out basin below a spring or a reservoir formed by a dam across a stream valley or depression below a spring. The dam may also be placed across a depression rather than a definite stream to catch diffuse surface water.

A spring-fed reservoir should be designed to maintain the water surface below the spring outlet. This eliminates the hazard of diverting the spring flow due to the increased head from the reservoir. When the spring flow is adequate to meet use requirements, surface waters should be diverted out of the reservoir to reduce sedimentation and to reduce spillway requirements.

Off-Stream Reservoir. The off-stream or by-pass reservoir is constructed adjacent to a continuously flowing stream. An intake, through either a pipe or an open channel, diverts water from the stream into the reservoir. Controls on the intake permit a reduction in sedimentation, particularly if all flood water can be diverted from the reservoir. Proper location and diking are essential to protect against stream-overflow damage.

11.5 Water Harvesting

The term "water harvesting" has come to be applied to any watershed manipulation carried out to increase surface runoff. In many arid areas, the majority of precipitation that infiltrates into the soil is lost for use either through direct evaporation or through transpiration by economically useless vegetation. For example, in the Colorado River Basin less than 6 percent of the precipitation appears as streamflow. Vegetation management has been shown to be effective as a means of increasing streamflow, and it may be anticipated that it will receive increasing application.

More immediate in application is the practice of water harvesting by catchments. Catchments are the areas of concrete, sheet metal, asphalt, or otherwise treated or waterproofed soil specifically constructed to catch and collect precipitation. Successful water harvesting requires attention not only to the collection of water but also to the conveyance and storage of the water collected. Catchment treatment may vary from soil smoothing and removal of vegetation to the application of plastic film or aluminum foil. Simple soil smoothing and vegetation removal have increased runoff by a factor of three in some tests.

Complete sealing will increase runoff to essentially 100 percent. Water harvesting is being applied to develop water supplies for wildlife, livestock, and occasional domestic use.

11.6 Ground Water

The classification of the earth's crust as a reservoir for water storage and movement of water is shown in Fig. 11.3. The profile of the earth is divided into two primary zones: the zone of rock fractures and the zone of rock flowage. In the rock-flowage zone, the water is in a chemically combined state and is not available. In the zone of rock fracture, interstitial water is contained in the pores of the soil or in the interstices of gravel and rock formations. This zone containing interstitial water is divided into the zone of aeration and the zone of saturation. As here considered, ground water occurs only in the zone of saturation. However, perched water tables, shown in Fig. 11.3, are often encountered in the zone of aeration. Intermediate water is gravitational water that is free to move to the

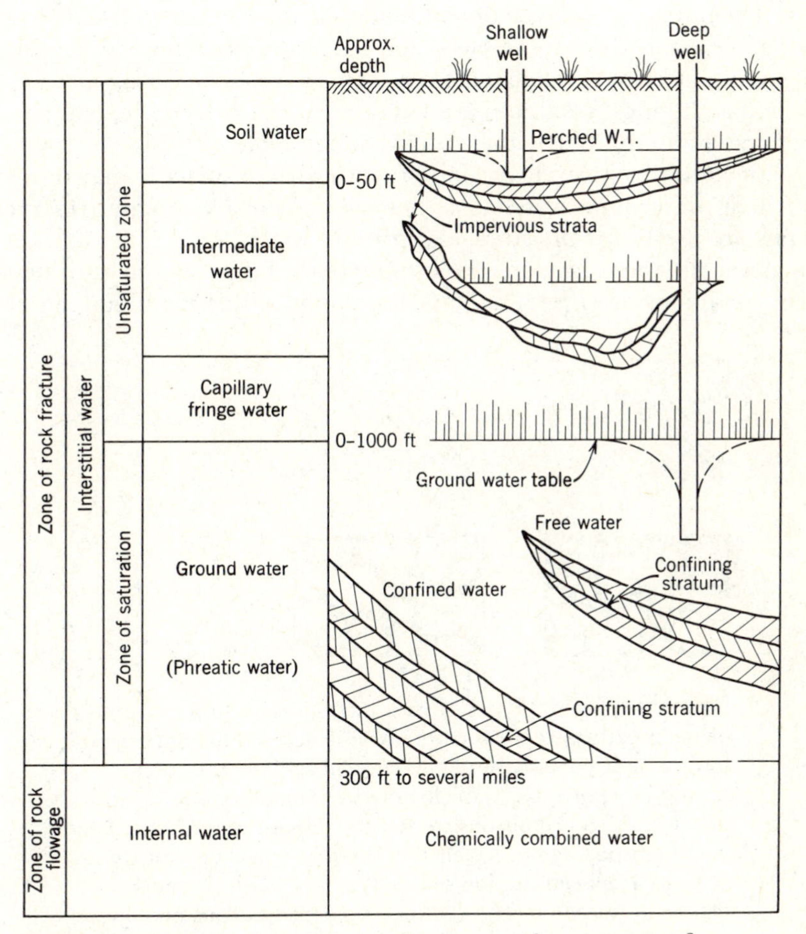

Figure 11.3. Classification of the earth's crust and occurrence of subsurface water.

ground water table. Whereas the ground water table is at the soil surface near lakes, swamps, and continuously flowing streams, it may be several hundred feet deep in drier regions. Ground water is often referred to as phreatic (a Greek term meaning "well") water. Capillary-fringe water and intermediate water exist above the ground water table and are present above perched water tables as well.

Formations from which ground water is derived in the zone of saturation have considerably different characteristics than the soil near the surface. The various types of deposits that furnish water supplies are shown in Fig. 11.4. Total porosity and permeability as indicated by the size and shape of the pores is shown in Table 11.1 Except in clay, porosity is generally a good indicator of specific yield and permeability. Usually, uncemented sand, gravel, fractured limestone, and rock formations are good water-bearing deposits.

Wells may be classed as gravity, artesian, or a combination of artesian and gravity, depending on the type of aquifer supplying the water. Gravity wells are those that penetrate the water table where water is not confined under pressure.

Gravity water may be obtained from wells, springs, and dugout reservoirs. Wells, by far the most common, are either shallow or deep, depending on the ground water depth. When the ground water table reaches the soil surface because the underlying strata are impervious, springs or seeps may develop. In areas where the ground water table is near the surface, dugout reservoirs or open pits are useful for the storage of ground water.

Whenever the water level in a hole rises above the top of the saturated zone, artesian conditions are present. Thus, according to this definition, an artesian

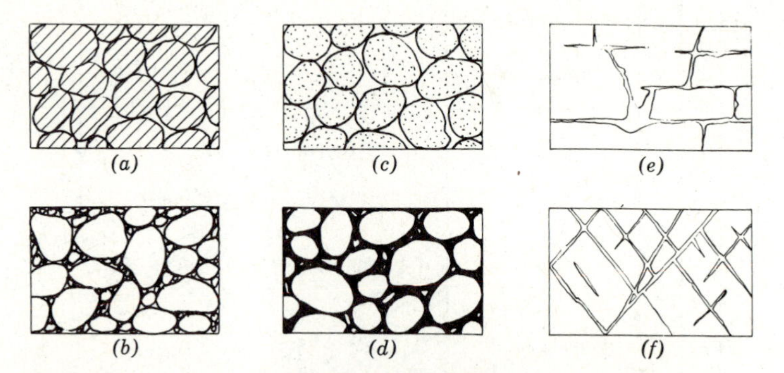

Figure 11.4. Several types of rock interstices and the relationship of rock texture to porosity. (*a*) Well-sorted sedimentary deposit having high porosity. (*b*) Poorly sorted sedimentary deposit having low porosity. (*c*) Well-sorted sedimentary deposit with pebbles which are porous so that the deposit as a whole has a very high porosity. (*d*) Well-sorted sedimentary deposit whose porosity has been diminished by the deposition of mineral matter in the interstices. (*e*) Rock rendered porous by solution. (*f*) Rock rendered porous by fracturing. (*Redrawn from Meinzer, 1923.*)

Table 11.1. Approximate Characteristics of Ground Water Aquifers

Soil Material	*Total Porosity (%)*	*Specific Yield (%)*	*Relative Permeability*
Dense limestone or shale	5	2	1
Sandstone	15	8	700
Gravel	25	22	5000
Sand	35	25	800
Clay	45	3	1

well is not necessarily a flowing well. The conditions under which artesian flow takes place are shown in Fig. 11.5. For artesian flow, the following conditions must be present: (1) pervious stratum with an intake area, (2) impervious strata below and above the water-bearing formation, (3) inclination of the strata, (4) source of water for recharge, and (5) the absence of a free outlet for the water-bearing formation, at a lower elevation. The level to which water rises in a pipe placed in the water-bearing formation is known as the piezometric head or (in three dimensions) the piezometric surface. Where the pervious stratum outcrops at the surface, artesian springs may develop. These are similar to water-table springs, except that the flow is under pressure.

11.7 Hydraulics of Wells

A cross section of a well installed in homogeneous soil overlying an impervious formation is shown in Fig. 11.6. Under static conditions, the water level will rise to the water table. When pumping begins, the water level in the well is lowered, thus removing free water from the surrounding soil. The water table around a well assumes the general form of an inverted cone, although it is not a true cone. The distance from the well to where the static water table is not lowered by drawdown is known as the radius of influence. The water level at the edge of the well will be slightly higher than in the well because of friction losses through the perforated casing. For a given rate of pumping, the water table surrounding a well in time nearly reaches a stable condition. The radius of influence may be predicted from the physical characteristics of the aquifer. A guide for such predictions is given in Table 11.2.

For a given well there is a definite relationship between drawdown and discharge. For thick water-bearing aquifers or artesian formations, the discharge-drawdown relationship is nearly a straight line. The discharge-drawdown relationship shown in Fig. 11.7 can be obtained by pumping the well at various rates and plotting the drawdown against the discharge. Test pumping a well should be continued for a considerable length of time. Short pumping tests are often misleading. It has been found that even 24-hr tests are not long enough, and

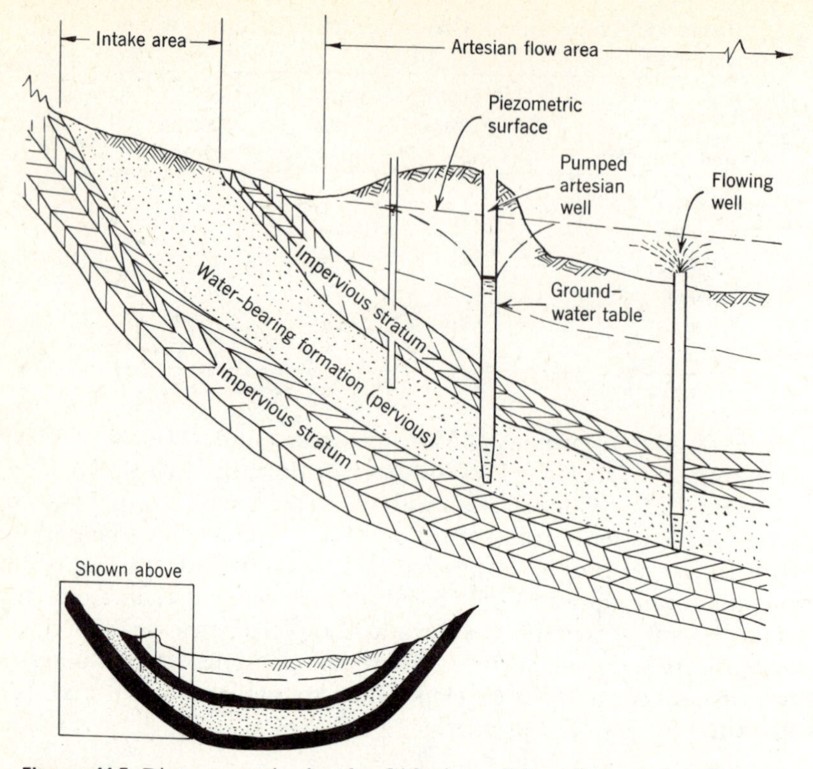

Figure 11.5. Diagrammatic sketch of ideal conditions for artesian flow.

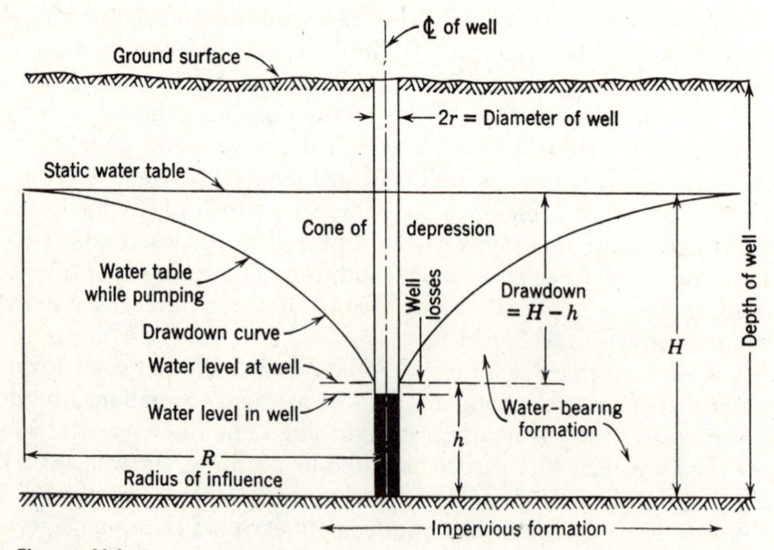

Figure 11.6. Cross section of a typical gravity well in homogeneous soil.

Table 11.2. Radius of Influence of Wells[a]

Soil Formation and Texture	*Radius of Influence (ft)*
Fine sand formations with some clay and silt	100–300
Fine to medium sand formations fairly clean and free from clay and silt	300–600
Coarse sand and fine gravel formations free from clay and silt	600–1000
Coarse sand and gravel, no clay or silt	1000–2000

[a]From Bennison, 1947.

30-day tests are more likely to indicate the true capacity of the well.

The rate of inflow into a gravity well (symbols illustrated in Fig. 11.6) is

$$q = \frac{3.14\ K\ (H^2 - h^2)}{\log_e (R/r)} \tag{11.2}$$

where q = rate of inflow in cfs,
K = hydraulic conductivity in feet per second,
H = height of static water level above the impermeable layer in feet,
h = height of water level in the well above the impermeable layer in feet,
R = radius of influence (Table 11.2) in feet,
r = radius of well casing or gravel envelope in feet.

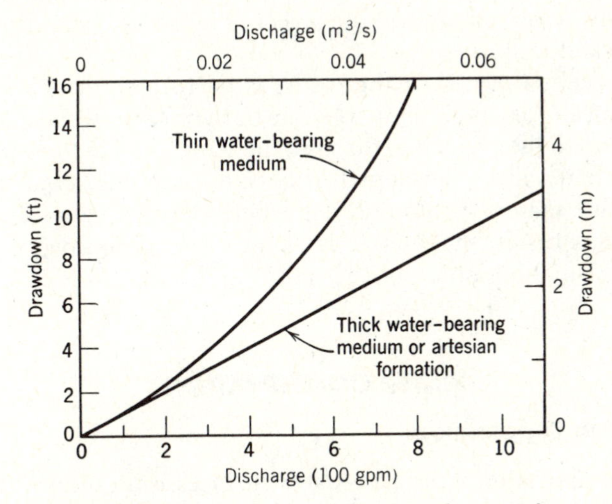

Figure 11.7. Relationship between drawdown and discharge of wells.

11.8 Construction of Wells

The three general types of wells are dug, driven, and drilled. The dug well is excavated deeper than the ground water level. Often it is lined with masonry, concrete, or steel to support the excavation. Because of the difficulty in digging below ground water level, dug wells do not penetrate the ground water to a depth sufficient to produce a high yield.

Wells up to 3 in. in diameter and 60 ft in depth may be constructed by driving a well point into unconsolidated material. A well point is a section of perforated pipe pointed for driving and connected to sections of plain pipe as it is driven to the desired depth. Sometimes the penetration of well points is aided by discharging a high-velocity jet of water at the tip of the point as it is driven.

Deeper and larger-diameter wells are drilled with cable-tool or rotary equipment. With cable tools, a heavy bit is repeatedly dropped onto material at the bottom of the well. Crushed material is removed periodically with a bailer. Cable-tool wells have been drilled to depths of 5000 ft. Deep wells are also drilled with rotary tools consisting of a bit rotated by a string of pipe. A mud slurry pumped through the drill pipe brings cuttings to the surface as it flows up the outside of the drill pipe. In unconsolidated materials, wells may be cased as drilling progresses. Casings may also be installed after drilling is completed.

Well casings are perforated where they pass through the water-yielding strata. In some situations, perforated casing may be formed in place by ripping or shooting holes through a solid casing. In most instances, better results are achieved by placing a corrosion-resistant screen at the water-bearing strata. Screens are made of brass, bronze, or special alloys to resist corrosion. Screen openings are selected to permit 50 to 70 percent of the particles in the aquifer to pass the screen. The open screen area should keep entrance velocities below 0.5 fps to minimize head loss. In aquifers of uniformly fine, unconsolidated material, a gravel pack may be placed around the screen. Figure 11.8 shows a cross section of a gravel-packed well.

After the screen is placed or the casing is perforated, a well should be developed by pumping at a high discharge rate or by surging with a plunger. These practices develop higher velocities through the screen and in the aquifer adjacent to the screen than will be developed in normal pumping from the well. This action brings fine materials into the well, where they are removed by pumping and bailing. As a result, the aquifer is opened for freer flow of water and is stabilized for normal operation of the well.

WATER CONSERVATION

11.9 Evaporation Suppression

Reduction of evaporation from free-water surfaces is an important water-conservation measure. Two broad approaches are employed, reduction of the free-water surface area and protection of free-water surfaces.

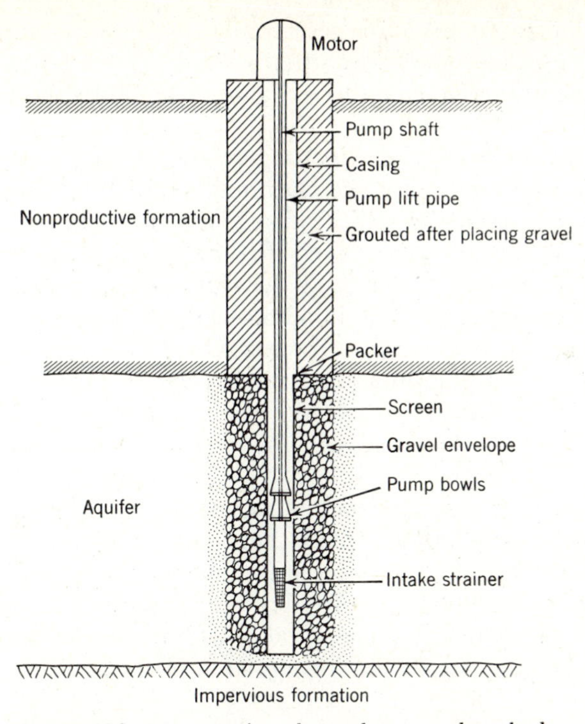

Figure 11.8. Cross section through a gravel-packed well. (*Modified from Linsley and Franzini, 1979.*)

Reduction of the free-water surface is accomplished by minimizing the surface-area-to-volume ratio of reservoirs. Storage of water in natural ground water reservoirs rather than in surface reservoirs also reduces evaporation losses.

Protection of free-water surfaces to reduce evaporation has been uneconomical except in special situations. Methods under investigation include monomolecular films, floating membranes, and floating particles. Each of these is designed to reduce the exposure of the free-water surface. Monomolecular films are made of fatty alcohols, such as hexadecanol. The film must be applied continuously to maintain it against wind action and biological deterioration. Floating membranes of butyl rubber are practical for small reservoirs. Floating particles or blocks of material such as perlite, foamed wax, or light concrete reduce the free-water surface while leaving animals access for drinking. The various methods have all been shown to be effective in reducing evaporation in small reservoirs.

11.10 Artificial Recharge

Ground water reservoirs are supplied by water percolating to them from the surface. Under natural conditions only a small fraction of rainfall reaches the ground water. Since ground water reservoirs provide evaporation-free storage and since surface runoff waters are often wasted, a logical water-conservation measure is to attempt to increase the recharge of ground water reservoirs from surface runoff.

The general methods of artificial recharge are basin, furrow or ditch, flooding, pit and shaft, or injection wells. The basin method of spreading water consists of a series of small basins formed by dikes or banks. The dikes often follow contour lines, and they are so arranged that the water flows from one basin to the next. In the furrow or ditch method, the water flows along a series of parallel ditches placed close together. The flooding method consists of ponding a thin layer of water over the land surface. Pits, shafts, and injection wells, as methods of recharge, are used primarily in municipal areas and industrial centers. Regardless of the method, it is desirable to recharge water that is relatively free of sediment. It is not unusual to use a combination of several methods for recharge purposes.

In artificial recharge basins, it is often necessary to periodically remove layers of accumulated sediment and to replace them with sand. Soil conditioners, such as organic residues, grasses, or chemical treatments, are effective in increasing infiltration rates. Some waters require desilting in settling basins or the use of chemical deflocculants as well as biological control treatment before they can be recharged without clogging the infiltration area or the aquifer.

11.11 Control of Seepage

Water conveyance losses from canals and ditches can be greatly reduced through reduction or elimination of seepage. Concrete linings are frequently placed in irrigation canals and ditches. Asphalt, fiberglass-reinforced asphalt, and plastic linings are also used in canals and reservoirs. Certain additives are successful in canals, reservoirs, and ditches for seepage reduction. One group of chemical additives reduces infiltration capacity by causing soil particles to swell and become hydrophilic. Fine clays are sometimes added to surface soils. These swell and seal soil pores to reduce infiltration rates.

11.12 Phreatophyte Control

The term "phreatophyte" includes plants that habitually obtain their water supply from the zone of saturation or from the overlying capillary fringe. Examples are tamarisk, cottonwood, willow, and mesquite. Eighty species of phreatophytes have been identified. Phreatophytes cover approximately 17 million acres in the western United States and use an estimated 25 million ac-ft of water annually. Phreatophyte growth is largely concentrated along the lower valleys of major rivers. Consumptive use of water by phreatophytes varies with species, climate, and depth to the ground water table. Under high water table conditions, water used by phreatophytes will approach open-pan evaporation. Maximum use of water by tamarisk in the Rio Grande Valley of New Mexico has been measured at 10.8 ft per year with a 2-ft depth to the water table.

Control of phreatophytes thus offers a great potential for water conservation. Control can be effected either by chemical or mechanical means; however, the cost of control has limited its application. To protect treated areas, replacement vegetation having relatively low water-use requirements must be established and maintained. Channelization is the most effective means of salvaging water that would otherwise be lost to phreatophytes. Through channelizaton and drainage, ground water can be lowered in phreatophyte-infested areas and more of the flow can be conveyed to downstream reservoirs. The accompanying lower water table greatly reduces the consumptive use by phreatophytes.

WATER RIGHTS

Two basic divergent doctrines regarding the right to use water exist, namely, riparian and appropriation. They are recognized either separately or as a combination of both doctrines in different states. Both doctrines apply only to surface water in natural watercourses and to water in well-defined underground streams. A comparison of the salient features of these two doctrines is shown in Table 11.3. In the future, adjudicated water rights based on highest-value use will become increasingly important.

Water rights doctrine for percolating ground water varies greatly from state to state. The riparian doctrine for well-defined underground aquifers generally applies to the eastern states, and to some western states. Otherwise, either the appropriation or correlative rights doctrine applies. Under the correlative rights

Table 11.3. Comparison of Water Rights Laws[a]

Characteristic	*Riparian*	*Doctrine of Appropriation*
Acquisition of water right	By ownership of riparian land	By permit from state (state ownership)
Quantity of water	Reasonable use	Restricted to that allowed by permit
Types of use allowed	Domestic, livestock, etc., but not precisely defined	Some beneficial use required
Loss of water by nonuse	No	Yes, but continued use not always required
Location where water may be used	On riparian land, but some exceptions	Anywhere, unless specified in permit

[a]Generally applicable only for surface water in well-defined channels and for water in well-defined underground aquifers. Some state laws on ground water deviate from the above.

doctrine the landowner's use of ground water not only must be reasonable in consideration of the similar rights of others, but it must be correlated with the uses of others in times of shortage.

11.13 Riparian Doctrine

The riparian doctrine, which is a principle of English common law, recognizes the right of a riparian owner to make reasonable use of the stream's flow, provided the water is used on riparian land. Riparian land is that which is contiguous to a stream or other body of surface water. The right of land ownership also includes the right of access to and use of the water, and this right is not lost by nonuse. Reasonable use of water generally implies that the landowner may use all that he or she needs for drinking, for household purposes, and for watering livestock. Where large herds of stock are watered or where irrigation is practiced, the riparian owner is not permitted to exhaust the remainder of the stream; the owner may use only his or her equitable share of the flow in relation to the needs of others similarly situated. Since few eastern states have statutory laws governing water rights, this doctrine is based mostly on court decisions.

11.4 Doctrine of Prior Appropriation

The doctrine of prior appropriation is based on the priority of development and use, that is, the first person to develop and put water to beneficial use has the prior right to continue use. The right of appropriation is acquired mainly by filing a claim in accordance with the laws of the state. The water must be put to some beneficial use, but the appropriator has the right to all water required to satisfy his or her needs at the given time and place. This principle assumes that it is better to let individuals, prior in time, take all the water rather than to distribute inadequate amounts to several owners. Appropriated water rights are not limited to riparian land and may be lost by nonuse or abandonment.

This doctrine is recognized in most of the 17 western states, although in some it is in combination with the riparian doctrine. State water rights laws differ on specific details and they may change from time to time, making generalizatons that apply from state to state difficult.

WATER MEASUREMENT

Effective use of water requires that flow rates and volumes be measured and expressed quantitatively. Water measurement is based upon application of the formula

$$q = av \tag{11.3}$$

where q = flow rate, volume per unit time in cubic feet per second
a = cross-sectional area of flow in square feet
v = mean velocity of flow in feet per second

Flow measurement thus involves determination of mean velocity and area of flow. Some techniques make each of these determinations separately and use them directly in Eq. 11.3. In others the calibration of the measurement device gives the flow directly. Measurement of flow volume requires integration of the flow rate over the time period involved.

Units of volume and of volume per unit time are basically required in water measurement. In agriculture, the units of volume commonly used in the United States are the gallon, cubic foot, and acre-foot. Common units of rate of flow in English units are the gallon per minute, cubic foot per second, and Miner's inch. The Miner's inch is defined by state legislation and varies from state to state.

11.15 Floats

A crude estimate of the velocity of a stream may be made by determining the velocity of an object floating with the current. A straight uniform section of stream several hundred feet long should be selected and marked by stakes or range poles on the bank. The time required for an object floating on the surface to traverse the marked course is measured and the velocity calculated. The average surface velocity is determined by averaging float velocities measured at a number of distances from the bank. Mean velocity of the stream is often taken as 0.6 to 0.9 of the average surface velocity.

Floats consisting of a weight attached to a floating buoy are sometimes used to measure directly mean velocity. The weight is submerged to the depth of mean velocity, and the buoy marks its travel downstream. The float method has the advantage of giving an estimate of velocity with a minimum of equipment. The method is obviously lacking in precision.

Example 11.1. Determine the discharge in an irrigation canal 10 ft wide with an average depth of 2.4 ft. Time trial runs for surface floats to travel 250 ft were 96, 91, 93, 95, and 90 sec. Assume that the average velocity of flow is 80 percent of the surface (float) velocity.

Solution. The average time of flow for the five trials is 93 sec. The cross-sectional area of the canal is $2.4 \times 10 = 24$ sq ft. Substituting in Eq. 11.3,

$$q = 0.80 \times 24 \times \frac{250}{93} = 51.6 \text{ cfs} \quad (1.46 \text{ m}^3\text{/s})$$

11.16 Impeller Meters

Instruments employing an impeller that rotates at a speed proportional to the velocity of flowing water are often used for velocity determination. These instruments are commonly applied to the determination of flow in open channels, and in such applications are called current meters. A measurement of velocity is made at a number of points in the channel cross section, and an average velocity is calculated. A detailed discussion of the use of current meters is given in Schwab et al. (1981).

Impeller meters are also used for velocity measurement in closed conduits. In this application, the cross-sectional area of flow remains constant with time, and the meters are calibrated to read directly in cumulative volume or in flow rate. Figures 11.9 and 11.10 show typical installations of impeller meters in closed conduits.

11.17 Slope Area

The basic equations for velocity of open-channel flow may be applied to streamflow measurements. The equation most commonly accepted is the Manning formula, discussed in Chapter 7. A nomograph for calculating the velocity from the Manning formula is given in Fig. 7.3. Application of the formula to estimation of flow in open channels requires measurement of the slope of the water surface and measurement of the properties of the cross section of flow. The reach of the channel selected should be uniform and if possible as much as 1000 ft long.

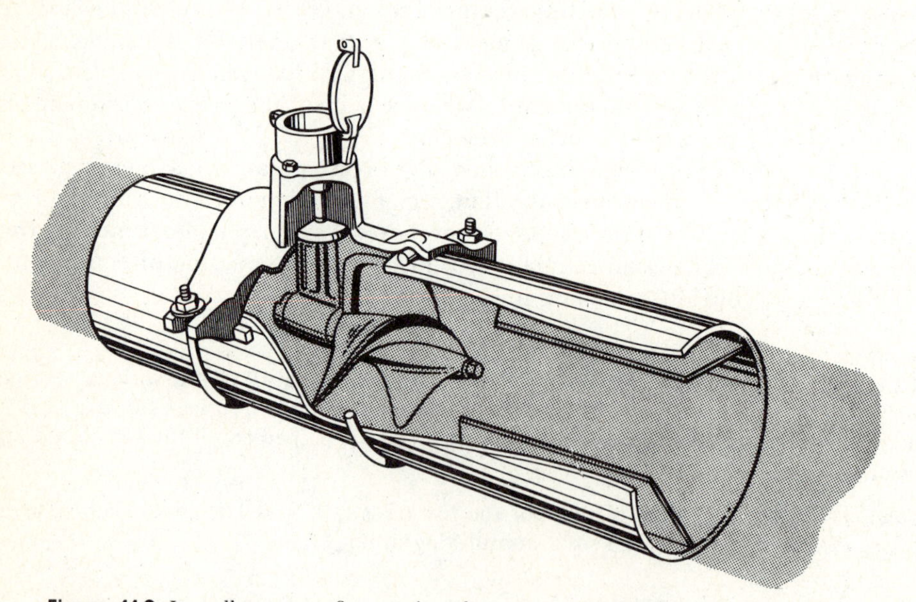

Figure 11.9. Impeller meter for use in a low-pressure pipeline. (*Courtesy Sparling Division, Hersey-Sparling Meter Company, El Monte, California.*)

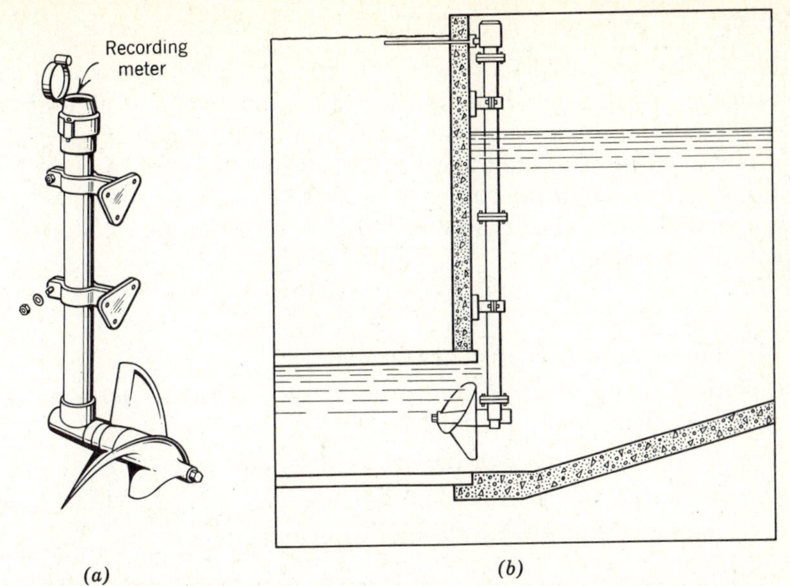

Figure 11.10. Impeller meter for use in open channels. (*a*) Basic meter assembly. (*b*) Installation with recorder at an inverted siphon. (*Courtesy Sparling Division, Hersey-Sparling Meter Company, El Monte, California.*)

The value of the roughness coefficient must be estimated, and this is difficult to do accurately. Table 7.2 gives values of n that are helpful in arriving at such estimates.

The slope area method is sometimes used in estimating the discharge of past flood peaks. The cross-sectional area and flow gradient are measured from high-water marks along the channel. This method gives only a rough approximation of the peak flow.

Example 11.2. Determine the discharge rate of a concrete-lined 2.00-ft width irrigation canal, rectangular in cross section having a uniform depth of flow of 1.10 ft. By instrument survey the difference in elevation of the water surface at points 300 ft apart was 0.21 ft.

Solution. The cross-sectional area of flow, $a = 1.10 \times 2.00 = 2.20$ sq ft and the wetted perimeter, $p = 1.10 + 2.00 + 1.10 = 4.20$ ft. The slope of the water surface, $s = (0.21/300) = 0.0007$. From Table 7.2, select a roughness coefficient, $n = 0.015$ for concrete. Substituting in Eq. 7.1 (the Manning formula),

$$v = \frac{1.49}{0.015} \times \left(\frac{2.20}{4.20}\right)^{2/3} \times 0.0007^{1/2}$$
$$= 99.33 \times 0.650 \times 0.0265$$
$$= 1.71 \text{ fps} \quad \text{(see Fig.7.3)}$$

Substituting in Eq. 11.3,

$$q = 2.20 \times 1.71 = 3.76 \text{ cfs} \quad (0.11 \text{ m}^3\text{/s})$$

11.18 Orifices

An orifice is usually a circular opening through which water flows. When the thickness of the orifice is more than two or three diameters, as in a thick wall, the orifice becomes a short tube. The orifice is a good measuring device, since the velocity is a function of the water head. As water flows through the orifice, head is lost by friction and the contraction of the jet with free outflow reduces the cross-sectional area to less than the orifice. The discharge of an orifice with a thin plate and free flow is

$$q = Ca(2gh)^{1/2} \tag{11.4}$$

where q = discharge in cubic feet per second
C = discharge coefficient, the product of a velocity coefficient and a coefficient of contraction, usually 0.6
a = cross-sectional area of the orifice in square feet
g = acceleration of gravity, 32.2 ft/(sec)(sec)
h = static water head above center of orifice in feet

The coefficient C is dependent on the orifice configuration and must be determined for each orifice design. Figure 11.11 illustrates an end-cap orifice for the measurement of discharge from an irrigation pump. Discharges for orifices 6 in. or less in diameter are given in Table 10.5.

The head producing the flow through an orifice can be measured with a single tube manometer, as shown in Fig. 11.11. The height of water in the tube above the center of the orifice is the head shown in Eq. 11.4. Differential ma-

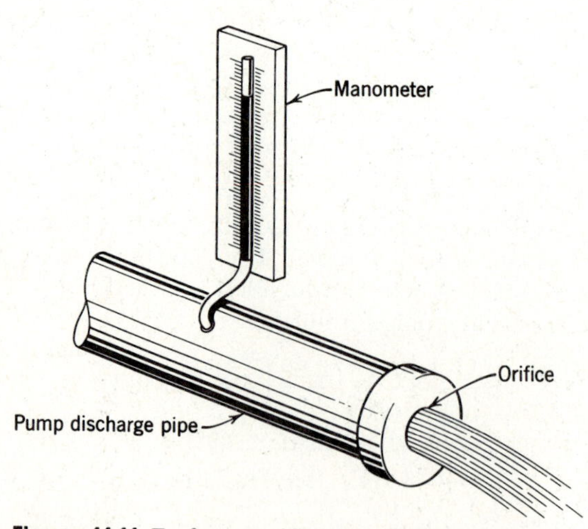

Figure 11.11. End-cap orifice installed for the measurement of flow from a pump discharge.

nometers using a gage liquid lighter or heavier than water are used for measuring the pressure difference in an elbow meter, across an orifice plate, in a pipe, and in other types of meters. Many types of commercial meters are available, usually provided with calibration curves for determining the flow rate.

11.19 Weirs and Flumes

For accurate measurement of flow in open channels, it is desirable to install structures of known hydraulic characteristics. They have a consistent relationship between head and discharge. The action of weirs and flumes in open channels is analogous to that of orifices and tubes for closed conduits.

Weirs. A weir consists of a barrier placed in a stream to constrict the flow and cause it to fall over a crest. The basic equation for flow through such a structure is given in Table 10.2 as well as the discharge capacity for broadcrested weirs up to 20 ft in width and flow depths up to 3 ft. A large weir on a drop spillway channel stabilization structure is shown in Fig. 10.2. Weir openings may be rectangular, trapezoidal, or triangular in cross section, or they may take special shapes to give desired head-discharge relationships. Consult standard hydraulic handbooks and references, such as King and Brater (1963) and Parshall (1950)

Figure 11.12. Rectangular weir for the measurement of flow in a small stream.

for the detailed discussion of weirs. A typical temporary weir for measuring stream flow is illustrated in Fig. 11.12.

Flumes. Specially shaped and stabilized channel sections may also be used to measure flow. Such a section is termed a flume. Flumes are generally less inclined to catch floating debris and sediment than are weirs, and for this reason they are particularly suited to measurement of runoff. One common type of measuring device is the Parshall flume, illustrated in Fig. 11.13. The Parshall flume has the advantage of requiring a very low head loss for operation. Discharge tables for all sizes of flumes are available in Parshall (1950). When the head at H_a (see Fig. 11.13) is less than 0.7 H_a, the flow for flumes of 1 to 8 ft throat width is

$$Q = 4WH_a^{1.522W^{0.026}} \tag{11.5}$$

When H_b is greater than 0.7 H_a, both H_a and H_b must be considered in determining the discharge, and reference should be made to the calibration tables presented by Parshall (1950).

11.20 Other Flow Measurement Methods

Velocity determinations may be made in open channels or closed conduits with pitot tubes (King and Brater, 1963). A pitot tube used in a closed conduit may

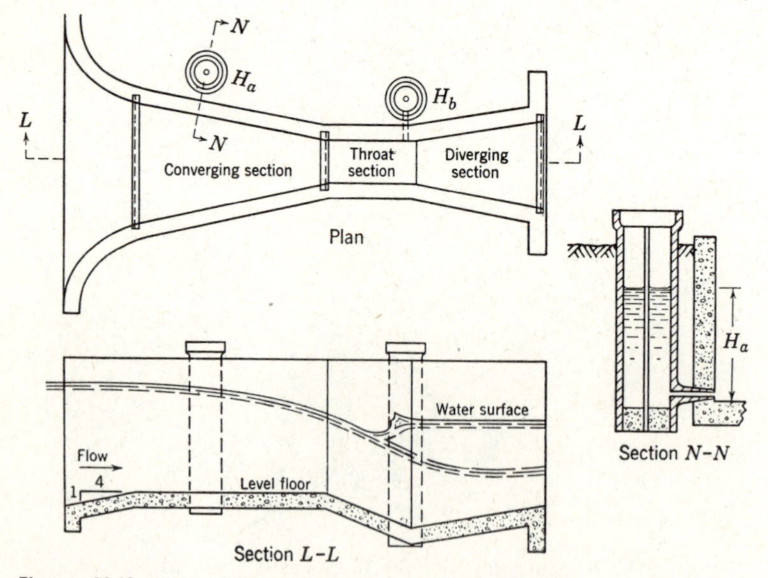

Figure 11.13. Parshall measuring flume. (*Redrawn from Parshall, 1950.*)

be calibrated to read directly the flow rate from one velocity measurement. One such device, designed for insertion through the wall of a pipe, is known as a Cox meter.

Pipe elbows offer an opportunity for measurement of flow. Water flowing through an elbow exerts different centrifugal forces at the inside and outside radii of the elbow, as shown in Fig. 11.14. The resulting difference in pressure may be measured through a differential manometer. Discharge through the elbow is a function of the square root of the pressure differential with the coefficient determined by calibration for each size of elbow.

Where a horizontal pipe discharges to the atmosphere, the trajectory of the jet is a function of the velocity of discharge. Measurement of the coordinates of one point on the surface of the jet with the reference origin at the upper surface of the jet at the point of discharge makes possible calculation of the flow. The laws controlling the velocity and displacement of falling bodies are applied to calculation of the velocity of the jet. Hansen et al. (1980) discuss the method in more detail.

For open channels the vane or pendulum-type meter is growing in use. The vane that extends into the water from the surface is of variable width. Discharge is measured by calibrating the angular displacement of the vane. Another method is the injection of a fluorescent dye or substance that changes the electrical

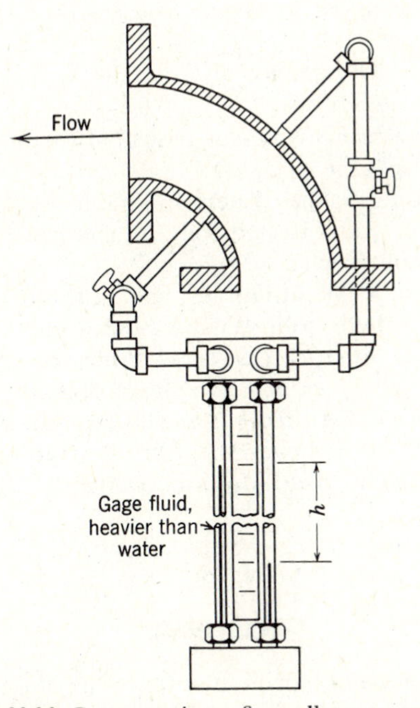

Figure 11.14. Cross section of an elbow meter with a differential manometer. *Source*: Lansford (1936).

conductivity of the stream. It is detected with suitable equipment downstream so as to determine the velocity of flow.

REFERENCES

Ayers, R. S., and D. W. Westcot (1976) "Water Quality for Agriculture, Irrigation and Drainage," Paper No. 29, Food and Agriculture Organization of United Nations, Rome.

Bennison, E. W. (1947) *Ground Water, Its Development, Uses and Conservation*, E. E. Johnson, Inc., St. Paul, Minn.

Bear, J. (1979) *Hydraulics of Groundwater*, McGraw-Hill Book Co., New York.

Bouwer, H. (1978) *Groundwater Hydrology*. McGraw-Hill Book Co., New York.

Garrity, T. A., Jr., and E. T. Nitzschke, Jr. (1967) *Water Law Atlas*, State Bureau of Mines and Mineral Resources and New Mexico Institute Mining and Technology, Cir. 95, Socorro, N.M.

Hagan, R. M., H. R. Haise, and T. W. Edminster (1967) "Irrigation of Agricultural Lands," Monograph No. 11, *Am. Soc. Agronomy*, The Society, Madison, Wis.

Hansen, V. E., O. W. Israelsen, and G. E. Stringham (1980) *Irrigation Principles and Practices*, 4th ed., John Wiley & Sons, New York.

Johnson, E. E., Inc. (1966) *Ground Water and Wells*, E. E. Johnson, Inc., St. Paul, Minn.

King, H. W., and E. F. Brater (1976) *Handbook of Hydraulics*, 6th ed., McGraw-Hill Book Co., New York.

Langsford, W. M. (1936) "The Use of an Elbow in a Pipe Line for Determining the Rate of Flow in the Pipe," *Univ. Ill. Eng. Exp. Sta. Bull. 289*.

Linsley, R. K., and J. B. Franzini (1979) *Water-Resources Engineering*, McGraw-Hill Book Co., New York.

Meinzer, O. E. (1923) "The Occurrence of Groundwater in the United States," U.S. Geological Survey, *Water Supply Paper 489*.

Parshall, R. L. (1950) "Measuring Water in Irrigation Channels with Parshall Flumes and Small Weirs," *U.S. Dept. Agr. Cir. 843*.

Rhoades, J. D. (1983) "Using Saline Waters for Irrigation," Paper presented at the International Workshop on Salt-Affected Soils of Latin America, Maracay, Venezuela, International Society of Soil Science.

Schwab, G. O., R. K. Frevert, T. W. Edminster, and K. K. Barnes (1981) *Soil and Water Conservation Engineering*, 3rd ed., John Wiley & Sons, New York.

Soil Conservation Society of America (1971) *A Primer on Agricultural Pollution*, The Society, Ankeny, IA. (See also *Jour. Soil and Water Conservation* **26** (2):44–65, 1971.)

U.S. Department Agriculture (1954) *Diagnosis and Improvement of Saline and Alkali Soils*, U.S. Dept. Agr. Handb. 60, U.S. Government Printing Office, Washington, D.C.

U.S. Bureau Reclamation (1953) *Water Measurement Manual*, U.S. Government Printing Office, Washington, D.C.

PROBLEMS

11.1 What is the classification of irrigation water having the following characteristics: concentration of Na, Ca, and Mg is 22, 3, and 1.5 milliequivalents per liter, respectively, and the electrical conductivity is 200 micromhos per cm at 25°C? What problems might arise in using this water on fine-textured soil? Consult the U.S. Deptartment of Agriculture Handbook 60.

11.2 Determine the discharge of a stream having a cross-sectional area of 200 ft^2 (18.6 m^2) by the float method. Trial runs for surface floats to travel 300 ft (91.4 m) were 122, 128, 123, 124, and 128 sec.

11.3 Determine the discharge of a stream having a cross-sectional area of 100 ft^2 (9.3 m^2) and a wetted perimeter of 30 ft (9.14 m) using the slope-area method. The channel has some weeds and stones and straight banks and is flowing at full stage. The difference in elevation of the water surface at points 400 ft (122 m) apart is 0.28 ft (0.085 m).

11.4 Determine the capacity of a Parshall flume having a throat width W of 1.25 ft (0.38 m) for H_a = 1.30 ft (0.40 m) and H_b = 0.90 ft (0.27 m).

11.5 An irrigation reservoir has a storage capacity of 80 ac-ft (9.864 ha-m). If the irrigation requirement for the crop is 24 in. (610 mm) and the seepage and evaporation losses are 60 percent of the stored water, how many acres (hectares) can be irrigated?

11.6 Compute the flow rate into a gravity well 24 in. (610 mm) in diameter if the depth of the water-bearing stratum is 80 ft (24.38 m), the drawdown is 30 ft (9.14 m), soil hydraulic conductivity is 3 iph (76 mm/h), and the radius of influence is 600 ft (183 m).

11.7 Compute the flow rate through a 3-in. (76-mm) diameter orifice in the end of a pipe if the pressure is 35 psi (81 ft or 24.7 m of head) and the orifice coefficient C = 0.6.

CHAPTER 12

FARM RESERVOIRS

A farm reservoir is a multiple-use conservation structure that may, depending on its location and size, store water for irrigation, livestock, spray water, fish production, recreation, fire protection, or any combination of these uses. A reservoir is known by a variety of local names, for instance, it may be called a pond, tank, or lake. In general, reservoirs are constructed by excavating or digging out a depression, thus forming a dugout reservoir, by constructing a dam in a natural ravine to form an on-stream reservoir, or by diverting the runoff into an off-stream pond. Surface-water development is discussed further in Chapter 11. Reservoirs are often built for their aesthetic and recreational value. Water for livestock is perhaps their most common use. With the proper filtering and purification equipment, water is suitable for domestic use. Ponds provide little, if any, flood reduction downstream, but they reduce the amount and rate of runoff immediately below the structure and help stabilize the channel both above and below the dam.

12.1 Site Selection

Many factors need to be considered in selecting the site, and seldom will all of them be optimum or the most desirable at a given location. Adequate storage capacity with the least amount of earth movement is normally the most important consideration.

Topography of the Pond Site. The dam should be located where the water depth will be at least the minimum depth indicated in Fig. 12.1 over 15 to 25 percent of the water area at normal water level. Making the dam center line perpendicular to the contour lines will result in the least amount of fill volume. Fill material is usually obtained from the inundated area to give the desired depth of water. Channel slope above the dam should range from 4 to 8 percent. Steep, narrow channels will give a small surface area and volume, while flat, wide channels produce a large surface area and volume, with too much water at shallow depths. Shallow depths can be eliminated by steepening the side slopes along the water line or by excavation. On large ponds, wave damage can be

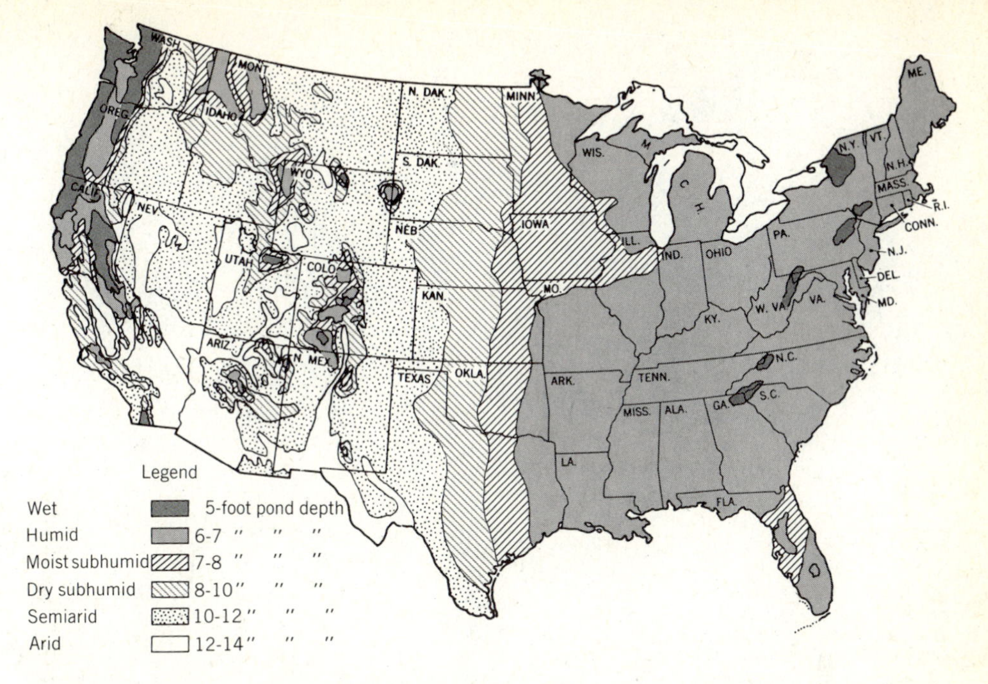

Figure 12.1. Recommended minimum depths for water in farm ponds. *Source:* U.S. Soil Conservation Service (1982).

reduced by placing the dam so that prevailing winds do not strike directly against the upstream face; by constructing a berm; or by using riprap. Topography is not important for dugout or upground ponds, as they can be built in flat areas.

Soil and Underlying Strata. Foundation material under the dam and under the water surface should be impermeable, and suitable fill soil should be available near the dam. Fill soil should generally have no more than 20 percent gravel, 20 to 50 percent sand, less than 30 percent silt, and 15 to 25 percent clay. Topsoil should be eliminated entirely in the fill. Deep peat, sand, gravel, or marl should be avoided as foundation material. Horizontal strata of sand and gravel or fractured rock at shallow depth may result in high seepage losses. Soil conditions prior to construction must be carefully evaluated because the control of seepage after construction is much more difficult.

Watershed Area. The size of the contributing watershed should be large enough to supply the desired amount of water, but not so large as to produce rapid sedimentation in the pond and to create an erosion problem in the flood spillway at the dam. The minimum size of watershed in acres for each acre-foot of storage capacity in the pond is shown in Fig. 12.2. (An acre-foot is a volume equal to 1 ft depth over 1 acre or 43,560 cu ft.) Watersheds larger than two to three times the minimum area in Fig. 12.2 may result in a short reservoir life

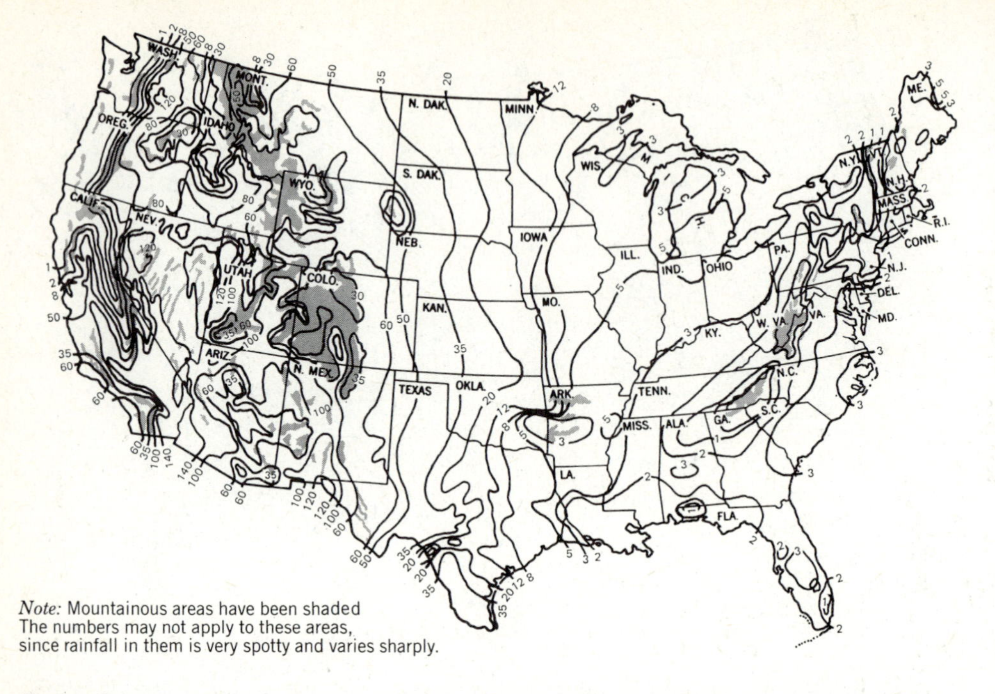

Figure 12.2. Size of drainage area in acres to impound 1 ac-ft of water storage. *Source:* Hamilton and Jepson (1940).

or an expensive flood spillway. These problems are less serious where the watershed has low runoff-producing characteristics, such as flat topography, grass or forest cover, and permeable soils. Completely cultivated watersheds should be avoided, but well-sodded grass waterways and grass cover around the pond will greatly reduce sediment inflow. Locations near farm buildings are desirable for fire protection, but the pollution hazard may be greater. Water from barnyards and other sources of pollution should be diverted away from the pond.

Flood Spillway Location. The most economical spillway is a grassed waterway around the end of the dam or across an adjacent ridge and into another natural channel. Excavated soil from the spillway can be placed in the dam. Thus, a large spillway will generally have little effect on the cost. Steep slopes along the waterway should be avoided where possible, to reduce the danger of erosion. Where good grass sod is already established, the flow may be allowed to spread out without a well-shaped channel. Concrete or permanent structures in the dam for carrying flood flows are normally not recommended because they are too costly.

Location of Water Use. The site should be as near to point of use as possible. For livestock in a pasture, a number of sites may be available, but for recreation or aesthetic uses, the number of choices may be more limited. Pumping from remote sites is often not practical, but where the pond is at a higher elevation than the point of use, gravity flow is possible.

12.2 Water-Storage Requirements

The storage capacity of a pond will depend on the water needs, evaporation from the water surface, seepage into the soil or through the dam, storage allowed for sedimentation, and the amount of carry-over from one year to the next. Several feet of depth are needed for fish to survive through the winter months. Water needs for domestic uses, livestock, spraying, irrigation, and fire protection may be estimated from Table 12.1. Evaporation from a pond-water surface can be estimated by multiplying local pan evaporation values published by the Weather Service by a factor of about 0.7. Evaporation can be reduced by selecting a site having a small surface area and deep depth. Seepage losses depend on the soil and construction techniques. For this reason they are difficult to predict. Studies have shown that ponds with large ratios of storage volume to watershed area provide a reliable supply even in dry years. Thus, within reasonable limits, maximum capacity for a given site is desirable. Minimum storage is usually computed by estimating the total annual needs and allowing 40 to 60 percent of the total storage for seepage, evaporation, sediment storage, and other non-usable requirements.

Table 12.1. Water Requirements for Farm Uses

Type of Use	*Gallons per Day*[a]	*Average acre-feet per Year*[b]
Household, all purposes per person	50–100	0.08
Dry cow or steer per 1000-lb weight	9–18	0.015
Milking cow per 1000-lb weight, including milkhouse and barn sanitation	18–40	0.032
Swine per 100-lb weight	1–1.5	0.002
Sheep per 100-lb weight	1–1.5	0.002
Chickens per 100 head	6–9	0.008
Turkeys per 100 head	10–15	0.014
Horses or mules per 1000-lb weight	8–12	0.011
Orchard spraying	1 gal per year of tree age per application	
Irrigation (humid region) 1–1.5 ac-ft per acre per season		
(arid region) 1–5 ac-ft per acre per season		

[a]Values for air temperatures of 50° F and 90° F, respectively.
[b]Acre-feet per year = 0.00112 × gallons per day.
Source: Midwest Plan Service (1979), and other sources.

12.3 Earth Dam Design

The design of the dam should be based on the most economical use of the available materials adjacent to the site. The most common type is a homogeneous dam with a core extending down to an impervious stratum (see Fig. 12.3). When good fill material is limited, it is placed in the core or center section of the dam. Borings at the dam site should be taken at least 5 ft below the lowest level in the reservoir. Where soil or geologic conditions indicate possible seepage or foundation hazards, deeper explorations should be made. The height of the dam is determined from estimated storage requirements plus an allowance for flood storage and freeboard.

Flood Storage Depth. The depth of water measured from the crest of the trickle spillway (normal water level) to the bottom of the flood spillway is for storage of flood water. This storage is provided so that the flood spillway does not have to carry the runoff from small storms. Flood storage depth may be selected from Table 12.2 for a pond with a given surface area and with a given return period peak runoff rate from the watershed.

Freeboard. Freeboard is usually 2 ft, which is the depth from the bottom of the flood spillway to top of the dam. It includes an allowance of 0.5 ft for frost action and a depth of flow in the flood spillway of 1.0 ft, with the remaining 0.5 ft height for wave action. Freeboard is necessary to assure that the dam does

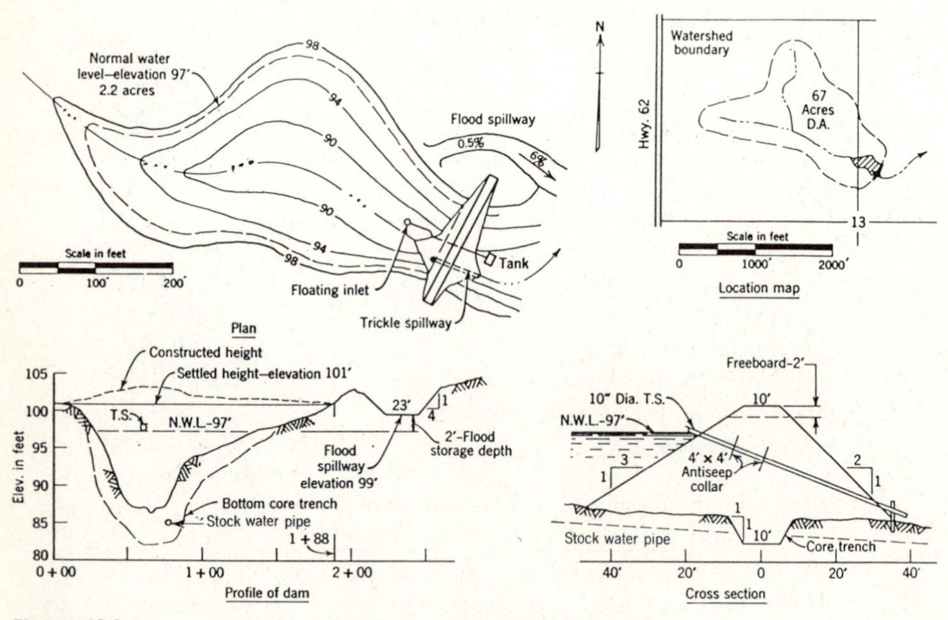

Figure 12.3. Farm pond plan and layout for Example 12.1.

Table 12.2. Flood Storage Depth for Farm Ponds (in Feet)

Watershed Peak Runoff Rate (cfs)	*Water Surface Area at Normal Water Level (acres)*				
	0.5	*1*	*2*	*3*	*5*
15 or less	1.0	1.0	—	—	—
15 to 25	1.5	1.0	0.5	—	—
25 to 35	2.0	1.5	1.0	0.5	—
35 to 45	2.0	2.0	1.0	0.5	0.5
45 to 60	2.0	2.0	1.5	1.0	1.0
60 to 80	—	2.5	2.0	1.5	1.0
80 to 100	—	—	2.5	2.0	1.5

not overtop. Where the dam height is over 15 ft and the length of the water surface is greater than 400 ft, the freeboard should be increased.

Side Slopes. For dams less than 50 ft in height with average soil, the side slopes should be not steeper than 3 : 1 (horizontal to vertical) on the upstream face and 2 : 1 on the downstream side. For coarse or uncompactable soils the side slopes should be flatter to assure stability. Upstream side slopes should be flatter than those on the downstream side because saturated soil is less stable than the unsaturated downstream slope. For ease of mowing with machinery, the downstream slope may be reduced to 2.5 : 1 or flatter.

Top Width. The minimum top width for dams up to 10 ft in height should be about 8 ft. The width should be increased about 0.5 ft for each additional foot of dam height. Where the top is to serve as a roadway, this minimum should be increased to 12 ft to provide a 2-ft shoulder for safety.

Settlement Allowance. Earth fills compacted in thin layers at optimum moisture content on an unyielding foundation will settle less than 1 percent. Since these conditions are usually not met, an allowance of 5 to 10 percent of the settled height should be added to the top of the fill during construction. On dams with the highest fill in the center, the top of the dam just after construction will have a slight crown sloping to either end (see Fig. 12.3).

12.4 Trickle Spillway

This spillway is a pipe or other permanent-type structure through the dam that will maintain the normal water level at the elevation of the entrance opening and will carry a small flow to a safe outlet below the dam. Pipe diameter may be selected from Table 12.3. The purpose of the trickle spillway is to carry any long-duration flow, such as that from a tile or from small storms. In this way the trickle spillway keeps the amount of soil moisture in the bottom of the flood spillway at an optimum level for good plant growth. The size of the trickle

Table 12.3. Trickle-Pipe Spillway Diameter (in Inches)[a]

Watershed Peak Runoff Rate (cfs)	*Water Surface Area at Normal Water Level (acres)*				
	0.5	*1*	*2*	*3*	*5*
15 or less[b]	6	6	—	—	—
15 to 25	8	6	6	—	—
25 to 35	8	8	6	6	—
35 to 45	10	10	8	6	6
45 to 60	10	10	10	8	6
60 to 80	—	12	10	10	8
80 to 100	—	12	12	12	10

[a]For corrugated metal pipes. For smooth wall pipes, use the next smaller diameter.

[b]Trickle spillway is not essential for watersheds less than 10 acres with no ground water flow.

spillway should be large enough to carry the long-duration flow, but it should not be designed to handle flood flows. The capacities of various types of pipe spillways are given in Chapter 10. A pipe with a hood inlet entrance or a drop-inlet pipe spillway is suitable for the trickle spillways (Fig. 12.4). The location of a typical spillway is shown in Fig. 12.3.

Any durable material (such as steel, corrugated metal, or concrete) that cannot be damaged by settling, heavy loads, and the like is suitable for the trickle spillway. Where the fill material surrounding the pipe is not specially compacted, concrete or metal antiseep collars should be attached and sealed to the pipe. The number and size of the collars should be sufficient to increase the length of the creep distance at least 10 percent. The creep distance is the length along the pipe within the dam, measured from the upstream side of the dam to the point of exit on the backslope of the dam. The increase in creep distance for an antiseep collar is measured from the pipe outward to the edge of the collar (see Fig. 12.4*a*). For example, the increase in creep distance for a 4 by 4 ft collar on a 12-in. diameter pipe would be twice the distance from the pipe or $2(2 - 0.5) = 3.0$ ft. All pipes should be centered in the antiseep collar, which should be located near the middle or slightly upstream in the dam. Soil should be well tamped in layers under and around the pipe and around the collar. Antiseep collars are usually not required for dams less than 10 ft in height.

The trickle spillway may, in some cases, serve as a drain for cleaning, draining, or restocking the pond with fish. The drain valve of one such spillway, shown in Fig. 12.4*b*, can be operated from above the water surface. Such inlets should have a screen or debris shield to prevent clogging the pipe.

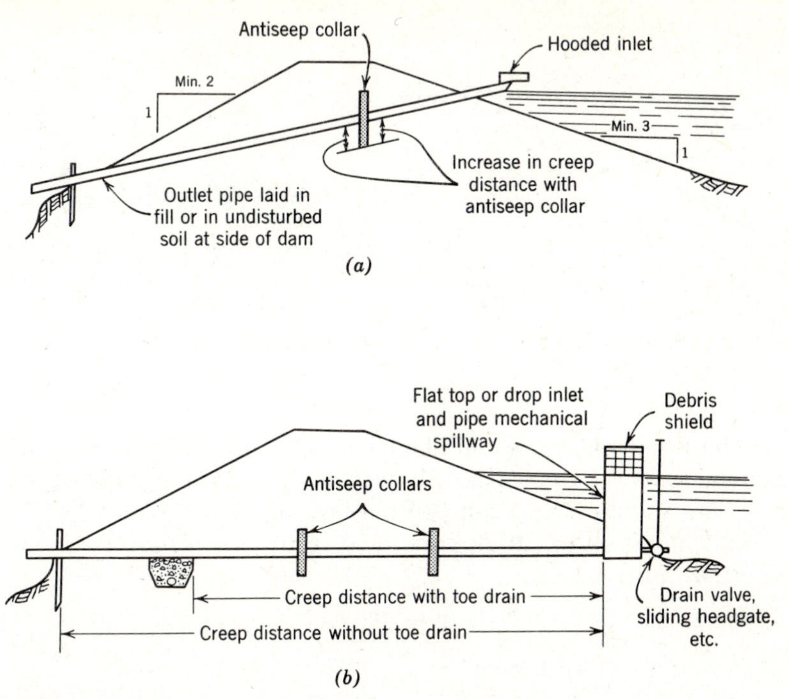

Figure 12.4. Types of trickle spillways and antiseep collars. (*a*) Hooded inlet and pipe. (*b*) Vertical riser and drainpipe combined. *Source*: Schwab et al. (1981).

12.5 Water Pipe and Drain

A steel pipe $1\frac{1}{4}$ in. in diameter through the dam is recommended for supplying water to livestock or for other uses. Two 2 by 2 ft antiseep collars are usually sufficient because of better compaction around a small pipe. The location of a typical water pipe is shown in Fig. 12.5. Such a pipe may be used to drain the pond. All valves should be frost-proof and should be equipped to operate above the water surface. Water from the pipe may be supplied to livestock by a tank and float valve. To obtain the best quality of water in the pond a floating inlet as shown in Fig. 12.5 should be attached to the water pipe.

12.6 Flood Spillway

All ponds filled by direct surface runoff from the watershed above should have a flood (emergency) spillway that will safely bypass floods that exceed the temporary storage capacity of the pond. The design capacity of the flood spillway should be the peak runoff rate for a 25-year return period storm (see Chapter 5). A return period of 50 or 100 years should be selected where potential damages

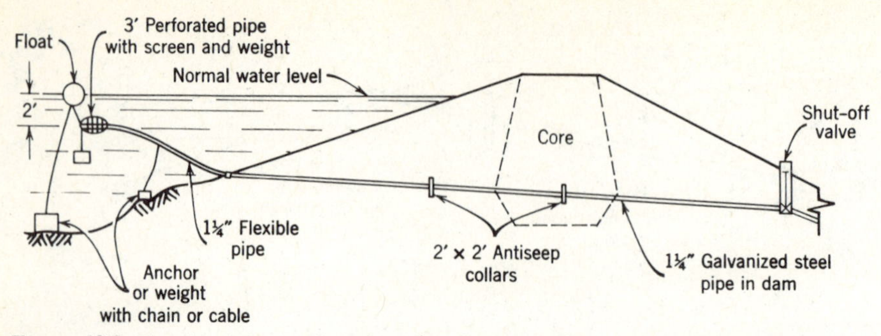

Figure 12.5. Floating inlet and outlet pipe for removing water from a pond.

are high. The flood spillway should be trapezoidal in cross section with a minimum bottom width of about 8 ft. The upper end of the spillway around the end of the dam should have a slope of not less than 0.5 percent. The width of this portion of the spillway can be computed from the weir formula. Assuming a flow depth of 1 ft,

$$W = q/3.2 \tag{12.1}$$

where W = bottom width in feet

q = runoff rate for 25-year return period storm in cubic feet per second

The upper entrance to the flood spillway should be flared out at a 45° angle on each side, to allow smooth flow of water into the channel. The lower end of the spillway normally has a much steeper slope, depending on the topography, and should be designed as a grassed waterway as described in Chapter 7. The outlet end of the spillway should extend well below the dam or into an adjacent waterway.

12.7 Seepage Control

Seepage control can best be achieved by proper design and construction. Seepage may occur through the dam or into the profile where the pond area is underlaid by rock, sand, and gravel strata or solution channels. Knowledge of the soil conditions can best be obtained by soil borings at the site.

Where a limited amount of good fill soil is available, it should be placed in the center section of the dam and in the core trench, which should extend down to an impermeable layer. The core trench should have a minimum width of about 8 ft and side slopes of 1 : 1 or flatter (see Fig. 12.3). Before the fill is placed, the topsoil or other permeable material under the dam should be removed. Maximum density of the fill should be obtained by compaction at the optimum moisture content. Soil is near optimum if, when compressed in the hand, it forms a ball that will not fall apart and from which free water cannot be squeezed. As previously discussed, soil should be carefully compacted under and around pipe or other structures through the dam. The core of good material should be as wide as possible. The next-best fill material should be placed upstream in the face of the dam and the poorest in the downstream section. Steel, concrete, wood, and other materials are suitable as thin-section diaphragms in the dam, but they are generally too costly for ponds.

Sealing of the reservoir area above the dam to prevent seepage may be accomplished by (1) compaction of the area to be covered by water, either with the soil in place or with a blanket of locally occurring heavy clay, (2) dispersement of clay soil with chemicals, (3) application of swelling clays, such as bentonite, and (4) lining with plastic, butyl, or asphaltic materials. Where rock, gravel, or other strata are exposed, the area should be covered in layers 4 to 8 in. thick and compacted with a sheepsfoot roller. The minimum depth should be 2 ft for water 8 ft deep, and it should be greater as the water depth increases. Where the existing soil is suitable, it should be ripped to a depth of 1 ft and then compacted. Where soils do not meet the criteria described in Section 12.1, bentonite or other high-swelling clays should be thoroughly mixed into the soil and then compacted. Soil should have 10 to 15 percent sand to provide the necessary soil strength. Common chemicals for dispersing the soil include salt, soda ash, sodium polyphosphates, and other salts. Effectiveness depends on the chemical and mineralogical composition of the soil. For this reason laboratory testing is necessary to determine the rate and kind of dispersant. Thorough pulverizing and mixing of the soil and chemicals with a rotary tiller or disk is also important.

12.8 Pond Design

The design procedure for a farm pond is described in the following example:

Example 12.1 Design a pond in northeastern Missouri to provide water for 200 steers, for irrigating a 0.5-acre garden, and for a family of 6 persons. A suitable site for the dam has a drainage area of 67 acres, and the estimated design runoff for a 25-year return period storm is 75 cfs. Seepage and evaporation losses from the pond are estimated as 50 percent of the storage capacity and sediment storage allowance as 10 percent.

Solution. From Table 12.1 the annual water requirements are

200 steers (200 × 0.015)	3.0 ac-ft
Irrigation water (0.5 × 1.5)	0.75
Domestic use (6 × 0.08)	0.48
Total	4.23 ac-ft

Storage requirements allowing for seepage loss, evaporation, and sediment storage of 60 percent (50 + 10) are

$$S = 4.23 \times \frac{100}{100 - 60} = 10.6 \text{ ac-ft}$$

From Fig. 12.2, read 6.5 acres of drainage area for each acre-foot of storage. Required watershed size is (10.6 × 6.5)69 acres. The selected site is satisfactory.

From a contour map of the reservoir area and a field investigation of the soils, a dam site was selected as shown in Fig. 12.3. By measuring the area within each contour line with a planimeter, the water-level height was determined as shown in Table 12.4. The area was measured to the center line of the dam. This procedure is sufficiently accurate, because most of the soil in the dam is usually taken from the reservoir area. (An approximate volume may be obtained by multiplying the water-surface area by 0.4 times the maximum water depth.) Storage at this site could be increased by raising the water level or by moving the dam further downstream. If by so doing, sufficient storage could not be obtained, another site would have to be selected.

By interpolation between contours at elevations of 94 and 98 ft, 10.5 ac-ft of storage is available at an elevation of 97 ft, which is selected as the normal water surface. For this site, the following specifications can be determined:

1. Crest elevation for trickle spillway (Table 12.4)	97.0 ft
2. Flood storage depth for 75 cfs and 2.2-ac surface area from Table 12.2 (interpolate as 1.9, round to 2.0 ft)	2.0
Elevation of flood spillway	**99.0 ft**
3. Freeboard (wave height 0.5, frost depth 0.5, and water flow depth 1.0ft)	2.0
Elevation of top of dam (settled height)	**101.0 ft**
4. Maximum allowance for settlement at station 0 + 72 (10% × 13.7)	1.4 ft
5. Top width of dam [8 + 0.5 × (13.7 − 10)] =	10 ft
6. Select dam side slopes upstream 3 : 1, downstream 2 : 1	

Table 12.4. Volume of Pond-Water Storage[a]

Contour Elevation (ft)	*Area Within Contour Line to Center Line of Dam (acres)*[b]	*Average Area (acres)*	*Contour Interval (ft)*	*Volume of Storage (ac-ft)*	*Accumulated Storage Volume (ac-ft)*
88	0.1				0
		0.4	2	0.8	
90	0.7				0.8
		1.0	4	4.0	
94	1.3				4.8
		1.9	4	7.6	
98	2.5				12.4
To elevation 97, surface area = 2.2 ac and storage volume = 12.4 − (7.6/4) = 10.5 ac-ft					

[a]Computed for the pond in Fig. 12.3 and Example 12.1. The number of contour lines was reduced to simplify computations.

[b]Estimated from the map in Fig. 12.3.

7. Trickle spillway diameter and length for 2.2-ac surface area and 75 cfs from Table 12.3 — 10 in.; 60 ft (estimate from Fig. 12.3)
8. Antiseep collar on trickle spillway, required increase in creep distance = 10 percent × 60 = 6 ft [creep distance for 4-ft collar = (4 − 1)2 = 6.0 ft] (outside pipe diameter assumed 1 ft) — two; 4 × 4 ft
9. Livestock water pipe, diameter, and length under dam — $1\frac{1}{4}$ in.; 100 ft
10. Volume of fill in dam and core trench (Fig. 12.6 and Table 12.5) — 1646 yd^3
11. Flood spillway width at control section (level) around end of dam from Eq. 12.1 W = 75/3.2 (flow depth 1.0 ft) — 23 ft
12. Flood spillway width below dam for maximum slope of 6 percent, maximum velocity 6 fps, 4 : 1 side slopes, retardance D, trapezoidal shape cross section (nomographs in Chapter 7) — 0.6 ft depth; 21 ft bottom width

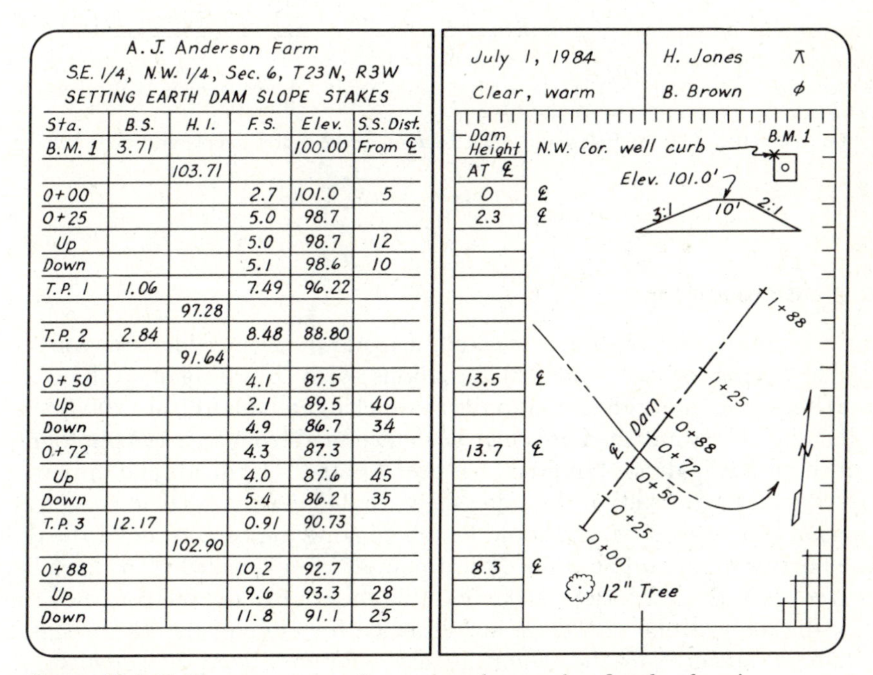

A. J. Anderson Farm
S.E. 1/4, N.W. 1/4, Sec. 6, T23 N, R3W
SETTING EARTH DAM SLOPE STAKES

Sta.	B.S.	H.I.	F.S.	Elev.	S.S. Dist. From ℄
B.M. 1	3.71			100.00	
		103.71			
0+00			2.7	101.0	5
0+25			5.0	98.7	
Up			5.0	98.7	12
Down			5.1	98.6	10
T.P. 1	1.06		7.49	96.22	
		97.28			
T.P. 2	2.84		8.48	88.80	
		91.64			
0+50			4.1	87.5	
Up			2.1	89.5	40
Down			4.9	86.7	34
0+72			4.3	87.3	
Up			4.0	87.6	45
Down			5.4	86.2	35
T.P. 3	12.17		0.91	90.73	
		102.90			
0+88			10.2	92.7	
Up			9.6	93.3	28
Down			11.8	91.1	25

July 1, 1984 | H. Jones ⊼
Clear, warm | B. Brown ϕ

Dam Height AT ℄	
0	℄
2.3	℄
13.5	℄
13.7	℄
8.3	℄

Figure 12.6. Field survey notes for setting slope stakes for the dam in Example 12.1. (Data for stations 1 + 25 and 1 + 88 are not shown.)

Table 12.5. Volume of an Earth Dam[a]

Station along Dam	*Height[c] of Dam (ft)*	*Cross-Sectional Area (ft²)[b]*	*Average Cross-Sectional Area (ft²)*	*Length of Section (ft)*	*Volume of Section (ft³)*
0 + 00	0	0			
			18	25	450
0 + 25	2.3	36			
			313	25	7,825
0 + 50	13.5	590			
			598	22	13,156
0 + 72	13.7	606			
			430	16	6,880
0 + 88	8.3	255			
			184	37	6,808
1 + 25	5.0	113			
			57	63	3,591
1 + 88	0	0		Volume in dam =	38,710
Volume in core trench (estimated average depth 3.3 ft and length 140 ft)					
				[(3.3 × 3.3) + (3 × 10)] 140 =	5,725
				Total earth fill volume =	44,435 ft³
				or (44,435/27) =	1,646 yd³

[a]Computed for the pond in Fig. 12.3 and Example 12.1.

[b]Cross-sectional area = $2.5h^2 + 10h$, where h is the dam height. This equation is valid only for 3 : 1 and 2 : 1 side slopes and for 10-ft top width.

[c]Height of dam (101 − ground elevation at each station at the center line, from Fig. 12.6).

12.9 Field Procedure

After a tentative site has been selected, a contour map of the reservoir area should be prepared (see Chapter 4). Elevations normally are obtained along cross section lines taken perpendicular to the channel at intervals of 50 to 200 ft above the dam site. The watershed area may be determined from an aerial photograph on which the water-divide boundary has been located. The divide and the runoff-producing characteristics of the watershed are determined by field survey. Soil borings to the desired depth should be taken along the center line of the dam, in the borrow area, and at other critical points. After detailed plans have been prepared, as in Fig. 12.3, slope stakes locating the edge of the fill from the center line of the dam can be set. These stakes locate the edge of the fill for starting the dam. Distances from the center line are computed from the equation,

$$d = (E_t - E)s + w/2 \tag{12.2}$$

where d = distance from center line to edge of fill in feet
E_t = Elevation of top of dam (settled height) in feet
E = elevation of the upstream or downstream slope stake in feet
s = side slope ratio of the dam
w = top width of dam in feet

For the pond in Example 12.1, the surveying notes are shown in Fig. 12.6 and the slope-stake distances are computed in Fig. 12.7 for station 0 + 50. Where the ground is nearly level, such as at station 0 + 25 in Fig. 12.6, distances can be computed directly from Eq. 12.2, since elevations at the slope stakes are the same as at the center line. Where the ground is sloping (perpendicular to the dam center line), slope stakes must be located by trial and error. The procedure illustrated in Fig. 12.7 is as follows: Point A is the first trial point obtained by substituting, in Eq. 12.2, the elevation of the ground at the center line, for which $d = (101 - 87.5)3 + 5 = 45.5$ ft. The actual elevation at A is 90.5, for which $d = (101 - 90.5)3 + 5 = 36.5$ ft. This is the distance to the second trial point B. The actual elevation of B is 89.3, for which $d = (101 - 89.3)3 + 5 = 40.1$ ft. This is the location of the third point C, which has an elevation of 89.5 ft. Since this elevation is within 0.2 ft of that at point B, point C is close enough to the correct distance of 39.5 ft shown in Fig. 12.7. Usually one or two trials are sufficient. Failure to set side slopes in this manner will result in a variation of slope on the face of the dam. If the stake had been set at the first trial point A in Fig. 12.7, the side slope would have been too flat. The downstream slope stakes are set in the same manner described above.

The shore line should be staked at about 50-ft intervals at the normal water level. This line will indicate where trees should be removed and where the water

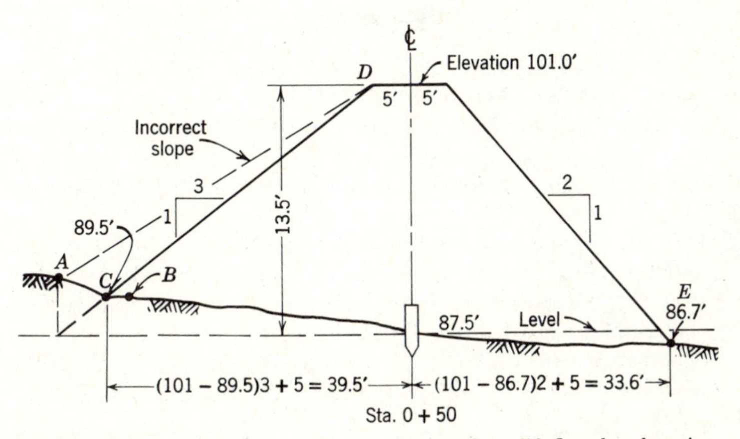

Figure 12.7. Setting slope stakes at station 0 + 50 for the dam in Example 12.1.

depth will be shallow. As construction proceeds, the water-supply pipe, trickle spillway, and flood spillway should be staked out.

12.10 Construction and Maintenance

All trees, stumps, and shrubs should be removed from the dam site and from the area to be inundated. Sod and topsoil should be removed and stockpiled. Before the placement of fill in the dam, the existing soil should be thoroughly plowed and disked and be near optimum moisture content. The fill material should be placed in 4- to 8-in. layers, evenly, over the entire dam. The sheepsfoot roller is best suited for compacting the fill. Heavy hauling equipment should use varied travel paths so as to avoid overcompaction. Hand-operated pneumatic or motor tampers are best for compacting around pipes and antiseep collars. Along the shore line, except at the dam, the slope should be increased to 2 : 1 to a minimum depth of 4 ft. The elimination of shallow water will minimize the growth of cattails and other water weeds.

Slopes and other areas above the water line where the subsoil has been exposed should be covered with about 6 in. of topsoil, fertilized, and seeded with a suitable grass. The entire pond area should be fenced to prevent damage to the dam, spillways, and banks. Sedimentation can be reduced by protecting waterways with grass and by establishing adequate erosion-control practices on the watershed.

Normally, little maintenance is required, but the pond should be inspected occasionally for evidence of seepage on the downstream face of the dam, piping, wave action, and damage by animals or humans. Weed growth and algae in the pond can be controlled with suitable chemicals. Trees should not be allowed to grow near the dam.

REFERENCES

Hamilton, C. L., and H. G. Jepson (1940) "Stock-Water Developments; Wells, Springs, and Ponds," *U.S. Dept. Agr. Farmers' Bull. 1859*.

Midwest Plan Service (1979) *Private Water Systems,* MWPS-14, Iowa State University, Ames, Iowa.

Neely, W. W. et al. (1965) "Warm-Water Ponds for Fishing," *U.S. Dept. Agr. Farmers' Bull 2210*.

Schwab, G. O., R. K. Frevert, T. W. Edminster, and K. K. Barnes (1981) *Soil and Water Conservation Engineering*, 3rd ed., John Wiley & Sons, New York.

Sewell, J. I. (1968) "Laboratory and Field Tests of Pond Sealing by Chemical Treatment," *Tennessee Agr. Exp. Sta. Bull. 437*.

Sisson, D. R., and R. H. Austin (1960) "Farm Ponds," *Purdue Agr. Ext. Bull. 369*.

U.S. Bureau Reclamation (1977) *Design of Small Dams*, 2nd ed., U.S. Government Printing Office, Washington, D.C.

U.S. Soil Conservation Service (1982) *Ponds—Planning, Design, Construction, U.S. Dept. Agr. Handb. 590*, U.S. Government Printing Office, Washington, D.C.

U.S. Soil Conservation Service (1971) *Ponds for Water Supply and Recreation*, U.S. Dept. Agr. Handb. 387, U.S. Government Printing Office, Washington, D.C.

PROBLEMS

12.1 Determine the storage requirements for a pond that is to provide sufficient water for 100 milk cows and 200 hogs, and spray water for 1000 apple trees with an average age of 10 years and with 4 applications per year. Seepage and evaporation losses are estimated to be 54 percent of the stored water.

12.2 Determine the minimum-size watershed area above an on-stream pond at your present location that will hold 11 ac-ft of water.

12.3 Determine the usable water storage for a pond at your present location that is to be filled with runoff from a 25-acre watershed. Assume seepage and evaporation loss as 60 percent. If the volume of runoff is not adequate, what measures could be taken to secure additional water? Assume that all runoff is required to fill the pond.

12.4 Determine the elevation of the normal water level to store 8.1 ac-ft of water in a pond. From a contour map of the reservoir area, the following area of each contour line to the center line of the dam was obtained: contour elevation 40, 0.2 acre; 42, 0.6; 44, 0.8; 46, 1.0; 48, 1.5; 50, 1.7. All storage below elevation 40 is to be reserved for sediment.

12.5 The elevation of the lowest point along a dam is 36.0 ft. The normal water level is to be 10 ft above this elevation, and the water surface area is 1.2 acre. The peak runoff from the watershed is 28 cfs for a 25-year return period storm. Determine the elevation of the trickle spillway, the flood spillway, and the top of the dam; the diameter for the trickle spillway; the top width of the dam; and the constructed height before and after settlement. Assume a settlement allowance of 10 percent of the settled height.

12.6 The height of fill for a dam at station 1 + 00 along the center line is 10 ft, and at 2 + 00 the height is 12 ft. Compute the volume of fill in cubic yards between these two stations, assuming that the top width of the dam is 10 ft and that the base of the dam is level. Side slopes for the dam are 3 : 1 and 2 : 1.

12.7 Write a simplified equation for the volume of fill in cubic yards per foot of length for an earth dam where the height of the dam at the center line is h, the top width is w, and the side slopes are 3 : 1 and 2 : 1. Assume that the base of the dam perpendicular to the center line is level.

12.8 Determine the width of the flood spillway at the level control section around the end of the dam with a maximum depth of flow of 1 ft (0.3 m). The runoff from a 25-year return period storm is 64 cfs (1.81 m^3/s). Determine the minimum bottom width of the lower end of this grassed spillway if the slope is 12 percent. Assume a maximum velocity of 7 fps (2.13 m/s) and retardance class D for the grass. See Chapter 7.

12.9 How far from the center line of the dam should the up- and down-stream slope stakes be set if the dam is 15 ft (4.57 m) high at the center line and the top width is 12 ft (3.66 m)? Assume that the ground is level perpendicular to the center line of the dam.

12.10 Solve Problem 12.9 if the ground slope is a uniform 10 percent toward the downstream side.

12.11 Locate and design a farm pond from a topographic map and related data supplied by your instructor.

CHAPTER 13

SURFACE DRAINAGE

Surface drainage is the removal of excess water using constructed open ditches, field ditches, land grading, and related structures. In this text the application is for land that has insufficient natural slope to provide adequate drainage for good agricultural production. In humid areas many thousands of miles of main outlet ditches have been built mostly by organized drainage districts. In arid regions under irrigation, drainage ditches are necessary to remove water required for leaching undesirable salts from the soil and to dispose of excess rainfall. Outlet ditches must be large enough to carry floodwater and of sufficient depth to provide outlets for tile drains. The design of outlet ditches is beyond the scope of this book, but the principles of grassed waterway design discussed in Chapter 7 are applicable. The primary differences are that the side slopes are steeper and the roughness coefficients in the Manning equation are normally lower for drainage ditches. The design of irrigation canals is basically the same as for drainage channels.

In this chapter surface drainage will be limited to field-size areas, and will include broad shallow surface drains that carry runoff to the point of entrance to outlet ditches. Land grading, which results in a continuous land slope toward the field ditches, is an important part of a surface drainage system. Land grading is essential for surface irrigation. It is the same as land leveling or land shaping, common terminology in irrigated regions.

Surface drainage systems must be suitable for mechanized operations on various types of topography, such as ponded areas, flat fields, and gently sloping land. Similar drains are needed to collect wastewater from surface irrigation systems. Ponded areas (potholes) are frequently found in glaciated regions, where erosion has not formed natural outlets. Flat or nearly level land having impermeable subsoils frequently requires surface drainage. Claypan, fragipan, or heavy-textured soils are examples. Flat land is defined as that with slopes less than 2 percent most of which is less than 1 percent.

From an extensive survey, Gain (1964) estimated that over 100 million acres in the eastern United States and about 8 million acres in the eastern provinces of Canada would benefit from surface drainage. The location and intensity of these drainage problem areas are shown in Fig. 13.1.

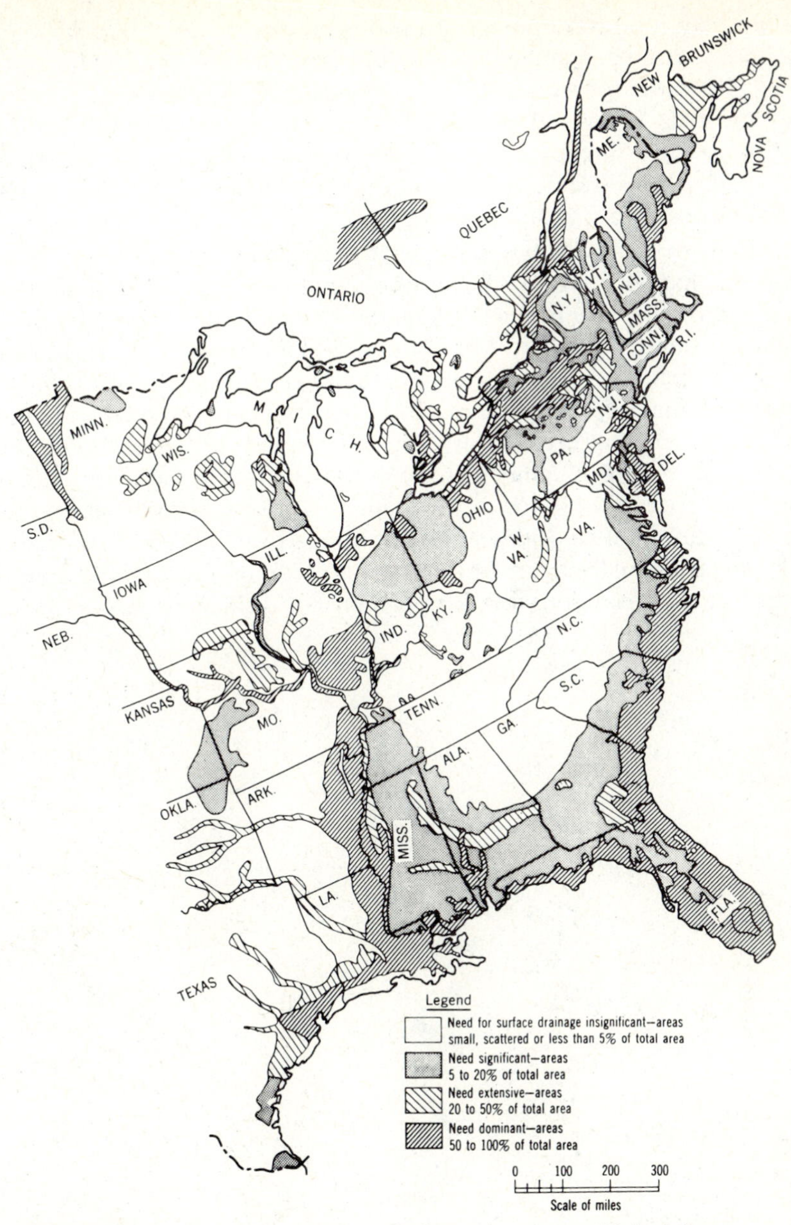

Figure 13.1. Lands needing surface drainage in eastern United States and Canada. *Source*: Gain (1964).

13.1 Random Field Ditches

Random potholes or depressional areas are common in glaciated regions and in other areas where the land is nearly level. Water accumulates in these shallow depressions, causing crop damage. Improper tillage or deadfurrows may also result in poor surface drainage. A random field-ditch system is illustrated in Fig. 13.2. The outlet for such a system may be a natural channel, a constructed ditch, or the field below, allowing the water to spread out where no distinct channel exists. Field ditches connecting depressions are normally less than 3 ft in depth. Such drains should follow routes that provide minimum cuts and the least interference with farming operations.

The design of field ditches is similar to the design of grass waterways, as discussed in Chapter 7. Where farming operations cross the channel, the side slopes should be flat; that is, 8 : 1 or greater for depths of 1 ft or less and 10 : 1 or greater for depths over 2 ft. Minimum side slopes of 4 : 1 are possible if the field is farmed parallel to the drain. The depth is determined primarily by the topography of the area, outlet conditions, and the capacity of the channel. A minimum cross-sectional area of 5 ft^2 is recommended. The grade in the channel should be such that the channel will not erode or fill with sediment. Maximum velocities vary from 2.5 fps for sandy loam and 3.0 fps for silt loam to 5.0 fps for clay soil. Minimum velocities should be about 1 to 2 fps for flow depths less than 3 ft. For triangular channels, the maximum grade for sandy loam is about 0.2 percent and for clay soils it is 1.0 percent. For drainage areas larger than 5

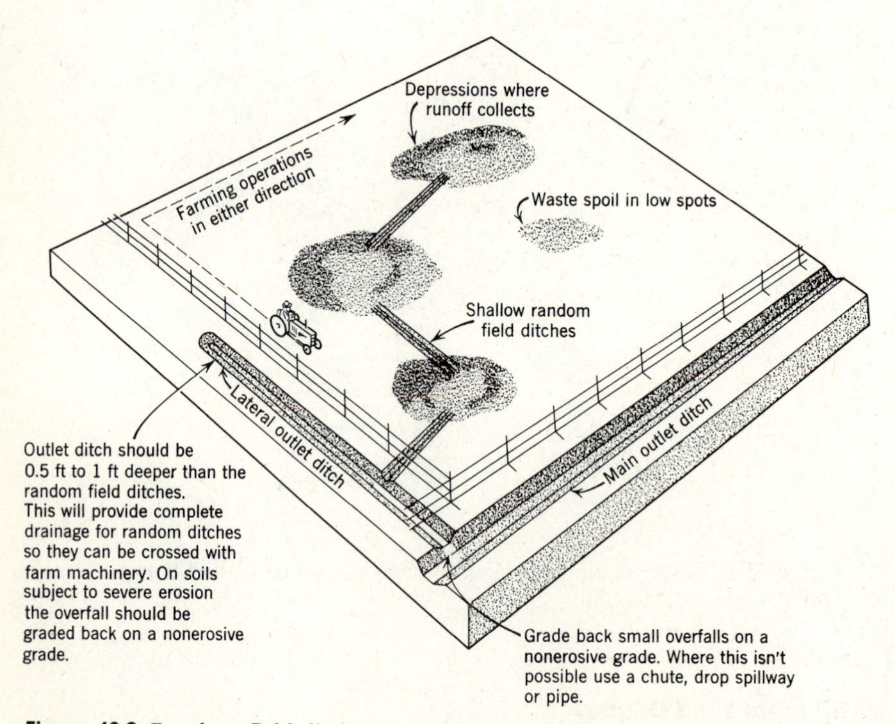

Figure 13.2. Random field-ditch system for surface drainage. (*Redrawn from Beauchamp, 1952.*)

acres, the discharge capacity of the channel should be based on the runoff for a 10-year return period storm. Because of crop damage from flooding, surface water should be removed within a period of about 24 h.

The plan, profile, and cross section of a random field ditch is shown in Fig. 13.3. The field-survey notes and a method for computing the cuts and volume of excavation are illustrated in Fig. 13.4. The detailed procedure is explained in Example 13.1. In staking out the drain in Fig. 13.3, elevations of the ground surface were taken on 50-ft stations starting from the bottom of the pothole and going over the adjacent ridge to such a point as will provide natural

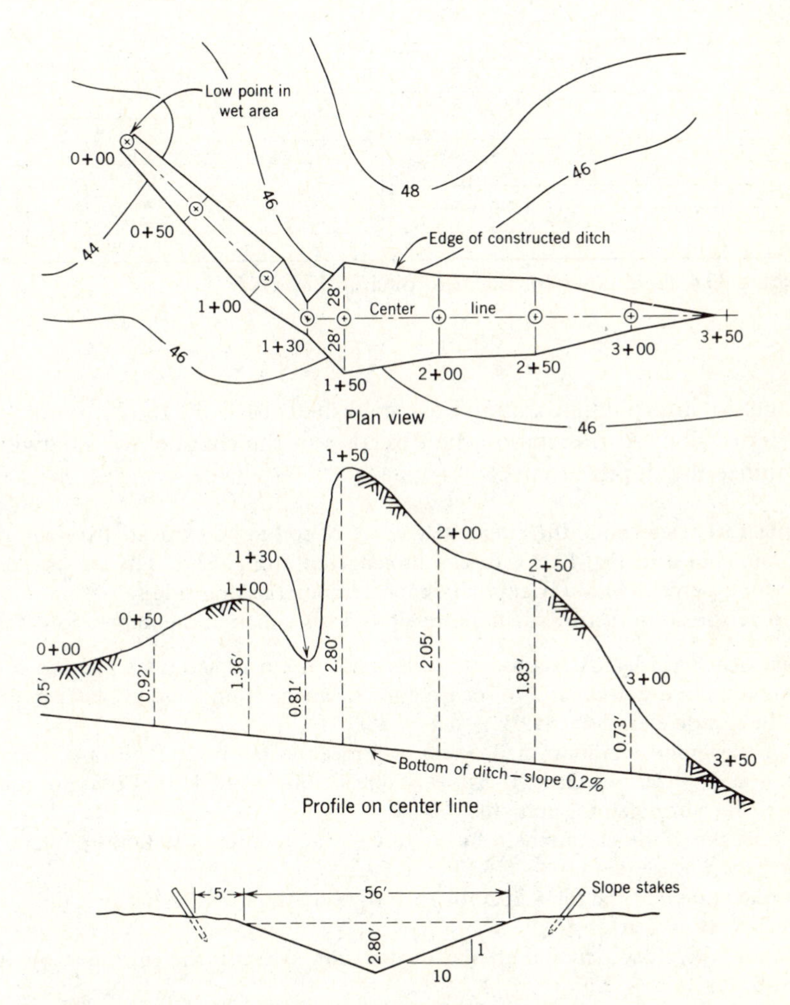

Figure 13.3. Plan, profile, and cross section of a random field ditch.

John Doe Farm
S.E. 1/4, Sec. 6, T81N, R6W
Surface Drain

May 10, 1984 — Clear, cool — J. Moe ⊼ — R. Timm φ

Sta.	B.S.	H.I.	F.S.	Hub Elev.	Grade Elev.	Cut	1/2 Top Width	X-Sect. Area	Avg. X-Sect. Area	Length ft.	Volume ft.
B.M. 1	1.36			50.00							
		51.36									
0+00			7.35	44.01	43.51	0.50	5.0	2.5			
									5.5	50	275
0+50			7.03	44.33	43.41	0.92	9.2	8.5			
									13.5	50	675
1+00			6.69	44.67	43.31	1.36	13.6	18.5	12.5	30	375
1+30			7.30	44.06	43.25	0.81	8.1	6.5	42.5	20	850
1+50			5.35	46.01	43.21	2.80	28.0	78.4			
									60.2	50	3,010
2+00			6.20	45.16	43.11	2.05	20.5	42.0			
									37.8	50	1,890
2+50			6.52	44.84	43.01	1.83	18.3	33.5			
									19.4	50	970
3+00			7.72	43.64	42.91	0.73	7.3	5.3			
									2.7	40	108
3+50			8.71	42.65	42.81	−0.16	0	0			
									TOTAL − 8,153		
B.M. 1			1.35	50.01				8153/27 = 302 cu. yds.			
			Error	0.01							

Figure 13.4. Field notes for the field ditch in Fig. 13.3.

drainage. An intermediate station was established at 1 + 30, the center of another small depression. At this station the direction of the channel was changed so as to minimize the depth of cuts.

Example 13.1. Determine the cuts and volume of soil to be excavated for the random field drain shown in Fig. 13.3 with elevations given in Fig. 13.4. The soil will permit a grade of 0.2 percent with a triangular-shaped channel having side slopes of 10 : 1. The soil is to be spread in nearby small depressions, rather than in the depression at 0 + 00.

Solution. See Fig. 13.4. At station 0 + 00, allow a cut of 0.5 ft to provide for some sedimentation in the drain and to secure good drainage from the pothole. Compute and record the grade elevation, 44.01 − 0.5 = 43.51 ft.

From the grade elevation at 0 + 00, subtract the fall in 50 ft to obtain the grade elevation at 0 + 50, which is 43.51 − (0.002 × 50) = 43.41 ft. Compute the grade elevations for all remaining stations.

Subtract the grade elevation from the hub elevation at each station to obtain the cut (e.g. at station 0 + 00, 44.01 − 43.51 = 0.50).

Compute one half the top width for each station, which is the cut times the side slope ratio (0.5 × 10 = 5 ft).

Compute the cross-sectional area at each station, which is the cut times one half the top width (0.5 × 5 = 2.5).

Obtain the average cross-sectional area for each pair of adjacent stations (2.5 + 8.5)/2 = 5.5.

Compute the volume of excavation between each pair of adjacent stations, which is the length of the section times the average cross-sectional area ($50 \times 5.5 = 275$). Note that near the last station, the grade line would intersect the ground surface at about 3 + 40 rather than 3 + 50. Add the volumes of each segment to obtain the total excavation.

Surface water in large depressions may also be removed through tile drains or discharged into a drainage well. Such wells are feasible only in special situations where the true water table is low and where an impervious layer is underlaid by pervious sand, gravel, or rock formations. They may contaminate ground water supplies. With tile outlets, the water may flow by gravity or it may be raised by a pump to a higher level. Most pumps for this purpose are of the propeller type, which discharge several hundred gallons per minute with lifts from 5 to 10 ft. The design of such pumping installations is beyond the scope of this text.

As shown in Fig. 13.5, a surface inlet, sometimes called an open inlet, is an intake structure for the removal of surface water from potholes, road ditches, depressions, and farmsteads. They should be placed at the lowest point along fence rows or in land which is in permanent vegetation. Where the inlet is in a cultivated field, the area immediately around the intake should be kept in grass. To prevent the entrance of trash, a fence may be constructed around it. Galvanized metal pipe or a manhole constructed of brick or monolithic concrete is also satisfactory. Manholes with sediment basins are sometimes used as surface inlets. At the surface of the ground a concrete collar should extend around the intake on the riser to prevent growth of vegetation and to hold it in place. On top of the riser a beehive cover or other suitable grate is necessary to prevent trash from entering the tile.

Where the surface inlet is connected to a main tile line, it is good practice to offset the surface inlet several feet from the main. Such construction may eliminate failure of the system if the surface inlet structure should become damaged.

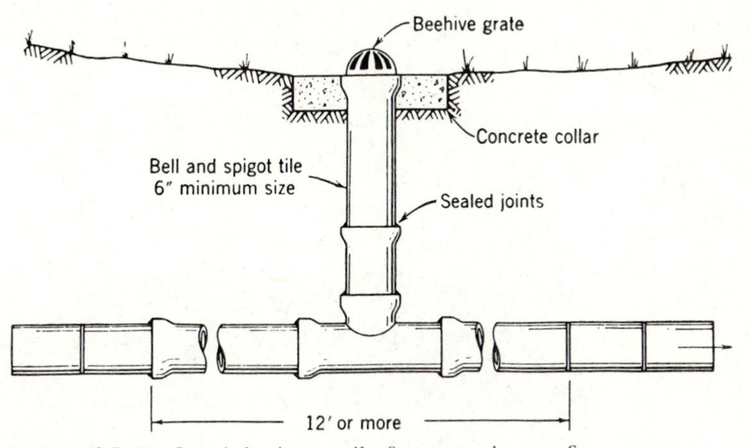

Figure 13.5. Surface inlet into a tile for removing surface water.

Where the quantity of water to be removed is small, (as in local depressions), blind inlets, also called "French drains," may be installed over the tile drain. These are constructed by backfilling the tile trench with various gradations of materials, such as gravel or coarse sand, or with corn cobs, straw, and similar substances. Although such inlets may not be permanently effective, they are economical to install and do not interfere with farming operations. As the voids in the backfill of the blind inlet become filled, its effectiveness is reduced. Since the soil surface has a tendency to seal, a narrow strip of dense sod-forming grass or other vegetation directly over the tile line will greatly increase the flow through the inlet. Where cultivation over the tile can be avoided, flow may be increased by extending the coarse material to the soil surface.

13.2 Bedding

Bedding is a method of surface drainage consisting of narrow-width plow lands in which the deadfurrows run parallel to the prevailing land slope (see Fig. 13.6).

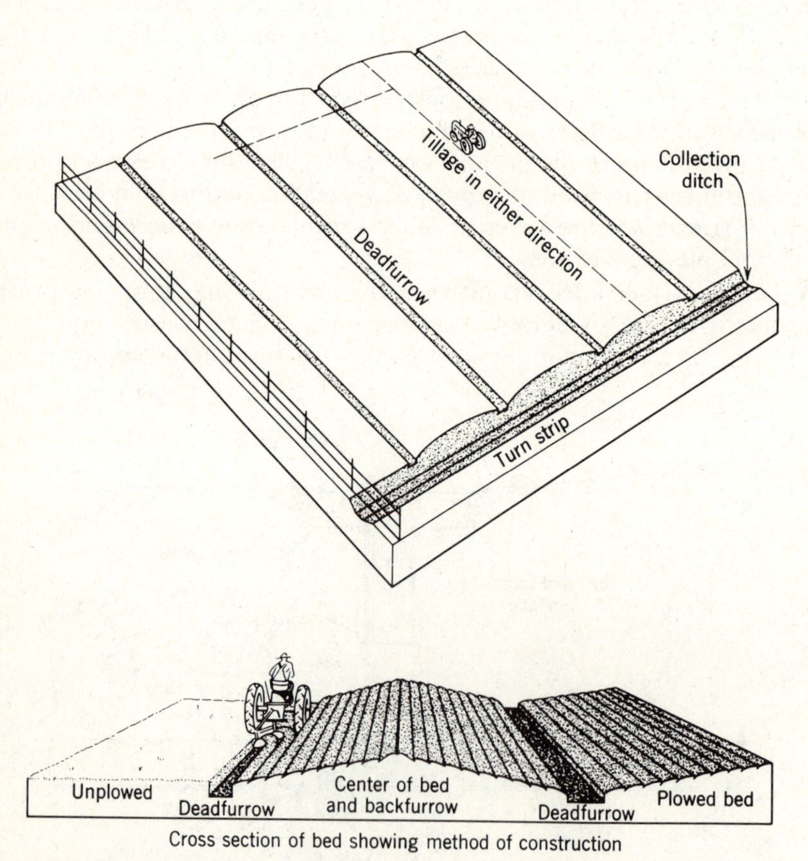

Figure 13.6. Bedding method of surface drainage for flat land.

The area between two adjacent deadfurrows is known as a bed. Bedding is most practicable on flat slopes of less than 1.5 percent where the soils are slowly permeable and subsurface drainage is not economical. Studies in southern Iowa showed that level land gave slightly better yields than bedding.

The design and layout of a bedding system involves the proper spacing of deadfurrows, depth of bed, and grade in the channel. The width of bed depends on the land slope, drainage characteristics of the soil, and the cropping system. Bed widths recommended by Beauchamp (1952) for the Corn Belt region vary from 23 to 37 ft for very slow internal drainage, from 44 to 51 ft for slow internal drainage, and from 58 to 93 ft for fair internal drainage. The depth of bed depends on the soil characteristics and tillage practices. In the bedded area the direction of farming may be parallel or normal to the deadfurrows. Tillage practices parallel to the beds have a tendency to retard water movement to the deadfurrows. Plowing is always parallel to the deadfurrows.

13.3 Parallel Field Ditch System

Parallel field ditches are similar to bedding except that the channels are spaced farther apart and have a greater capacity than the deadfurrows. This system is well adapted to flat, poorly drained soils with numerous small depressions which must be filled by land grading.

The design and layout are similar to that for bedding except that drains need not be equally spaced and the water may move in only one direction. The layout of such a field system is shown in Fig. 13.7. The size of the ditch may be varied, depending on grade, soil, and drainage area. The depth of the drain should be a minimum of 0.75 ft and have a minimum cross-sectional area of 5 ft^2. The side slopes should be 8 : 1 or flatter to facilitate crossing with farm machinery. As in bedding, plowing operations must be parallel to the channels, but planting, cultivating, and harvesting are normally perpendicular to them. The rows should have a continuous slope to the drains. The maximum length for rows having a continuous slope in one direction is 600 ft, allowing a maximum spacing of 1200 ft where the rows drain in both directions. In very flat land with little or no slope, some of the excavated soil may be used to provide the necessary grade. However, the length and grade of the rows should be limited so as to prevent damage by erosion. On highly erosive soils which are slowly permeable, the slope length should be reduced to 300 ft or less.

The cross section for field ditches may be V-shaped, trapezoidal, or parabolic. The W-ditch shown in Fig. 13.8 is essentially two parallel single ditches with a narrow spacing. All of the spoil is placed between the channels, making the cross section similar to that of a road. The advantages of the W-ditch are (1) it allows better row drainage because spoil does not have to be spread, (2) it may be used as a turn row, (3) it may serve as a field road, (4) it can be constructed and maintained with ordinary farm equipment, and (5) it may be seeded to grass or row crops. The disadvantages of the W-ditch are that (1) the spoil is not available for filling depressions, and that (2) a large area is occupied by the drains. The

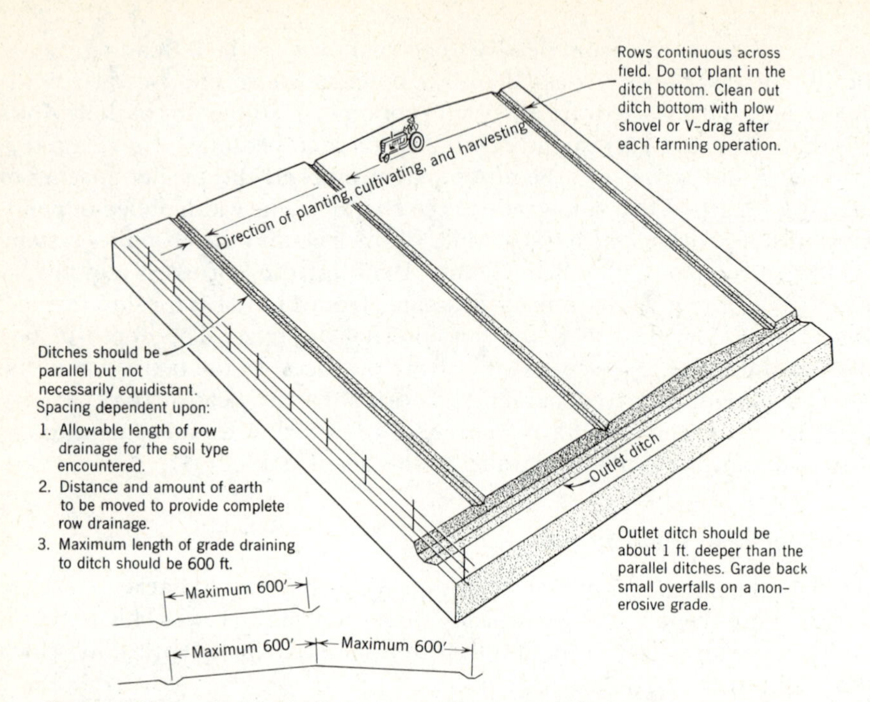

Figure 13.7. Parallel field-ditch system for surface drainage.

minimum width between the pair of drains varies from 15 to 50 ft. The W-ditch should be located so that the rows drain from both directions.

13.4 Parallel Lateral Ditch System

This system is similar to the parallel field ditch system except that the ditches are deeper. The minimum depth for these ditches is 2 ft, with side slopes steeper than 4 : 1. These channels are not designed for crossing with machinery. Such a system of ditches is suitable for controlling the water table and for subirrigation. In peat and muck soils, the side slopes may be vertical. Such ditches may be installed in peat and muck to obtain initial subsidence prior to tile installation. Any ditch will provide subsurface drainage as effectively as a tile of the same depth.

The layout of a ditch system is similar to that for the drains in Fig. 13.7. As in other methods of drainage on flat land, the surface must be smoothed and shaped so that water will move to the ditches. Farming operations must be parallel to the channels.

Parallel lateral ditches are common in the low lands of The Netherlands, West Germany, and England. Many of these lands are below sea level and because of

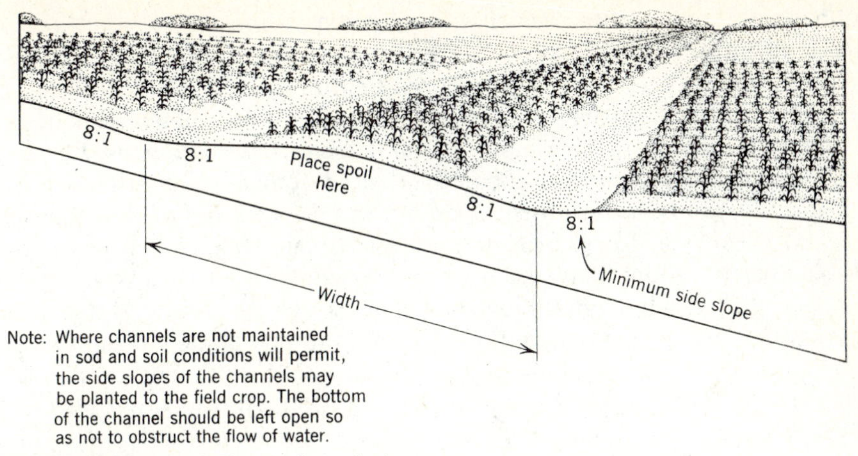

Figure 13.8. W-ditch for surface drainage.

their high water table are used mostly for pasture. Water may be in the channels for a large part of the year. Drainage water is removed entirely by pumping.

For water table control or subirrigation, the soil must be several feet deep and permeable, especially horizontally, and be underlaid with an impermeable material to prevent deep seepage. During dry seasons, the water level is held at the proper depth by suitably located control structures. In wet seasons the controls are removed so the ditches may provide surface drainage. In organic soils the water level is maintained from 1 to 4 ft below the surface to provide water for plant growth, to control subsidence, and to reduce fire and wind-erosion hazards.

13.5 Land Grading

Land grading is the operation of producing a plane land surface with a continuous slope. It is the same as land leveling, a term more common in irrigated areas. A field properly graded for surface drainage may also be suitable for surface irrigation. Land grading is normally a necessary operation in conjunction with the surface drainage systems previously described.

In grading the land, the high areas are cut down and the spoil is moved into the low sections. Reduced plant growth may occur on the fill areas, although the exposure of subsoil in the cuts is usually more serious. Precision in establishing the desired elevation is extremely important, because the slopes are usually a few tenths of a percent. A specified design slope is more important for irrigation than for drainage. A variable slope for drainage is not objectionable, provided it is continuous.

Cuts and fills are computed from an instrument survey with ground elevations taken to the nearest 0.1 ft on a 100-ft square grid for horizontal control. Elevations are taken at other critical points, such as highs and lows between grid stakes and in the outlet ditch.

The elevations are then recorded on a field map, as illustrated in Fig. 13.9. For ease of calculating earthwork volumes, the grid points along the bottom and left sides of the field were established 50 ft from these base lines. By plotting the ground profiles in the direction of the desired drainage as shown in Fig. 13.10, cuts and fills can be determined. A design slope for each line is established by trial and error, which will provide an approximate balance between cuts and fills as well as reduce haul distances to reasonable limits. Cuts and fills are estimated graphically from the original ground surface and the design profile. If cross-slope drainage is desired, profiles may also be plotted at right angles to the original lines, that is, along lines 1, 2, etc. in Fig. 13.9. The profiles shown in Fig. 13.10 are normally plotted directly on the map, which has been drawn on graph paper with 10 lines per inch. An approximate method of determining earthwork volume is to assume that the cut or fill at a grid point represents the average for 50 ft from the point in four directions (see Example 13.2). More accurate methods are described in Phelan (1961) and Schwab et al. (1981).

Example 13.2. Determine the volume of cuts and fills and the cut-fill ratio for the field in Fig. 13.9.

Solution. Using the cuts and fills from the profiles plotted in Fig. 13.10 and assuming that each value represents a full grid square, the volume of cuts is (start with line *A*, cut at *A*2 is 0.3, *A*3 is 0.1 and on line *B*, *B*2 is 0.4, etc.)

$$\begin{aligned} V_c &= (0.3+0.1+0.4+0.4+0.2+0.2+0.2+0.1+0.1+0.2+0.1+0.3+0.2) \\ &\quad \times (100 \times 100) \\ &= 28{,}000\text{ft}^3 \end{aligned}$$

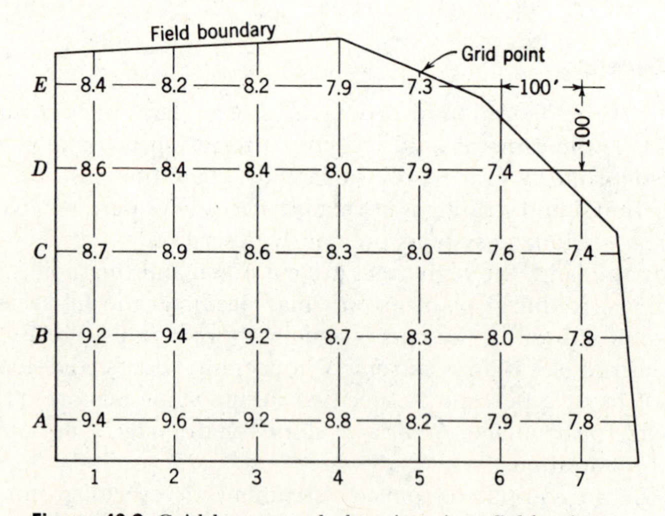

Figure 13.9. Grid layout and elevations in a field prior to land-grading operations.

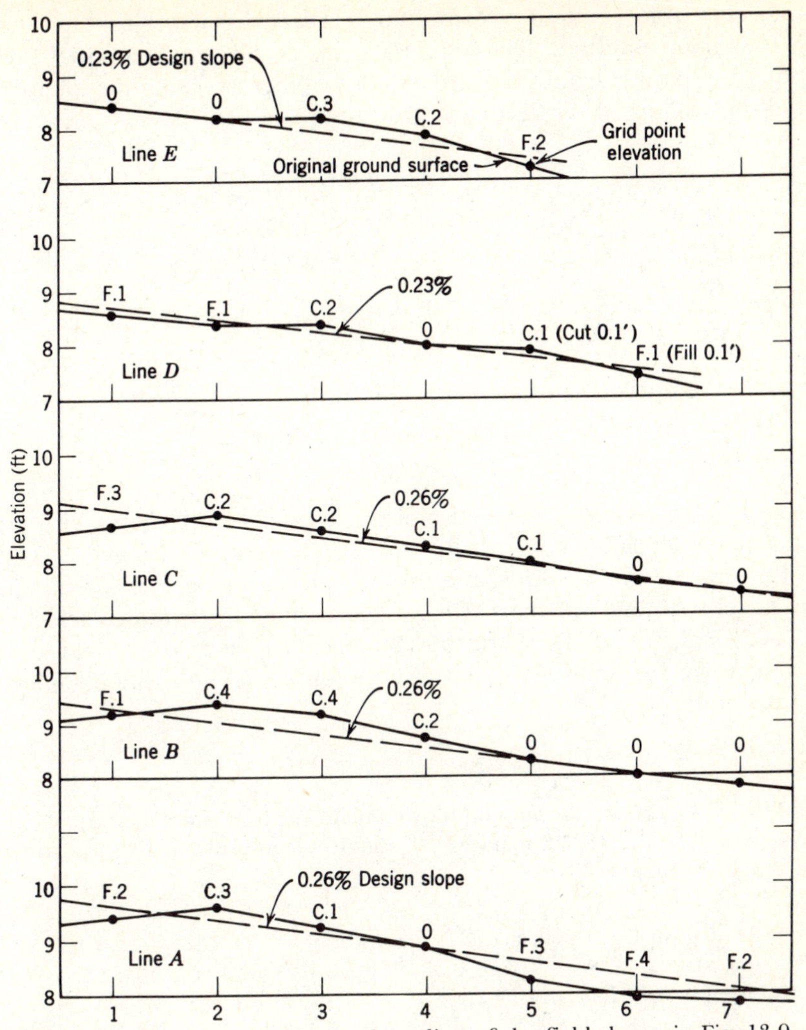

Figure 13.10. Cuts and fills for land grading of the field shown in Fig. 13.9.

The volume of fill is (start with line *A*, *A*1 is a fill of 0.2, *A*5 is 0.3, *A*6 is 0.4, *A*7 is 0.2, and on line *B B*1 is 0.1, etc.)

$$V_f = (0.2 + 0.3 + 0.4 + 0.2 + 0.1 + 0.3 + 0.1 + 0.1 + 0.1 + 0.2) \times (100 \times 100)$$
$$= 20{,}000 \text{ ft}^3$$

The cut-fill ratio is

$$C/F \text{ ratio} = 28{,}000/20{,}000 = 1.4$$

This ratio may be increased by lowering one or more of the design profiles or by changing the slope to increase cuts and decrease fills. The opposite changes will decrease the ratio.

A more accurate method of computing earthwork volume for land grading is the four-point method. The volume of cuts or fills are computed for each grid square separately, and they are added together to obtain the total. By this method the volume of cuts for each grid square is

$$V_c = \frac{L \times L(SC)^2}{4(SC + SF)} \tag{13.1}$$

and the volume of fills is

$$V_f = \frac{L \times L(SF)^2}{4(SC + SF)} \tag{13.2}$$

where L = length of one side of the grid in feet,
SC = sum of the cuts on a grid square in feet,
SF = sum of the fills on a grid square in feet.

The above equations apply to a rectangular grid provided the appropriate lengths are known as well as the cuts or fills for the four corners. If a grid has all cuts or all fills, the volume is simply the product of the average cut or fill and the grid area. The volume in cubic yards can be obtained by dividing each equation by 27. Although the total cut and fill volume calculations for a field would be time-consuming for hand computation, a solution can be easily obtained on a digital computer.

Example 13.3. Determine the cut and fill volume by the four-point method for the 100-ft grid square $A1$, $A2$, $B1$, $B2$ in Fig. 13.9 using the cuts and fills for this grid square shown in Fig. 13.10.

Solution. Read from Fig. 13.10, $F.2$, $C.3$, $F.1$ and $C.4$. $SC = 0.4 + 0.3 = 0.7$, $SF = 0.1 + 0.2 = 0.3$, and $SC + SF = 0.7 + 0.3 = 1.0$.

Substituting in Eqs. 13.1 and 13.2,

$$V_c = \frac{100 \times 100\ (0.7)^2}{4 \times 1.0} = 1225 \text{ cu ft}$$

$$V_f = \frac{100 \times 100\ (0.3)^2}{4 \times 1.0} = 225 \text{ cu ft}$$

Experience has shown that, in land grading or leveling, the cut-fill ratio should be greater than 1. Compaction from equipment in the cut area which reduces the volume and also compaction in the fill area which increases the fill volume needed are believed to be the principal reasons for this effect. Marr (1957) stated that on level ground between stakes the operator has an optical illusion of a dip in the middle, and therefore in filling, crowning often occurs. The ratio of cut to fill volume for various soil conditions is shown in Table 13.1. These should

Table 13.1. Suggested Cut-Fill Ratios for Land Grading

Soil Conditions	*Ratio: Volume of Cut to Volume of Fill*
Organic soils	
Cuts less than 0.4 ft	2.0
Cuts greater than 0.4 ft	1.7
Clay loam soil	1.2 to 1.45
Clay soils	1.3 to 1.4
Medium textured soils	1.2 to 1.25
Sandy soils, compacted	1.1 to 1.5
Bottomland soil in Mississippi and Arkansas	1.5

Source: Coote and Zwerman (1970) and other references.

be adjusted where local conditions dictate. Land grading costs are normally based on the volume of the cuts.

13.6 Land Smoothing

Land smoothing is the practice of removing small surface irregularities on the land surface after land grading with scrapers and other heavy equipment. A land plane or land leveler is required for land smoothing. They are made with blades up to 15 ft in width and lengths up to 90 ft. The blade on the land plane is mounted midway on the frame and is adjustable vertically so that the depth of cut and amount of soil carried can be regulated. The purpose of land smoothing is to make a uniform plane surface either for surface drainage or for the uniform distribution of irrigation water. The smoothing operation is ordinarily done in the field after grading without detailed surveys or plans.

A smoothing operation consists of a minimum of three passes with the leveler. The first two passes are made on opposite diagonals as shown in Fig. 13.11, and the last path is made in the direction of cultivation. The soil should be relatively free of crop residue or trash to allow a good distribution of the soil in the depressions. Because fill areas continue to settle and farming with large equipment may produce surface irregularities, land smoothing may be required annually or every few years. Scheduling the operation when soil and crop conditions are suitable is often a problem.

13.7 Construction

Prior to making an instrument survey or construction, heavy vegetation and residue should be removed or plowed under. For the construction of field ditches with shallow cuts to 1-ft depth, moldboard and disk plows, small scrapers, blade graders, and other light equipment are suitable. For depths up to 2.5 ft, such equipment as motor graders, scrapers, and heavy terracing machines are more

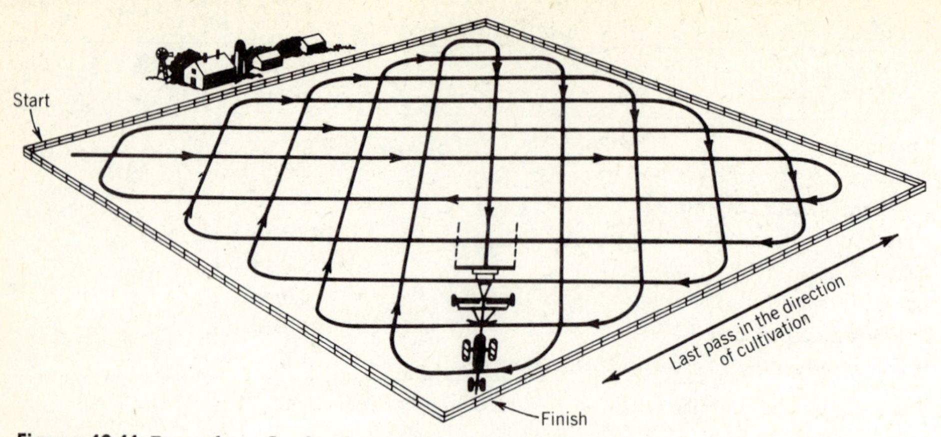

Figure 13.11. Procedure for land-smoothing operations. *Source:* Drablos and Moe (1984).

appropriate. Bulldozers equipped with push- or pull-back blades and carryall scrapers are suitable for deeper cuts.

Several methods of establishing the grade line of field ditches are shown in Fig. 13.12. The first method (*a*) is suitable for blade-type equipment. The cut may be easily checked after the first cut is made, but the reference hubs are later removed by construction. With any of these methods, the final grade elevations should be checked with a surveying instrument. With the method shown in Fig. 13.12*b*, the hub stakes are not removed during construction. To guide the operator, slope stakes may be set at each station, as illustrated in Fig. 13.3. with laser grade control equipment, stakes are not required, only an elevation at one point and the channel grade.

In land grading operations, the major earthwork is accomplished with heavy scrapers or pans equipped with laser grade control systems for accurate depth control. Some operators drive the equipment over the field empty, noting the elevations from the laser as they proceed. Grid points may be marked in the field with the cut or fill as noted in Fig. 13.13. The general direction of earth movement is noted by the long arrows. As shown in Fig. 13.14, the operator avoids the grid stakes, since these are marked with the cut or fill. After the grade is checked, a land plane removes these small islands and other minor irregularities on the surface.

13.8 Maintenance

Plowing parallel to field ditches is usually adequate to maintain the channel depth and shape. Deadfurrows placed in the channels with backfurrows on the ridges between drains will maintain and accentuate the drainage system. After the first few years, it may be advisable to shift the positions of the deadfurrows and backfurrows to avoid developing too deep a drain or too high a ridge.

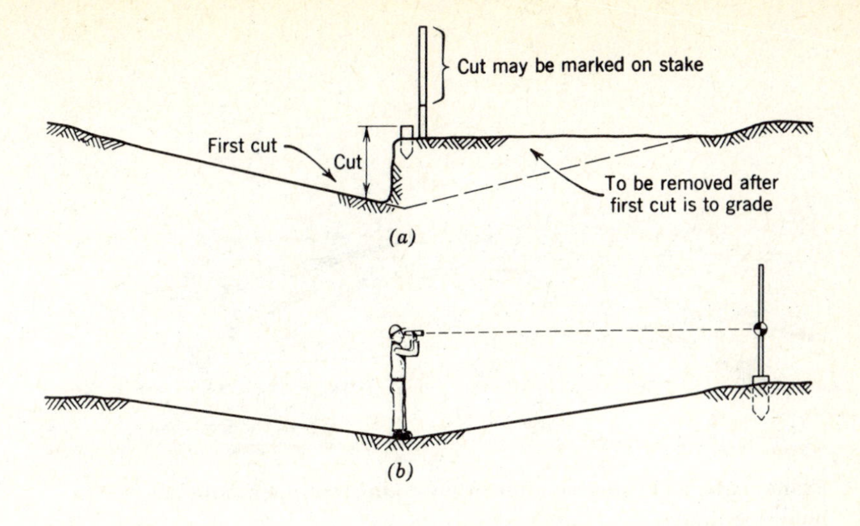

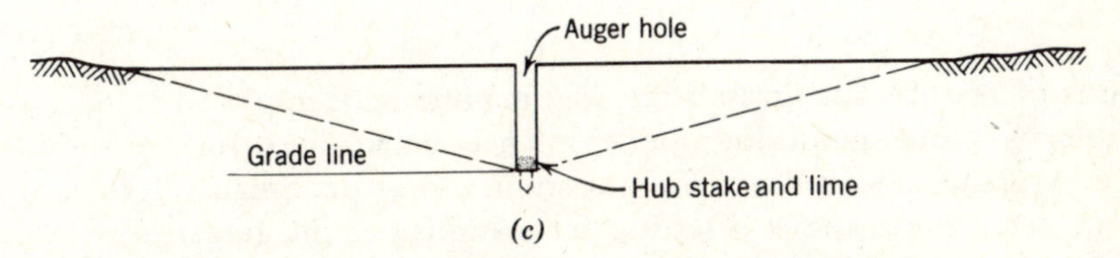

Figure 13.12. Methods for establishing grade for surface drains.

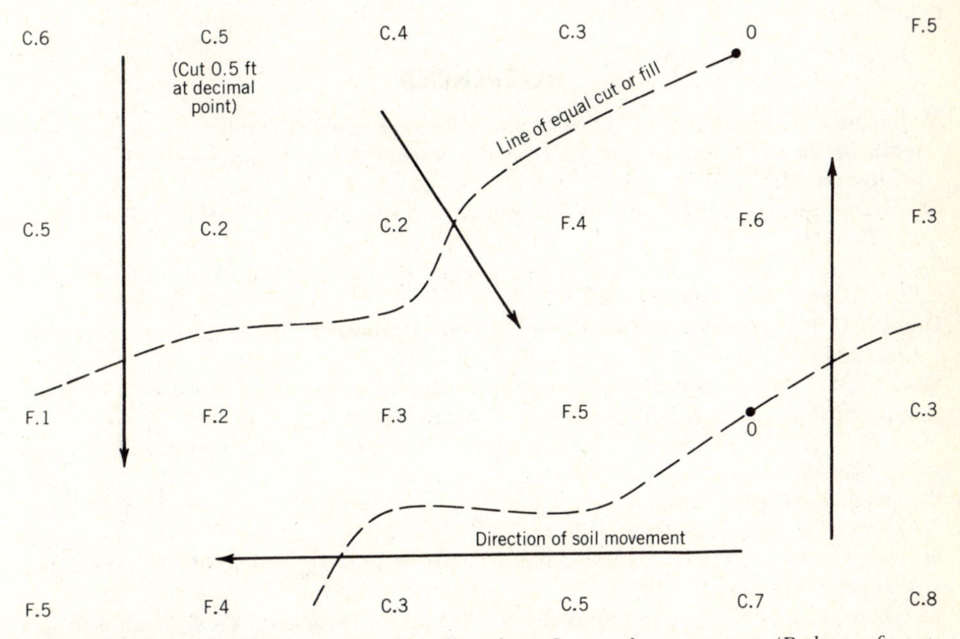

Figure 13.13. Cut and fill map showing directions for earth movement. (*Redrawn from Anderson, 1980.*)

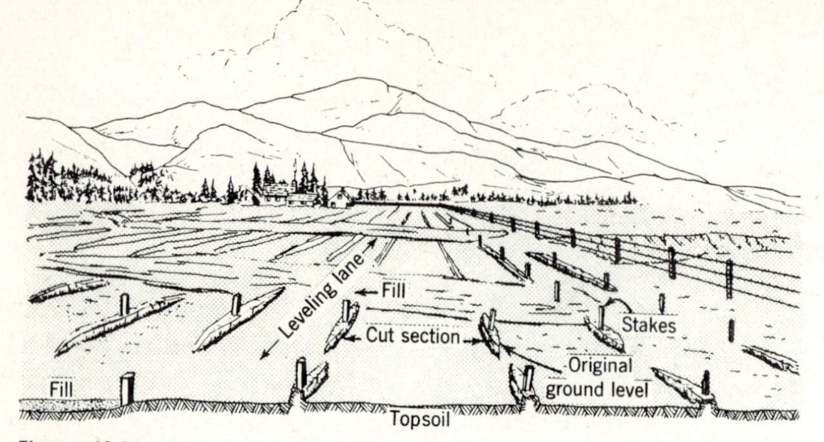

Figure 13.14. Field surface after heavy-equipment operation, but before land smoothing.

Methods of terrace and grass waterway maintenance apply equally as well to field ditches. Land-smoothing operations may be required for several years or even every year, since fill soil has a tendency to settle. Small ridges across the slope or small depressions resulting from improper tillage can usually be removed by land smoothing.

REFERENCES

Anderson, C. L. et al. (1980) "Land Shaping Requirements," Chapter 8 in M. E. Jensen (ed.), *Design and Operation of Farm Irrigation Systems*, ASAE Monograph, The Society, St. Joseph, Michigan.

Beauchamp, K. H. (1952) "Surface Drainage of Tight Soils in the Midwest," *Agr. Eng.* **33**: 208–212.

Coote, D. R., and P. J. Zwerman (1970) "Surface Drainage of Flat Lands in the Eastern U.S.," *Cornell University Ext. Bull. 1224*.

Drablos, C. J. W., and R. C. Moe (1984) "Illinois Drainage Guide," *University Illinois Ext. Cir. 1226*.

Gain, E. W. (1964) "Nature and Scope of Surface Drainage in Eastern United States and Canada," *Am. Soc. Agr. Eng. Trans.* **7** (2):167–169.

Marr, J. C. (1957) "Grading Land for Surface Irrigation," *California Agr. Exp. Sta. Cir. 438*, Revised.

Phelan, J. T. (1961) "Land Leveling and Grading," in Richey, C. B. et al., *Agricultural Engineers' Handbook*, McGraw-Hill Book Co., New York.

Schwab, G. O. et al. (1981) *Soil and Water Conservation Engineering*, 3rd ed., John Wiley & Sons, New York.

Zwerman, P. J. (1969) "Land Smoothing and Surface Drainage," *Cornell University Ext. Bull. 1214*.

PROBLEMS

13.1 Determine the volume of excavation in cubic yards for a surface ditch with 10 : 1 side slopes if the cuts at successive 50-ft (15-m) stations are 0.5, 2, 4, 3, and 0 ft (0.15, 0.6, 1.2, 0.9, 0 m). Use the average end area between stations to compute the volume.

13.2 Compute the volume of soil to be excavated in cubic yards for a random field ditch with 8 : 1 side slopes if the cuts at consecutive 50-ft (15-m) stations are 0.5, 1.0, 2.4, 1.8, 0.8, and 0 ft (0.15, 0.3, 0.7, 0.6, 0.2, 0 m).

13.3 If the depth of cut at station 0 + 00 in Fig. 13.4 is reduced to 0.1 ft (0.03 m) and the grade is increased to 0.3 percent, compute the cut at each station and the total volume of excavation.

13.4 From the survey notes below, determine the depth of cut at each station for a surface ditch that is to remove water from a pothole. The slope of the drain from 0 + 00 to 1 + 00 is 0.2 percent. Select a uniform slope from 1 + 00 to 2 + 00 so that the cut at 2 + 00 is zero. Compute the slope.

Sta.	*B.S.*	*F.S.*	*Elev.*	*Cut*
BM 1	4.6		50.0 ft	
0 + 00		6.2		0.2 (bottom of pothole)
0 + 50		5.5		
1 + 00		3.0		
1 + 50		4.6		
2 + 00		7.0		0

13.5 Raise the design profile on lines *B* and *C* in Fig. 13.10 so that the cut-fill ratio is about 1.2 rather than 1.4 as in Example 13.2. Compute the new volume of cut and volume of fill. If the contractor charges $1.50 per cubic yard of cut, what is the cost for grading the field?

13.6 Plot the ground elevations in Fig. 13.9 along the vertical lines, that is, on lines 1, 2, etc. By selecting nearly uniform design slopes, determine cuts and fills so as to give a cut-fill ratio of about 1.3. Compute the volume of cuts and fills.

13.7 Determine cuts and fills for a field assigned by the instructor for a given cut-fill ratio. Compute the volume of cuts and fills in cubic yards.

13.8 Compute the cut and fill volume in cubic yards (cubic meters) for a 100 × 100-ft (30 × 30-m) grid square if two cuts are 0.2 and 0.5 ft (0.06 and 0.15 m) and two fills are 0.4 and 0.2 ft (0.12 and 0.06 m).

13.9 Compute the volume of cuts, volume of fills, and the cut-fill ratio for the data in Fig. 13.13 using the four-point method. Assume the grid is 100 × 100 ft (30 × 30 m).

CHAPTER 14

SUBSURFACE DRAINAGE

Plants need air as well as moisture in their root zones. Excess water that is free to move to subsurface drains generally retards plant growth, because it fills soil voids and restricts proper aeration. For most cultivated crops, adequate surface drains are needed for flat or undulating topography to remove excess ponded water, and subsurface drains are required for soils with poor internal drainage and a high water table. Where soils do not have an impermeable layer below the root zone, internal drainage may be adequate and pipe drains are not needed. For maximum productivity of most crops, both surface and subsurface (pipe) drainage are essential regardless of whether drainage is man-made or occurs naturally. Surface drainage systems may give a greater return per dollar invested than a higher-cost subsurface system.

14.1 Benefits of Subsurface Drainage

Subsurface drainage increases crop yields by (1) removing gravitational or free water that is not directly available to plants, (2) increasing the volume of soil from which roots can obtain food, (3) increasing the movement and quantity of air in the soil, (4) providing conditions that permit the soil to warm up faster in the spring, (5) increasing the bacterial activity in the soil, which improves soil structure and makes plant food more readily available, (6) reducing soil erosion, since a well-drained soil has more capacity to hold rainfall, resulting in less runoff, and (7) removing toxic substances, such as sodium and other soluble salts that in high concentrations retard plant growth. Other benefits that may not result in increased yields include a reduction in time and labor for tillage and harvesting operations. With a crop such as corn, a delay in planting date will decrease yields. Planting in wet soils is likely to decrease plant stands. A delay in harvesting time will increase machine losses as well as grain damage.

As illustrated in Fig. 14.1, pipe drains permit deep root development by lowering the water table, especially during the spring months. A plant with a

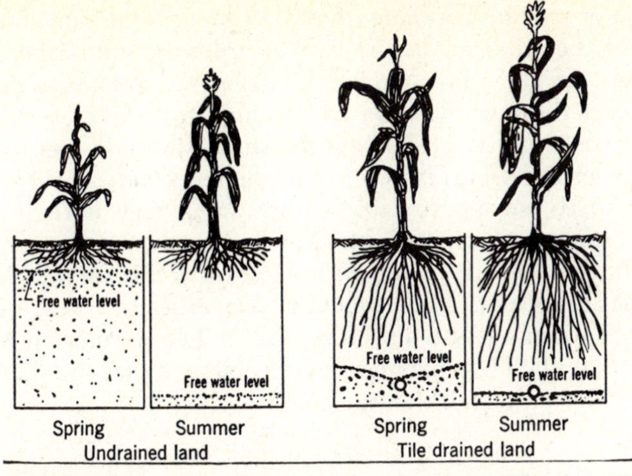

Figure 14.1. Root development of crops grown on drained and undrained land. (*Redrawn from Manson and Rost, 1951.*)

deep root system can withstand droughts better than shallow rooted plants because larger quantities of water are readily available when needed, as well as are more plant nutrients. A large root system is also desirable later in the season because the plant is larger and transpires more water.

14.2 Movement of Water into Pipe Drains

In a uniform soil saturated to the surface, the movement of water takes place along the paths indicated in Fig. 14.2. Since the flow lines are so drawn that the

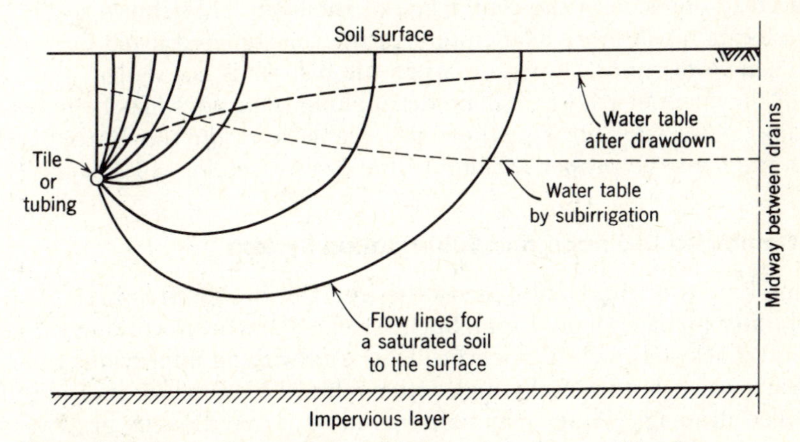

Figure 14.2. Water flow paths to a drain and the water table after drainage or after subirrigation.

quantity of water moving between any two of them is the same, the water table will be lower in the soil near the drains than from the soil farther away. This is evident because the flow lines are farther apart, the greater the distance from the drains. After the saturated soil has drained for a day or so, the resulting water table or free water level will have the shape shown in Fig. 14.2. The water level directly over the drains is lower than the level midway between them.

Where the soil is more permeable in the backfill trench, the water table will be lower than shown, directly over the tile. Soil permeability and spacing of the drains have the greatest effect on the rate of drop of the water table. Since water must move greater distances horizontally than vertically to reach the drain, the horizontal permeability is the more important. The permeability of most soils decreases with depth. This change in permeability affects the shape of the flow lines and the rate of drop of the water table.

14.3 Drainage Systems

The layout of random, herringbone, and gridiron systems is shown in Fig. 14.3*a*, *b*, and *c*. The topography to which each of these systems is adapted is indicated by the dashed contour lines. The random system is suitable where the field is not to be completely drained. It is flexible in layout, permitting the location of drains where they are most needed. As will be discussed in Section 14.15, considerable cost savings are possible with proper layout.

In most large installations several of the layout patterns will be found. The lines are normally located so that the land surface along the drains has some slope toward the outlet. Interceptor drains shown in Fig. 14.3*d* are laid out nearly on the contour and are placed so as to drain as much of the seepy area below as possible. Extensive soil borings may be required to locate the impermeable layer above which the ground water is flowing.

Where grass waterways or channels are wet from seepage, pipe drains should be placed to one side of the center line of the channel as shown in Fig. 14.3*b*. Such a location will more likely intercept the seepage and avoid erosion along the trench from runoff, especially when the backfill is not settled. Where both sides of the channel are wet, a main drain along both sides may be necessary to intercept the seepage. Locating the impermeable layer and the source of seepage is important for the proper location of the drains (see Fig. 14.3*d*).

14.4 Combined Drainage and Subirrigation System

In humid areas of the United States the trend in design practice is to develop a total water management system. Such systems have been working for years in peat and muck soils with high permeability and with an impervious layer below the drains or with a naturally high water table. This practice is also known as controlled drainage. Water may be pumped or it may flow by gravity into the drains to provide subirrigation during dry conditions. More recent applications have been to adapt the system to heavier mineral soils. For subirrigation the

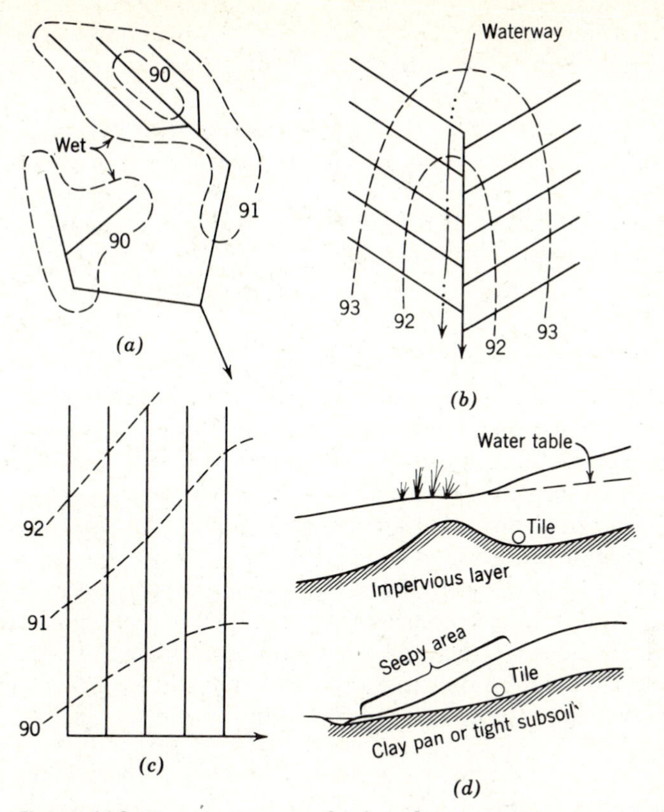

Figure 14.3. Common types of subsurface systems.

water table is higher at the drain than it is midway between drains, as shown in Fig. 14.2. Its shape is inverted from that for the drainage mode.

A water management simulation computer model DRAINMOD has been developed by Skaggs (1980) in North Carolina. It is a complex model involving more than 20 parameters. When it is properly calibrated, it will predict crop yields for any combination of conditions desired. DRAINMOD is based on a water balance in the soil profile. It uses climatological data on an hour-to-hour basis to predict the response of the water table, and the soil water regime above it, to various combinations of surface and subsurface water management practices. By simulating the performance of alternative systems over several years of record, an optimum system can be selected on the basis of probability. The model is composed of a number of separate components including methods to evaluate infiltration, subsurface drainage, surface drainage, evapotranspiration, subirrigation, and soil water distribution.

DRAINMOD can predict trafficability in the spring for tillage and planting operations, the time available for harvesting, the hydraulic loading capacities for land application of wastewater, and the plant response to wet and dry stress during the growing season. Plant response to drainage, adjusted for different stages of plant growth, is evaluated by the height and the duration of the water table, called SEW values.

The model has been validated using measured crop yields for several soils in North Carolina and Ohio. The computer program is operational on several university main frame computers, state Soil Conservation Service computers, and some microcomputers. It is too lengthy and detailed to be included in this book, but a user's manual is available.

14.5 Outlets

The failure of many drainage systems is caused by faulty outlets. As shown in Fig. 14.4*a*, corrugated metal or similar pipe with a flap gate to prevent entry of rodents is recommended. If there is a danger of flood water backing up into the drain, a flood gate may be installed in place of the flap gate.

Where the depth of the outlet ditch is inadequate, an automatically controlled pump with a small sump for storage, as shown in Fig. 14.4*b*, can be installed. Before considering such an installation, deepening of the outlet ditch should be investigated. Pump outlets require operation and maintenance costs that must be compared to those for the ditch excavation and maintenance. Where outlet ditches must pass through the land of others, right-of-way costs and legal problems often necessitate the construction of pump outlets.

14.6 Depth and Spacing

Because of variable soil and climatic factors, it is difficult to make specific recommendations for the proper depth and spacing of drains. The depth and spacing of drains apply only to laterals and have nothing to do with the depth of the mains or submains. Depths for mains are governed by outlet conditions and topography of the drainage area.

Deep drains are more effective than shallow ones in soils that do not have a claypan or tight subsoil at a depth less than the depth of the drain. Two feet of soil over the top of the drain is a minimum. Unless there is a tight subsoil, drains should be placed at optimum depth to provide minimum drainage cost. This depth is always measured to the bottom of the drain.

There is a definite relationship between depth and spacing of drains. For soils with a uniform profile, the deeper the drains the wider the spacing and vice versa. Average depth and spacing for soils of different texture are given in Table 14.1. Local field practices and recommendations of the state experiment stations should be followed. Most states have drainage guides that recommend depth and spacings by soil types and by type of crops to be grown.

Depth and spacing for humid areas are based on the rate of drop of the water table, while in arid irrigated land the maintenance of a low water table to prevent the upward movement of toxic salts and the removal of excess salts are the

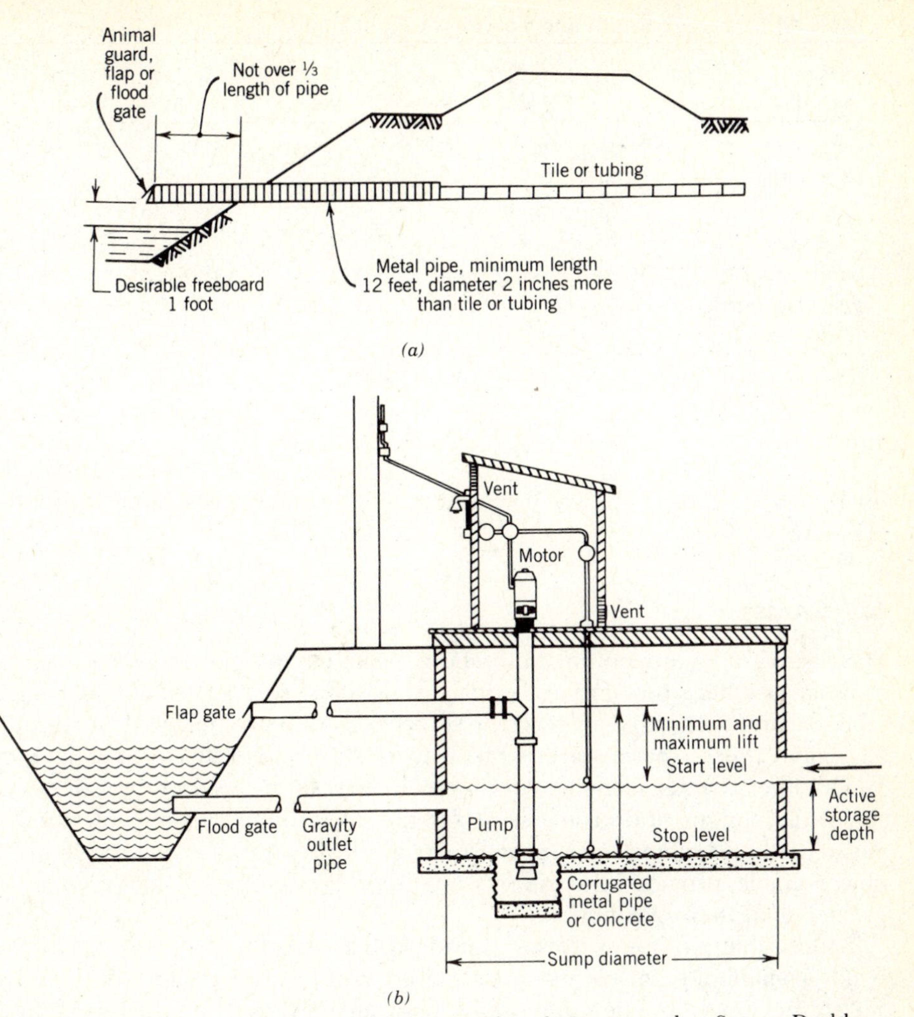

Figure 14.4. Outlets for subsurface drains. (*a*) Gravity. (*b*) Pump outlet. *Source:* Drablos and Moe (1984).

criteria for depth and spacing recommendations. Because soils in irrigated regions are generally more permeable than in humid areas, spacing of drains is wider (see Table 14.1). The greater depth of drains is necessary to provide a lower water table than is usually required in humid regions. Mathematical expressions have been developed for computing drain spacings for either of the criteria described above. Spacings can be computed by using the simulation model DRAINMOD, previously described.

For special crops that produce a high annual income, such as tobacco, fruit, and vegetable crops, the drain spacings given in Table 14.1 should be reduced as much as 25 to 50 percent. In humid regions, pipe drainage is most beneficial

Table 14.1. Average Depth and Spacing of Pipe Drains

Soil	*Hydraulic Conductivity*		*Spacing*	*Depth*
	Class[a]	*iph*	*(ft)*	*(ft)*
Clay	Very slow	0.05	30–50	3.0–3.5
Clay loam	Slow	0.05– 0.2	40–70	3.0–3.5
Average loam	Moderately slow	0.2 – 0.8	60–100	3.5–4.0
Fine sandy loam	Moderate	0.8 – 2.5	100–120	4.0–4.5
Sandy loam	Moderately rapid	2.5 – 5.0	100–200	4.0–5.0
Peat and muck	Rapid	5.0 –10.0	100–300	4.0–5.0
Irrigated soils	Variable		150–600	5.0–8.0

[a]Hydraulic conductivity as classified by O'Neal, 1949.

for row crops during the spring months. Pipe drainage is seldom justified for grass crops.

14.7 Grades

Maximum grades are limiting only where drains are designed for near maximum capacity or where they are embedded in unstable soil. Pipe embedded in fine sand or other unstable material may become undermined and settle out of alignment unless special care is taken to provide joints that fit snugly against one another. Under extreme conditions it may be necessary to install bell and spigot tile, tongue and groove concrete tile, metal pipe, or nonperforated corrugated plastic tubing. On mains steep grades up to 2 or 3 percent are not objectionable, provided the capacity at all points nearer the outlet is equal to or greater than the pipe above.

A desirable minimum working grade is 0.2 percent. Where sufficient slope is not available, the grade may be reduced to 0.1 percent where fine sand and silt are not present. The minimum velocity at full flow where fine sand and silt are present should be 1.5 fps as computed from the Manning equation (Chapter 7).

14.8 Drainage Rates

Pipe drains are designed to remove water at a given rate, called the *drainage coefficient*, which is the depth of water in inches to be removed in a 24-hour period from the drainage area. In humid regions the drainage coefficient depends largely on the rainfall, but varies with the soil, type of crop, and degree of surface drainage. Recommended drainage coefficients are shown in Table 14.2. In irrigated areas the discharge from drains may vary from about 10 to 50 percent of the water applied. Since not all of the area is irrigated at the same time, the drainage area to be used to calculate drain flow is not the same as the

Table 14.2. Drainage Coefficients for Pipe Drains in Humid Regions[a]

Crops and Degree of Surface Drainage	*Drainage Coefficient in ipd*	
	Mineral Soil[b]	*Organic Soil*
Field Crops		
Normal[c]	$\frac{3}{8} - \frac{1}{2}$	$\frac{1}{2} - \frac{3}{4}$
With blind inlets	$\frac{3}{8} - \frac{3}{4}$	$\frac{1}{2} - 1$
With surface inlets	$\frac{3}{4} - 1$	$\frac{3}{4} - 1\frac{1}{2}$
Truck Crops		
Normal[c]	$\frac{1}{2} - \frac{3}{4}$	$\frac{3}{4} - 1\frac{1}{2}$
With blind inlets	$\frac{1}{2} - 1$	$\frac{3}{4} - 2$
With surface inlets	$1 - 1\frac{1}{2}$	$2 - 4$

[a]From U.S. Soil Conservation Service, 1973.

[b]These values may vary, depending on special soil and crop conditions. Where available, local recommendations should be followed.

[c]Adequate surface drainage must be provided.

entire drained area, but is estimated from the area irrigated. Because of a wide range in flow, the drainage coefficient should be based on local recommendations.

14.9 Drainage Area

The drainage area is that actually drained by the pipe. Where surface water is to be removed by the drains, the watershed area is the drainage area even though it may not be entirely pipe drained. For a single drain, as in a random system, the width drained is the approximate pipe spacing for the soil.

14.10 Size of Drains

The size of drains depends upon the kind of pipe, the drainage area, the drainage coefficient, and the grade in the drain. By profile leveling the slope of the soil surface and the depth of the outlet can be determined. From these data the grade in the drain can be selected. Three-inches is the minimum recommended size; however, 4- and 6-in. are considered minimum sizes in many areas. The minimum size for perforated tubing or pipes could be reduced because misalignment at cracks or joints is not a problem. Six-inch pipe mains are often the minimum diameter recommended, as the greater capacity will reduce the duration of back flooding in the laterals when peak flows exceeded the design rate.

Pipe size can be determined from Fig. 14.5, which is based on the Manning velocity equation discussed in Chapter 7. For example, if the drainage area is 15 acres, the drainage coefficient (D. C.) is $\frac{3}{8}$ in., and the slope is 0.2 percent, the required corrugated plastic tubing size is 8 in. diameter, as noted by the dot on Figure 14.5*a*. This dot is located at the intersection of the vertical line through 0.2 percent grade and the horizontal line through 15 acres in the $\frac{3}{8}$-in. column. If a tubing size is required for a D.C. of $\frac{3}{16}$ in. or one half the $\frac{3}{8}$ rate, reduce the

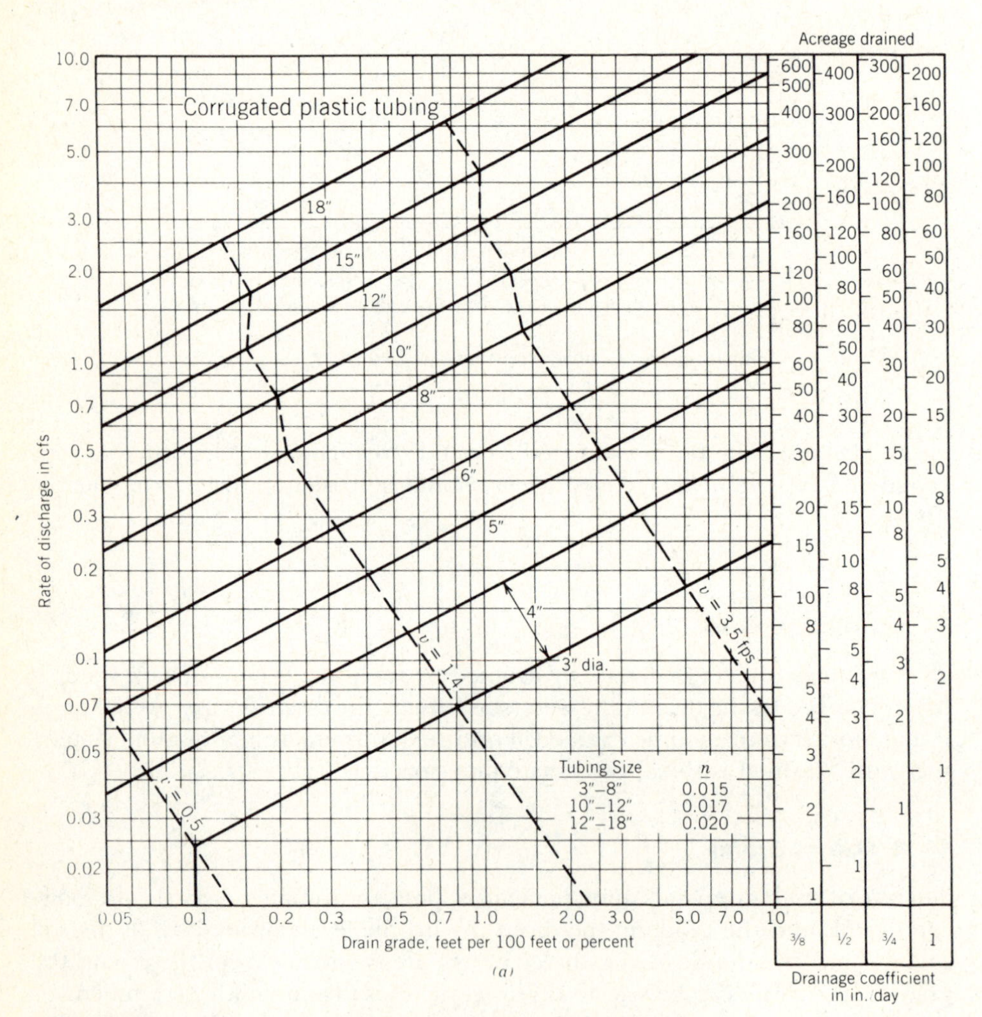

Figure 14.5. Pipe-size nomographs. (*a*) Corrugated plastic tubing.

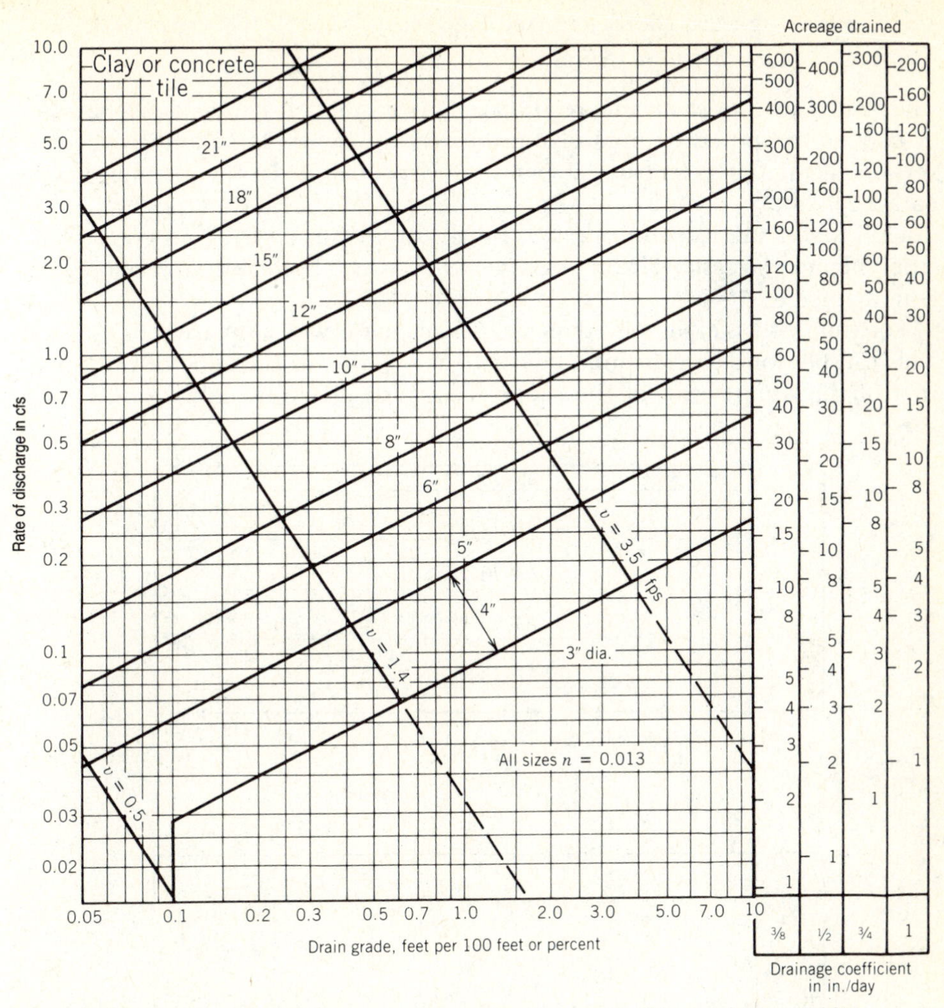

(Figure 14.5. Continued) (*b*) Clay or concrete tile. *Source*: Drablos and Moe (1984).

15 acres to one half, or 7.5 acres, and read 5-in. tubing at the 0.2 percent grade. The discharge rate for the 15 acres at $\frac{3}{8}$ in. is about 0.24 cfs as can be read from the left-hand scale in Fig. 14.5*a*. The 8-in. pipe will run only partly full, since its maximum capacity at full flow is about 0.48 cfs. This flow can be read on the left-hand scale at the intersection of the diagonal line above the 8-in. size and the 0.2 percent vertical line.

The pipe diameter may be computed directly from the Manning equation, from which

$$d = 16(qn/s^{1/2})^{3/8} \tag{14.1}$$

where d = pipe diameter in inches
q = required rate of discharge in cubic feet per second

n = roughness coefficient (see Fig. 14.5)
s = grade or slope of the drain in feet per foot

The rate of discharge may be read directly from the left-hand side of Fig. 14.5 or may be computed by multiplying the acreage by the D. C. and by the conversion factor 0.042 (cubic feet per second per acre). In the example in the paragraph above, $q = 15 \times 0.375 \times 0.042 = 0.24$ cfs. Substituting in Eq. 14.1 would give 6.2 in. diameter required for the drain. Since this size is greater than 6 in., the next largest available size is selected, which in the example above is an 8-in. diameter.

Equation 14.1 is easily solved on a hand calculator with a square root function. This can be done by remembering that the $\frac{3}{8}$ power is the cube of the square root of the square root of the square root. For example, $(256)^{3/8} = 2^3 = 8$.

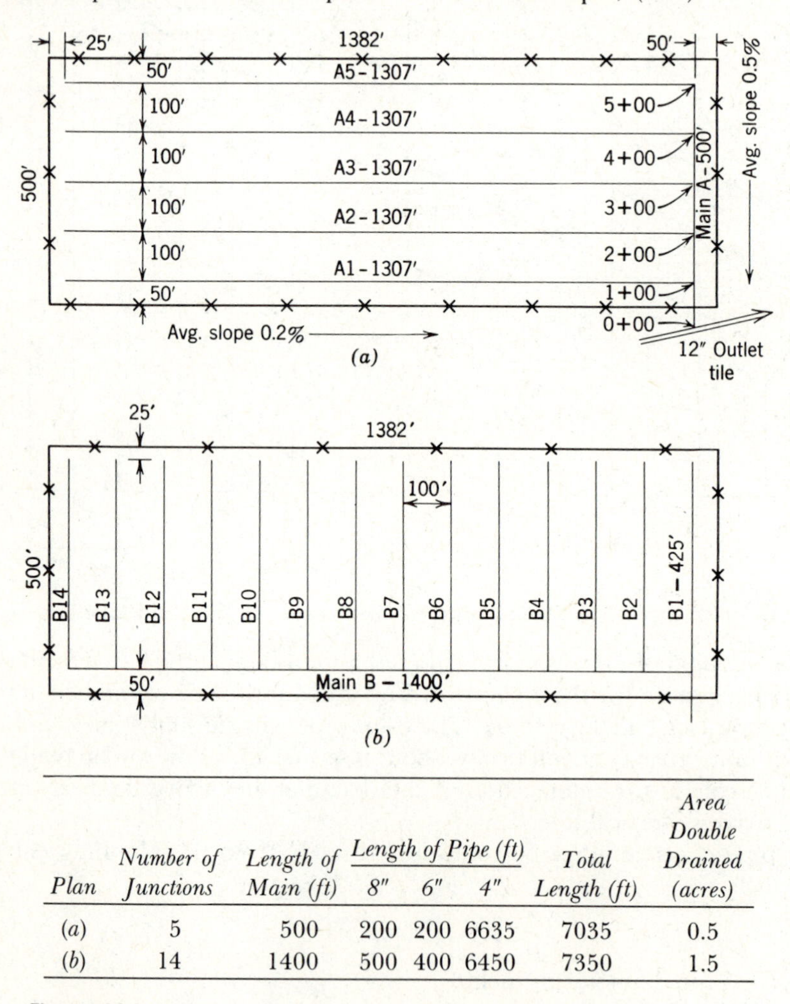

Plan	*Number of Junctions*	*Length of Main (ft)*	*Length of Pipe (ft)* 8"	6"	4"	*Total Length (ft)*	*Area Double Drained (acres)*
(*a*)	5	500	200	200	6635	7035	0.5
(*b*)	14	1400	500	400	6450	7350	1.5

Figure 14.6. A comparison of main and lateral layouts for subsurface drainage systems.

Taking the three square roots of 256 first will give 2, which is then cubed to get 8.

Example 14.1. Determine the depth, spacing, drainage coefficient, and type of system to provide uniform drainage for a field of average loam soil in Georgia where field crops are to be grown. No surface inlets are required. Size and shape of the field, general slope, and outlet are shown in Fig. 14.6*a*.

Solution. From Table 14.1 for average loam soil select a spacing of 100 ft and a depth of 4 ft (check with local recommendations). From Table 14.2 for normal surface drainage and for field crops the drainage coefficient for determining the tubing size is $\frac{1}{2}$ in. Because the field is long and has adequate slope toward one corner, the gridiron system (Fig. 14.6*a*) with long laterals is most suitable. Long laterals reduce the amount of large-size pipe in the main and the number of junctions. For efficiency, the main and lateral are placed 50 ft from fence lines. The upper ends of the lateral lines should extend to within $\frac{1}{4}$ of the spacing from fence lines or 25 ft.

Example 14.2. For the system described in Example 14.1 determine the amount of each size of pipe required. After staking and surveying was completed, the slopes were as follows: *A*1 and *A*2, 0.15 percent; *A*3, *A*4, and *A*5, 0.2 percent; and Main *A* 0.25 percent. Assume minimum size as 4-in.

Solution. The area drained by each lateral is (1307 × 100)/43,560 = 3.0 acres. Since the main is short and is partly drained by the laterals, the area drained by the main may be neglected. Determine the tubing size from Fig. 14.5*a* ($\frac{1}{2}$ in. D.C.) for each change in area drained or change in slope as shown below.

Line	*Area drained, (acres)*	*Slope (%)*	*Tubing size, (inches)*	*Length, (feet)*
*A*5	3.0	0.2	4	1307
*A*4	3.0	0.2	4	1307
*A*3	3.0	0.2	4	1307
*A*2	3.0	0.15	4	1307
*A*1	3.0	0.15	4	1307
Main *A*				
4+00 to 5+00	3.0	0.25	4	100
2+00 to 4+00	9.0 max	0.25	6	200
0+00 to 2+00	15.0 max	0.25	8	200

The length of 4-in. tubing required is 6635 ft; 6-in., 200 ft; and 8-in. 200 ft. Since the amount of 4- and 6-in. tubing is small, all 8-in. size might be more practical for Main A.

Example 14.3. From Fig. 14.7 determine the clay tile size for line *A*1. The soil is clay loam and all surface water from the 8-acre drainage area is to be removed through the surface inlet. Field crops are to be grown.

Solution. From Table 14.2 select $\frac{3}{8}$- and $\frac{3}{4}$-in. D.C. for normal conditions and surface inlets, respectively. The maximum drainage area for *A*1 above station 5 + 62 is the area

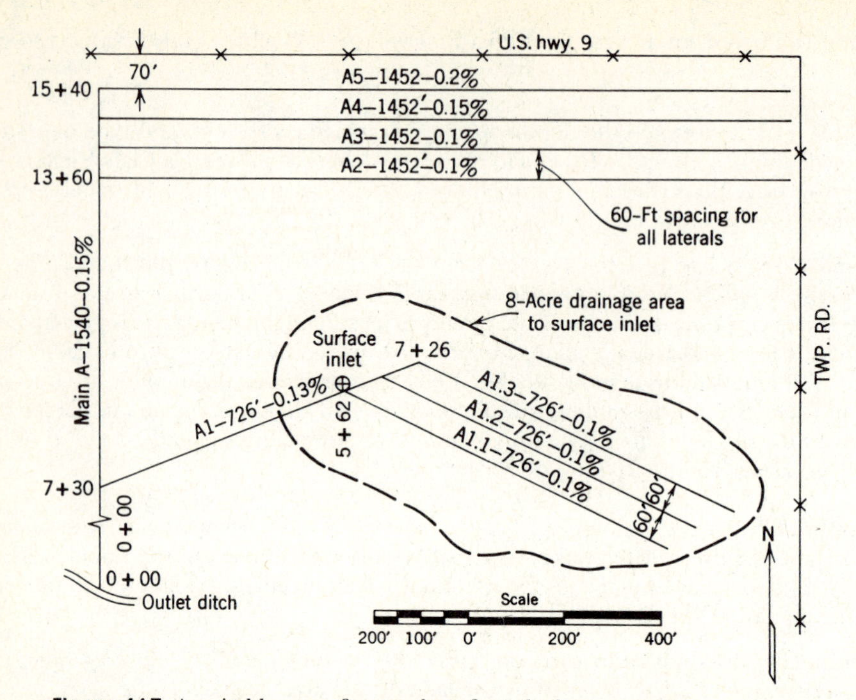

Figure 14.7. A suitable map for a subsurface drainage system.

drained by laterals *A*1.2 and *A*1.3, which is (2 × 60 × 726)/43,560 or 2.0 acres. From Fig. 14.5*b* read a minimum size of 4 in. The 8-acre drainage area to the surface inlet with a D.C. of $\frac{3}{4}$ in. is equivalent to 16 acres at $\frac{3}{8}$ in. The total drainage area for line *A*1 must include a 60-ft strip below the surface inlet, which is (562 × 60)/43,560 or 0.8 acre. From Fig. 14.5*b* read a tile size of 8-in. for 16.8 acres, slope of 0.13 percent, and D.C. of $\frac{3}{8}$ in. The total tile for *A*1 is 562 ft of 8-in. and 164 ft of 4-in.

Note that the area drained by the three laterals in the depression is not considered in determining the drainage area for line *A*1 below the surface inlet. To do so would result in using the same area twice. The assumption is made that the tile size will be adequate if the design is based on the D.C. for surface inlets.

14.11 Envelope Filters

For drainage of irrigated land in the West, gravel envelopes are placed completely around the drain. These envelopes are installed (1) to prevent the inflow of soil into the drain and (2) to increase the effective pipe diameter, which increases the inflow rate. Because of practical considerations and cost, the envelope is usually limited to one gradation of material, which is selected from local naturally occurring deposits. The minimum thickness of the filter is 3 in. In humid areas this practice is not necessary. However, initial backfilling with 6 to 12 in. of topsoil directly over the tile is desirable. Such a practice will prevent misalignment and increase inflow. So-called "blinding" can be done with corncobs, cinders, or other porous material.

For keeping out fine, uniform sands, thin filter materials, called "geotextiles" have been commercially developed. These materials are normally placed around the tubing at the extruding plant. These textiles may be made from nylon, polypropylene, and similar products in a variety of thicknesses and opening sizes. One type called a bean sock stretches giving variable sizes of openings. In Europe an organic material, coconut fiber, is shredded and wrapped uniformly around the tubing, but it deteriorates more rapidly than the synthetic fabrics. The opening size in the filter material should allow some of the fine soil particles to pass through so as not to plug the filter but to retain the larger particles that tend to deposit in the drain. The maximum opening size is related to the soil particle size distribution, but the relationship has not yet been fully established for all soils.

14.12 Accessories

As shown in Fig. 13.5, a surface inlet is an intake structure for the removal of surface water from potholes, road ditches, farmsteads, and the like. Surface drains (Chapter 13) are preferred, where possible, because of their greater capacity ro remove flood water. Surface inlets should be placed at the lowest point along fence rows or in land that is in permanent vegetation. Where the inlet is in a cultivated field, the area immediately around the intake should be kept in grass. The concrete collar will stabilize the riser and prevent vegetation from growing near the inlet.

To allow surface water to reach the drain more quickly, short lengths of the trench may be filled with various gradations of sand and gravel. The coarsest material is placed immediately over the pipe and the size is gradually reduced toward the surface. Short lengths backfilled in this manner are known as blind inlets. Such inlets have a tendency to seal at the soil surface and cannot be expected to remove large amounts of surface water. Keeping the area directly over the drain in permanent grass sod will increase the flow. Organic materials, such as corncobs, sawdust, and straw, are less satisfactory than more durable materials.

Relief pipes and breathers are small-size vertical risers extending from the drain line to the surface. The riser should be made of steel pipe or mortared bell and spigot tile and should be located at fence lines where they are not likely to be damaged.

Relief pipes serve to relieve the excess water pressure in the pipe during periods of high outflow, thus preventing blowouts. A relief pipe should be installed where a steep section of a main changes to a flat section, unless the capacity of the flat section exceeds the capacity of the steep section by 25 percent.

14.13 Selection of Drain Pipe

Drain pipes are made from clay or concrete in short lengths and of bituminous fiber, steel, or corrugated plastic perforated tubing. Good quality concrete tiles are very resistant to freezing and thawing but may be subject to deterioration in acid and alkaline soils. In these soils, concrete tiles should be used only if they are approved by local recommendations. Clay tiles are not affected by acid or alkaline soils. When concrete tiles are subjected to frequent alternate freezing and thawing conditions, they are safer to use although most clay tiles are resistant to frost damage.

Good clay or concrete drain tiles should have the following characteristics: (1) resistance to weathering and deterioration in the soil, (2) sufficient strength to support static and impact loads under conditions for which they are designed, (3) low water absorption, that is, a high density, (4) resistance to alternate freezing and thawing, (5) relative freedom from defects, such as cracks and ragged ends, and (6) uniformity in wall thickness and shape. Drain tiles that meet current specifications of the American Society for Testing Materials (ASTM) have the essential qualities listed above. Specifications have been prescribed for three classes of drain tile, namely, standard, extra-quality, and special-quality (concrete only) or heavy-duty (clay only). Standard-quality tile is satisfactory for drains of moderate size and depths found in most farm drainage work.

Corrugated plastic tubing, first manufactured in the United States about 1967, increased in usage to 80 percent in 1975 and to 95 percent of all agricultural drains by 1983. It is lightweight and unaffected by all soil chemicals. The weight of 4-in. diameter plastic tubing is only about $\frac{1}{25}$ of that for clay or concrete tile. Tubing is made in sizes varying from 1 to 24 in. in diameter and in black, white, red, yellow, as well as other colors. Small size tubing is coiled in lengths up to several thousand feet. The most common resins are high density polyeythlene (HDPE) and polyvinyl chloride (PVC). In the United States and Canada, HDPE has the greater share of the market, whereas in Europe PVC is more common. Tubing should meet ASTM Standards F405 or F667 for deflection and elongation. These standards specify either standard or heavy-duty quality.

14.14 Allowable Drain Pipe Depths

Frequently, deep cuts are necessary for mains, and the soil load on the pipe may cause failure. Allowable depths for standard and extra-quality clay or concrete tile are given in Table 14.3.

The width of the trench at the top of the tile governs the load regardless of the trench width above the top of the tile. Thus, a trench 2 ft wide at the top of the tile and 5 ft wide at the surface would result in a load on the tile equal to that from a trench 2 ft wide all the way to the surface. Under conditions to the left and below the heavy line in Table 14.3, tile may be installed at any depth without the danger of breakage. To reduce the possibility of failure, tiles should be selected that are uniform in quality. Where the depth of cut is greater than the allowable, the trench width may be reduced, or higher strength tile, corrugated metal pipe, and other more rigid materials may be installed.

Allowable depths for corrugated plastic tubing are given in Table 14.4. These depths are based on an allowable deflection of 20 percent of the nominal tubing

Table 14.3. Allowable Clay or Concrete Tile Depths to the Bottom of the Trench in Feet

Tile Size (in.)	*ASTM Class*	*Crushing Strengths (lbs/lin. ft)*	*Width of Trench at Top of Tile (in.)*						
			15	*18*	*21*	*24*	*27*	*30*	*36*
4 or 5	Standard	1200	[a]	8.0	7.3	7.3	7.3	7.3	7.3
	Extra-quality	1650	[a]	[a]	10.5	9.8	9.8	9.8	9.8
6	Standard	1200	[a]	8.9	6.3	6.3	6.3	6.3	6.3
	Extra-quality	1650	[a]	[a]	10.6	8.3	8.3	8.3	8.3
8	Standard	1200	[a]	9.2	6.4	5.4	5.4	5.4	5.4
	Extra-quality	1650	[a]	[a]	10.8	7.8	7.1	7.1	7.1
10	Standard	1200	[a]	9.5	6.6	5.4	4.8	4.8	4.8
	Extra-quality	1650	[a]	[a]	11.0	8.0	6.5	6.2	6.2
12	Standard	1200	[a]	9.7	6.8	5.6	5.0	4.7	4.7
	Extra-quality	1650	[a]	[a]	11.2	8.2	6.7	5.8	5.8
15	Extra-quality	1650	—	—	11.5	8.5	7.0	6.0	5.2
18	Extra-quality	1800	—	—	14.8	9.8	8.0	7.0	5.6
21	Extra-quality	2100	—	—	—	13.8	9.7	8.5	6.7

[a] Any depth is permissible at this width or less.

Assumptions: (1) Crushing strengths given are averages in pounds per linear foot based on sand-bearing method. Specifications for Drain Tile ASTM C4 and C412. (2) These values allow a safety factor of 1.5. (3) Loadings were computed for wet clay soil at 120 lb per cu ft. Somewhat greater depths are permissible for lighter soils. (4) Ordinary laying whereby the underside of the tile is well bedded on soil for 60°–90° of the circumference.

diameter, although tubing may withstand up to 30 percent deflection without collapsing. Good side support is an important factor in preventing deflection. For tubing installed with less than 2 ft of cover, gravel backfill should be placed around the tubing and to a depth equal to the pipe diameter above the tubing. For further information see Schwab et al. (1981).

14.15 Layout and Design Procedure

The first step in design is to make a preliminary survey to evaluate the feasibility of drainage. The kind of information needed is indicated in Fig. 14.8. Tile layout can be made directly in the field or on a topographic map. If the field is flat, several elevations should be taken with the level to locate the outlet and plan the type of system best suited to the area. As shown in Fig. 14.6, the total length of pipe required and the amount of large-size pipe can be reduced by using a layout having long laterals and a short main. This reduction is due to a smaller area that is drained both by the main and by the laterals. This double-drained area for the field in Fig. 14.6 is reduced from 1.5 acres with short laterals to 0.5 acre for the long-lateral layout (Fig. 14.6*a*). By joining the laterals to the main

Table 14.4. Allowable Corrugated Plastic Tubing Depths to the Bottom of the Trench in Feet

Nominal Tubing Diameter (in.)	*Tubing Quality (ASTM)*	*Trench Width at Top of Tubing (in.)*			
		12	*16*	*24*	*32 or more*
4	*Standard	12.7	7.0	5.5	5.2
	**Heavy-Duty	[a]	9.8	7.0	6.3
6	*Standard	10.1	6.9	5.4	5.2
	**Heavy-Duty	[a]	9.6	6.7	6.1
8	*Standard	10.3	7.1	5.6	5.3
	**Heavy-Duty	[a]	9.8	6.8	6.2
10	**	—	9.2	6.6	6.2
12	**	—	8.8	6.6	6.2
15	**	—	—	6.8	6.3

[a]Any depth is permissible at this width or less and for 8-inch trench width for all sizes.
*Pipe stiffness 13 psi at 20 percent deflection.
**Pipe stiffness 18 psi at 20 percent deflection.

Assumptions: (1) Tubing buried in loose, fine-textured soil with density of 109 lb/cu ft. (2) Modulus of soil reaction 50 psi. (3) Deflection lag factor 3.4. (4) Bedding angle 90°. (5) Vertical deflection 110 percent of horizontal deflection.

Note: Differences in commercial tubing from several manufacturers, including corrugation design and pipe stiffness, and soil conditions may change the assumptions and, therefore, the maximum depths may be more or less than stated above. These depths are based on limited research and should be used with CAUTION.

Source: Fenemor et al. (1979).

at an angle of 90°, the double-drained area is kept at a minimum. Such a layout also facilitates installation. Smaller angles are permissible provided the flow from the lateral is directed downstream.

After the final plan has been selected, the drain lines are then staked out by placing hub stakes and guard stakes (station markers) at each 50- or 100-ft station. The stakes must be offset about 5 ft from the center line of the trench so that they will not be removed by the excavation. The contractor can advise as to the offset distance and to which side of the trench to set the stakes.

After the drain line has been laid out, the elevations of the hub stakes are determined by profile leveling and the notes are recorded as shown in Fig. 14.9. This form of field notes is convenient for simple drainage systems as the cuts can be computed in the field and given to the contractor. On the left-hand side of Fig. 14.9 a blank line is provided between 100-ft station numbers. This method of spacing provides the proper scale for plotting the profile shown on the right-hand side of the page. The blank line also permits the recording of intermediate stations without confusing the data, for example, see station 2 + 50. The hub elevations in column five on the left-hand side of the page are then plotted on the right-hand side to determine the ground profile. The elevation of the outlet drain line or ditch is also plotted on the profile and the grade line for the pipe

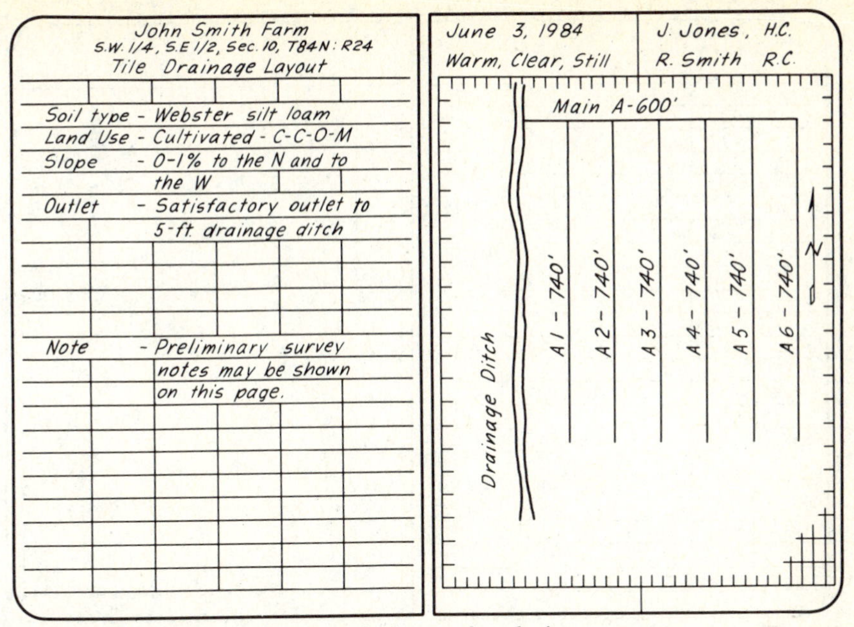

Figure 14.8. Field notes for a preliminary pipe-drainage survey.

is selected so that cuts at each station will not be too deep. The field notes for each lateral or main should be recorded on a separate page of the field book.

The grade elevation at 0 + 00 on the lateral is then determined from the grade elevation previously computed for the main at the junction. When the main and lateral are the same diameter, the grade elevation of the lateral should be 0.1 ft above the grade elevation of the main at the junction. In Fig. 14.9 a rise of 0.3 ft (93.50–93.20) was allowed for convenience. However, the pipes are connected in the normal manner and the extra fall is adjusted in the first few feet of the lateral. This additional fall at the junction causes the velocity to increase and reduces sedimentation.

Where the main is larger than the lateral, the grade line of the lateral must be higher so that the center line of the main and the center line of the lateral intersect. This difference in elevation can be computed by subtracting half the outside diameter of the lateral from half the outside diameter of the main.

As illustrated in Fig. 14.9, cuts are computed for each station by the subtraction of the grade elevation from the hub elevation. Grade elevations for each station above 0 + 00 are computed by adding the rise (distance times percent slope) to the grade elevation of the previous station. The contractor should be provided with station numbers, cuts, and grades for each line.

Where subsurface drains are to be installed with trenching machines or with drain plows equipped with laser grade control systems, the procedure of staking and determining cuts at 100-ft intervals is not required. Instead, it is necessary

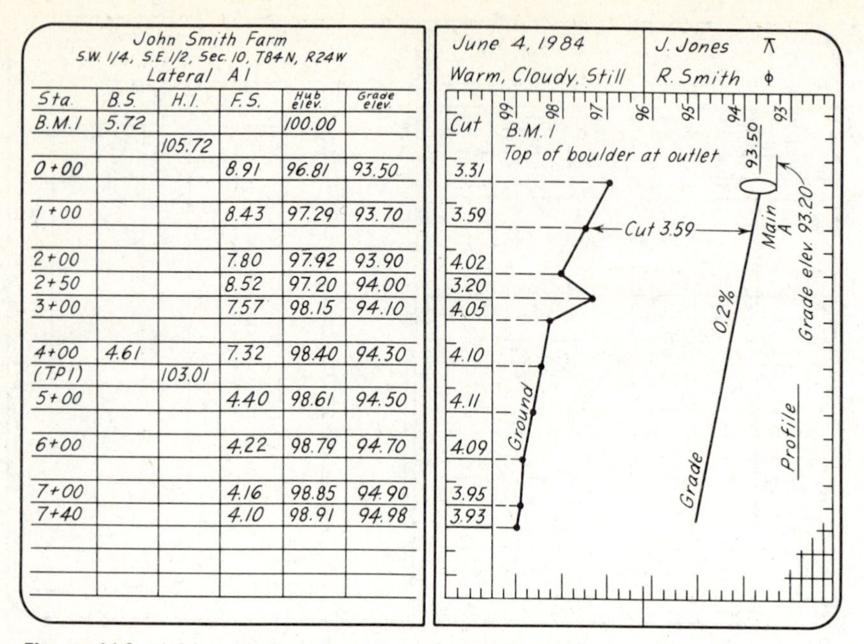

John Smith Farm
S.W. 1/4, S.E. 1/2, Sec. 10, T84N, R24W
Lateral A1

Sta.	B.S.	H.I.	F.S.	Hub elev.	Grade elev.
B.M.1	5.72			100.00	
		105.72			
0+00			8.91	96.81	93.50
1+00			8.43	97.29	93.70
2+00			7.80	97.92	93.90
2+50			8.52	97.20	94.00
3+00			7.57	98.15	94.10
4+00	4.61		7.32	98.40	94.30
(TP1)		103.01			
5+00			4.40	98.61	94.50
6+00			4.22	98.79	94.70
7+00			4.16	98.85	94.90
7+40			4.10	98.91	94.98

Figure 14.9. Field notes for recording cuts and grades.

to locate only the drain line and the points where changes in grade are required. The rotating laser beam from the command post can be set level or on a sloping plane for reference. A detector on the machine picks up the beam and automatically keeps the machine on the same grade as the plane produced by the laser beam. When a change in grade is desired, some machines are equipped with a "grade breaker," which will allow the machine to be controlled at a grade different from that of the laser beam. Sufficient survey points must be taken along the drain line to determine the desired grade and points of change to keep the cuts within reasonable limits.

Example 14.4. Determine the minimum grade elevation and cut at the outlet of a 4-in. diameter lateral to be connected to a 10-in. clay main if the hub stake elevation at the junction (0 + 00) is 96.81 as shown in Fig. 14.9.

Solution. The outside diameter (o.d.) of the 10-in. main is about 1.0 ft (o.d. in tenths of feet is numerically about equal to the inside tile diameter in inches for sizes up to 12 in.) and 0.4 ft for the 4-in. lateral. For the two lines to connect center to center

$$\text{rise in elevation} = \frac{1.0}{2} - \frac{0.4}{2} = 0.3 \text{ ft}$$

Minimum grade elevation of the 4-in. lateral is 93.20 + 0.3 = 93.50 ft, and the cut on the lateral is 96.81 − 93.50 = 3.31 ft, as shown in Fig. 14.9.

14.16 Installation and Maintenance

Wheel and endless-chain trenchers and drain plows (trenchless) are the most common installation machines. Backhoes require hand shaping of the trench bottom because accurate grade control devices have not been developed. By setting targets at each station or by using the laser plane of light at the desired grade (Fig. 14.10), a line parallel to the grade line is established from which the desired depth of cut and grade are maintained. The vertical distance (base height) from the sight bar or laser detector on the machine to the bottom of the digging mechanism is fixed. In Fig. 14.10 this base height is 10 ft. Subtracting the cut at each station from the base will give the height of the target above the hub stake. For example, at the first station in Fig. 14.10, the target height is 6.5 ft (10 − 3.5). The allowable variation from the true grade should be not more than 0.01 ft per in. of tile diameter in 100 ft.

The crack spacing between individual tiles (1-ft lengths) should be not more than $\frac{1}{8}$ in. for sandy soils and $\frac{1}{4}$ in. for clay. In unstable soils, tiles should be laid as close together as possible. For equivalent drainage, perforated tubing should have about 28 and 42 perforations per foot ($\frac{1}{4}$ in. diameter) to compare to $\frac{1}{8}$- and $\frac{1}{4}$-in. cracks, respectively. Perforating ordinary tile will give some increase in flow, but increasing the crack width could give the same effect.

Poor outlets and failure to make timely repairs are major maintenance problems. Drains that are designed and installed properly require little maintenance. Roots of brush and trees, particularly willow, elm, cottonwood, soft maple, and eucalyptus, will grow into the line and obstruct the flow, especially if the drain is fed by springs supplying water far into the dry season. Under such conditions this vegetation should be removed if it is within 100 ft of a line. If this is not convenient, a sealed tile or pipe may be installed. In a few areas minerals, such as iron and manganese, may be deposited in the openings, thus sealing the drains. Iron ochre is common in some areas, especially in peat and muck soils. It is a gelatinous material, caused by chemical oxidation or bacteria. Its removal or

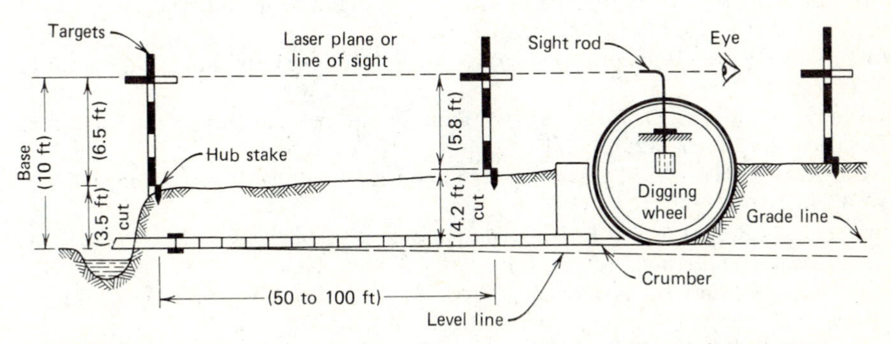

Figure 14.10. Establishing grade line with a trenching machine using targets or laser beam.

control is a difficult problem, for which there is not a completely satisfactory solution.

14.17 Mapping the Drainage System

After the drainage system has been installed, a suitable map should be made and filed with the deed to the property. As described in a previous chapter, this map may be made with a traverse table, or, if the system is simple, such as the one shown in Fig. 14.7, it may be made by measuring a few distances and angles with a steel tape. The drainage system can best be shown on an aerial photo or a topographic map. The owner of the farm should retain a copy of the map as it is useful in locating the lines at a later date. This map is likewise evidence that the drains have been installed and is especially valuable to future owners.

14.18 Estimating Cost of Drainage Systems

The principal items of cost for a subsurface drainage system are installation, engineering, pipe, and accessories, such as outlet tubes and surface inlets. Pipe prices vary from one region to another and with the distance from the plant. Most ditching-machine contractors charge a fixed price per unit length. Additional charges may be made for overcut below a specified depth, for large pipe and for backfilling. Engineering and supervision costs may vary from 5 to 10 percent of the total. In the Midwest the cost for materials is about the same as the trenching cost, but in California, where depths are 6 ft or more, installation is about twice the cost of the pipe.

REFERENCES

Drablos, C. J. W., and R. C. Moe (1984) "Illinois Drainage Guide," *University Illinois Ext. Cir. 1226*.

Fenemor, A. D., B. R. Bevier, and G. O. Schwab (1979) "Prediction of Deflection for Corrugated Plastic Tubing," *Trans. ASAE* **22**(6):1338–1342.

Grass, L. B. (1969) "Tile Clogging by Iron and Manganese in Imperial Valley, California," *J. Soil and Water Conservation* **24**(2):135–138.

Luthin, J. N. (1974) *Drainage Engineering*, 2nd ed., Kreiger Publishing Co., Huntington, N.Y.

Manson, P. W., and C. O. Rost (1951) "Farm Drainage—An Important Conservation Practice," *Agr. Engin.* **32**:325–327.

O'Neal, A. M. (1949) "Soil Characteristics Significant in Evaluating Permeability," *Soil Sci.* **67**:403–409.

Schwab, G. O., R. K. Frevert, K. K. Barnes, and T. W. Edminster (1981) *Soil and Water Conservation Engineering*, 3rd ed., John Wiley & Sons, New York.

Smedema, L. K., and D. W. Rycroft (1983) *Land Drainage*, Cornell University Press, Ithaca, N.Y.

Skaggs, R. W. (1980) "A Water Management Model for Artificially Drained Soils," *University North Carolina Water Resources Res. Inst. Tech. Bull. 267*.

U.S. Bureau Reclamation (1978) *Drainage Manual*, U.S. Government Printing Office, Washington, D.C.

U.S. Soil Conservation Service (1973) *Drainage of Agricultural Land*, Water Information Center, Inc., Port Washington, N.Y.

U.S. Soil Conservation Service (1979) *Engineering Field Manual* (Lithographed), Washington, D.C.

van Schilfgaarde, J. (ed.) (1974) "Drainage for Agriculture" Monograph 17, *Am. Soc. Agron.*, Madison, Wis.

PROBLEMS

14.1 A pipe drainage system draining 12 acres flows at design capacity for two days following a storm. If the system is designed using a D.C. of $\frac{1}{2}$ in., how many cubic feet of water will be removed during this period?

14.2 What size pipe (clay) is required to remove the surface water through a surface inlet if the runoff accumulates from 20 acres of mineral soil and the slope in the tile is 0.3 percent? Design for field crops in your area.

14.3 Determine the total drainage area in acres for the pipe system in Fig. 14.8, assuming that Main *A* drains only a 50-ft strip on one side of the main and the drain spacing is 100 ft.

14.4 Determine the plastic tubing size at the outlet of a 10-acre drainage system (*a*) if the D.C. is $\frac{3}{8}$ in., and the grade is 0.3 percent; (*b*) if the grade is reduced to 0.1 percent.

14.5 Determine the design drain flow from 10,000 ft of pipe at a spacing of 200 ft, following local recommendations for irrigated conditions. What size clay tile is required for the main at the outlet if the slope is 0.25 percent?

14.6 For the pipe system shown in Fig. 14.7, determine the size of clay tile required for Main *A* and lateral *A*1 if the D.C. is $\frac{3}{8}$ for normal conditions and $\frac{3}{4}$ in. for surface inlets. Assume that the location is in a humid area and that all surface water from the 8 acres must be removed through the surface inlet.

14.7 For the pipe system shown in Fig. 14.7, determine the length and size of all plastic tubing required. Assume that the drainage coefficients are to be the lowest recommended rate for truck crops. The field is in a humid area and (mineral) clay loam soil predominates. All surface water from the 8-acre depression is to be removed through the surface inlet.

14.8 Solve Problem 14.6, assuming that the surface inlet is not to be installed. Surface water is to be removed with a surface drain.

14.9 What is the maximum depth that a 10-in. standard-quality clay tile will withstand in a trench 24 in. wide? Extra-quality?

14.10 Six-inch tiles are to be installed in a trench 21 in. wide and 9.5 ft deep. Will standard-quality tile support the soil load? If not, what could be done to prevent breakage?

14.11 Using a topographic map supplied by your instructor, select a depth, a spacing, a drainage coefficient for the crop, and the layout for the subsurface drainage system. By estimating the grade from the elevations on the map, determine the size and length of plastic tubing required, including junctions and other accessories.

14.12 The elevation of a hub stake on a 10-in. tile main is 86.40. If the cut on the main at this station is 3.40 ft, what is the lowest grade elevation at which a 4-in. lateral should be connected? Assume that a 10 × 4-in. manufactured T-junction is to be installed and that the wall thickness of both sizes is 1 in.

14.13 If the sight bar on a trenching machine is mounted 9 ft above the bottom of the digging wheel, how high should the sighting target be set above the hub stake if the cut is 3.56 ft?

14.14 Using Fig. 14.9, compute the cuts for lateral *A*1 if the slope in the drain had been 0.3 percent up to station 3 + 00 and 0.16 percent above this point.

14.15 Estimate the cost of the drainage system in Fig. 14.6*a*. Assume the average cut for the main is to be 4.5 ft and for the laterals 4.0 ft or less. *Installation cost:* 4.0 ft or less, $30 per 100 ft; overcut, $1.00 per 0.1 ft per 100 ft. *Materials:* Pipe prices to be supplied by your instructor. Add 5 percent of length for breakage for clay or concrete, but no extra cost for plastic; pipe junctions at five times the cost per foot of pipe of the largest size. *Engineering:* 5 percent of total material and installation cost. Show bill of materials and compute the cost per acre.

14.16 What is the maximum depth at which 8-in. standard quality corrugated plastic tubing can be installed in a 24-in. wide trench to prevent a deflection not greater than 20 percent of the tubing diameter? What is the maximum depth at which heavy-duty tubing can be installed?

14.17 Four-inch diameter corrugated plastic tubing is to be installed in a trench 24 in. wide and 6.0 ft deep. Will standard quality tubing provide sufficient support to prevent a deflection of less than 20 percent? If not, what could be done to reduce the deflection?

14.18 The grade elevation at the outlet of a 600-ft lateral is 83.50 ft. The hub elevations at each 100-ft station starting at the outlet are 86.81, 87.29, 87.92, 88.15, 88.40, 88.61, and 88.79 ft. Select a uniform slope so as to provide an average cut of about 4 ft. Tabulate all grade elevations and compute the cuts for each station.

CHAPTER 15

IRRIGATION PRINCIPLES

Man's dependence upon irrigation can be traced to the earliest biblical references. Irrigation in very early times was practiced by the Egyptians, the Asians, and the Indians of North America. For the most part, water supplies were available to these people only during the periods of heavy runoff. Modern concepts of irrigation have been made possible only by the application of modern power sources to deep well pumps and by the storage of large quantities of water in reservoirs. Thus, by using either underground or surface reservoirs it is possible to bridge over the months and years and to even out water excesses and deficiencies.

Benefits from proper irrigation include (1) increased yields, (2) improved crop quality, particularly from vegetable and fruit crops, (3) controlled time of planting and harvesting so as to obtain a more favorable market price, (4) reduced damage from freezing temperatures, (5) reduced damage by control of high air temperatures, (6) increased efficiency of fertilizers and reduced cost of application, and (7) a stabilized farm income.

In semihumid and humid areas where irrigation primarily supplements natural precipitation, the prospective irrigator should answer positively the following questions before considering irrigation: (1) Is the water supply adequate and of good quality? (2) Is sufficient labor available to operate the irrigation system? (3) Is capital available to purchase the necessary equipment? (4) Will irrigation sufficiently increase crop yields over a period of years to justify the cost? Investment in alternative means of increasing production or conservation should also be considered if these practices are not already fully utilized. With limited capital such alternative practices might be more advantageous than irrigation. Not only is capital required for the irrigation system, but capital for additional fertilizer and seed is also needed. Higher fertility levels and higher planting rates are usually necessary in order to obtain maximum returns from the water applied.

15.1 Evapotranspiration

Estimates of both soil evaporation and transpiration from vegetation are necessary to determine the amount of water required in a given time. Evapotranspiration, a term used to express both of these losses, is simply the total water loss to the atmosphere. Although many methods have been developed to estimate evapotranspiration, the Blaney-Criddle equation is one of the simplest. The monthly evapotranspiration equation is

$$u = \frac{ktp}{100} = kf \tag{15.1}$$

where u = monthly evapotranspiration in inches

k = monthly evapotranspiration (consumptive use) coefficient (determined for each crop from experimental data)

t = mean monthly temperature in degrees Fahrenheit

p = monthly percent of daytime hours of year

$f = \frac{(tp)}{100}$ = monthly evapotranspiration (consumptive use) factor

For the entire growing season the following equation is more convenient

$$U = KF = \Sigma kf = K\Sigma f \tag{15.2}$$

where U and K correspond to u and k in Eq. 15.1 and F = sum of the monthly evapotranspiration (consumptive use) factors f for the period. Mean monthly temperatures and percent of daytime hours for each month can be determined from Weather Service records or from other data for the locality. Evapotranspiration coefficients for irrigated crops are given in Table 15.1.

Modifications of this method and other more sophisticated approaches are discussed by Schwab et al. (1981), Jensen (1974), and the U.S. Soil Conservation Service (1970). A comparison of various approaches with the results of field research has been prepared by the U.S. Bureau of Reclamation (1983).

15.2 Irrigation Requirements

The irrigation requirement is the quantity of water, exclusive of precipitation, to be supplied by artificial means. Irrigation requirements are dependent not only on evapotranspiration but also on the water application efficiency, the water supplied by percolation, by capillary movement from the ground water table, and by effective rainfall. Effective rainfall is the water that actually enters the soil profile. Estimates of effective rainfall may be taken from charts given in U.S. Soil Conservation Service (1970).

Example 15.1. Determine the evapotranspiration and irrigation requirements for wheat (small grain) at Plainview, Texas, if the water application efficiency is 65 percent.

Solution.

Month	t^a	p^b	f^c	*Rainfall*[a]
April	59.2	8.80	5.21	1.92
May	67.5	9.72	6.56	2.58
June	75.6	9.70	7.33	3.04
		Total	19.10	7.54

[a]Mean monthly temperature in degrees F and rainfall can be obtained from Weather Service records for the locality.
[b]Monthly percent of daytime hours of the year is available from Hansen et al. (1980) or U.S. Soil Conservation Service (1970).
[c]Monthly evapotranspiration factor ($tp/100$).

From Eq. 15.2 and Table 15.1, $K = 0.80$, $U = 0.80 \times 19.10 = 15.28$ in.

$$\text{irrigation water required} = \frac{(15.28 - 7.54)}{0.65} = 11.9 \text{ in.} \quad (302 \text{ mm})$$

15.3 Crop Needs

Water requirements and time of maximum demand vary with different crops. Although growing crops are continuously using water, the rate of transpiration depends on the kind of crop, the degree of maturity, and the atmospheric conditions, such as humidity, wind, and temperature. Where sufficient water is available, the moisture content should be maintained within the limits for optimum growth. The rate of growth at different soil moistures varies with different soils and crops. Some crops are able to withstand drought or high moisture content much better than others. During the later stages of maturity, the water needs are generally less than during the maximum growing period. When crops are ripening, irrigation is usually discontinued.

From soil moisture measurements or from the appearance of the soil or the crop, the irrigator is able to determine when and how much water should be applied. A recommended procedure involves taking soil samples from the root zone. With experience, it is possible to estimate the need for irrigation from the "feel" of the soil. Tensiometers or neutron probe sites located in representative areas of the field provide a more reliable basis for determining water needs. In practice, however, much irrigation water is applied on the basis of a routine schedule based on the experience of the farm manager.

Table 15.2 illustrates evapotranspiration estimates and water requirements for crops grown near Deming, New Mexico, based on the Blaney–Criddle method. It should be noted that a correction is made for expected effective rainfall in determining the field irrigation requirements. Table 15.3 presents the effect of soil texture on application rates and moisture-holding capacity. Table 15.4 gives approximate root depths and the daily moisture use for selected crops.

Table 15.1. Seasonal Evapotranspiration Crop Coefficients K for Irrigated Crops

Crop	*Length of Normal Growing Season or Period*[d]	*Evapotranspiration Coefficient* K[b]
Alfalfa	Between frosts	0.80 to 0.90
Bananas	Full year	0.80 to 1.00
Beans	3 months	0.60 to 0.70
Cocoa	Full year	0.70 to 0.80
Coffee	Full year	0.70 to 0.80
Corn (maize)	4 months	0.75 to 0.85
Cotton	7 months	0.60 to 0.70
Dates	Full year	0.65 to 0.80
Flax	7 to 8 months	0.70 to 0.80
Grains, small	3 months	0.75 to 0.85
Grain, sorghums	4 to 5 months	0.70 to 0.80
Oilseeds	3 to 5 months	0.65 to 0.75
Orchard crops:		
Avocado	Full year	0.50 to 0.55
Grapefruit	Full year	0.55 to 0.65
Orange and lemon	Full year	0.45 to 0.55
Walnuts	Between frosts	0.60 to 0.70
Deciduous	Between frosts	0.60 to 0.70
Pasture crops:		
Grass	Between frosts	0.75 to 0.85
Ladino whiteclover	Between frosts	0.80 to 0.85
Potatoes	3 to 5 months	0.65 to 0.75
Rice	3 to 5 months	1.00 to 1.10
Soybeans	140 days	0.65 to 0.70
Sugar beet	6 months	0.65 to 0.75
Sugarcane	Full year	0.80 to 0.90
Tobacco	4 months	0.70 to 0.80
Tomatoes	4 months	0.65 to 0.70
Truck crops, small	2 to 4 months	0.60 to 0.70
Vineyard	5 to 7 months	0.50 to 0.60

[a]Length of season depends largely on the variety and the time of year when the crop is grown. Annual crops grown during the winter period may take much longer than if they are grown in the summertime.

[b]The lower values of K for use in the Blaney-Criddle formula, $U = KF$, are for the more humid areas, and the higher values are for the more arid climates.

Source: U.S. Soil Conservation Service, 1970.

15.4 Seasonal Use of Water

To make maximum use of the available water, the irrigator should have a knowledge of the water requirements of crops at all times during the growing season.

Table 15.2. Seasonal Evapotranspiration and Irrigation Requirements For Crops Near Deming, New Mexico[a]

Crop	*Length of Growing Season (Days)*	*Evapo-transpiration Depth (in.)*	*Effective Rainfall Depth (in.)*	*ET Less Rainfall (in.)*	*Water Application Efficiency (%)*	*Irrigation Requirement Depth (in.)*
Alfalfa	197	36.0	6.0	30.0	70	42.9
Beans (dry)	92	13.2	4.0	9.2	65	14.1
Corn	137	23.1	5.3	17.8	65	27.4
Cotton	197	26.3	6.0	20.3	65	31.3
Grain (spring)	112	15.6	1.3	14.3	65	22.0
Sorghum	137	21.6	5.3	16.3	65	25.1

[a]Average frost-free period is April 15 to October 29. Irrigation prior to the frost-free period may be necessary for some crops.

Source: Jensen (1974).

It may be possible to select the crop to fit the water supply. Figure 15.1 shows evapotranspiration for two crops grown in the Salt River Valley of Arizona. Wheat, being a fairly short season crop, has the lowest seasonal use of 25.8 in. with the high water requirement occurring during March and April. Alfalfa, a long season crop, has a seasonal use of 74.3 in. and in this area grows during the entire year with the exception of December and January. Similar data are available in other regions.

15.5 Moisture Deficiency Recurrence

The duration and length of dry periods during the growing season in humid and semihumid areas largely determine the economic feasibility of irrigation.

Table 15.3. Effect of Soil Texture on Maximum Sprinkler Application Rates and Moisture-Holding Capacity

Soil Texture and Profile Conditions	*Moisture-Holding Capacity, (in./ft)*	*Maximum Sprinkler Application Rates with Vegetative Cover (in. per hr)*[a]	
		0% Slope	*10% Slope*
Sandy loam uniform to 6 ft	0.7–1.5	1.7	1.0
Sandy loam over compact subsoil	0.7–1.5	1.2	0.7
Silt loam uniform to 6 ft	1.5–2.0	1.0	0.6
Silt loam over compact subsoil	1.5–2.0	0.6	0.4
Clay to clay loam	2.0–2.5	0.2	0.1

[a]For bare soil or where plants are small reduce these rates to about one half.

Table 15.4. Root Depth and Peak Rate of Moisture Use for Certain Crops

Crop	*Root Depth (ft)*[a]	*Peak Rate of Moisture Use (in. per day)*		
		Climate		
		Cool	*Moderate*	*Hot*
Alfalfa	3.0–4.0	0.20	0.25	0.30
Beans	1.0–3.0	0.12	0.16	0.25
Corn	2.0–4.0	0.20	0.25	0.30
Pasture	1.5–2.5	0.20	0.25	0.30
Potatoes	1.0–2.0	0.14	0.20	0.25
Strawberries	1.0–1.5	0.12	0.16	0.25

[a]Depth above which most roots occur, unless restricted by hardpan or other such layers.

In the Northern Hemisphere the moisture deficiency during the months of June, July, and August is more serious than in earlier or later months.

15.6 The Soil Moisture Reservoir

In planning and managing irrigation it is helpful to think of the soil's capacity to store available moisture as the soil moisture reservoir. The reservoir is filled periodically by irrigations. It is slowly depleted by evapotranspiration. Water application in excess of the reservoir capacity is wasted unless it is used for leaching (Section 15.9). Irrigation must be scheduled to prevent the soil moisture reservoir from becoming so low as to inhibit plant growth.

Irrigation can raise the soil moisture to the field capacity. In either sprinkler or surface irrigation the infiltration capacity and the permeability of the soil will determine how fast water can be applied. In sprinkler irrigation the water should be applied at a rate lower than the infiltration capacity. In surface irrigation, the soil surface must be flooded to allow water to enter the soil.

In some cases where early spring runoff is available beyond that which may be stored in surface reservoirs, fields are sometimes flooded with surface runoff channeled through irrigation ditches in order to fill the soil moisture reservoir and to conserve other water supplies for use later in the season.

The lower limit of soil moisture for plant growth is the wilting point. The moisture-holding capacity of the soil is the difference between the field capacity and the wilting point, which is shown in Table 15.3 for soils of different texture. The calculation of the soil moisture-holding capacity is illustrated in Example 15.2.

Example 15.2. Determine the soil moisture-holding capacity for a sandy loam soil in inches depth if the field capacity is 33 percent, the wilting point is 23 percent, and the dry soil density is 70 lb/cu ft. The crop to be grown is alfalfa, which has a root depth of 3.0 ft (Table 15.4).

Solution. The total available soil moisture in the 3-ft root depth is

$$\frac{33 - 23}{100} \times 3.0 \times 70 = 21 \text{ lb/sq ft area}$$

By converting weight to depth (water weighs 62.4 lb/cu ft)

$$\text{Moisture depth} = \left(\frac{21}{62.4}\right) \times 12 = 4.0 \text{ in. (102 mm)}$$

15.7 Irrigation Scheduling

In many instances, irrigation districts, government agencies, or consulting engineering firms are providing irrigation water management services to farm operators. With information on the water-holding capacity of the soil and evapotranspiration rates, the time and amount of irrigation can be predicted. Computer programs may be utilized to facilitate the process. This service provides farmers with a basis for effective management of their irrigation activities.

The time between irrigations can be determined by estimating the allowable percentage soil moisture depletion, which is usually not greater than 40 to 60 percent of the water-holding capacity. Example 15.3 gives the procedure.

Example 15.3. Determine the depth of irrigation for potatoes grown in a clay loam soil in a moderate climate, assuming that irrigation is to begin when 45 percent of the soil moisture-holding capacity is depleted. If no rain, how many days later will a second irrigation be required?

Solution. Read from Table 15.3 a soil moisture-holding capacity for clay loam of 2.0 in./ft. For a root depth of 2.0 ft for potatoes given in Table 15.4, the depth of irrigation for 45 percent depletion is

$$2.0 \times 2.0 \times 0.45 = 1.8 \text{ in. (46 mm)}$$

The peak rate of moisture use from Table 15.4 is 0.2 in. per day for a moderate climate.
Time between irrigations = 1.8/0.2 = 9 days.

15.8 Salinity

The presence of soluble salts in the root zone can be a serious problem especially in arid regions. In subhumid regions, where irrigation is provided on a supplemental basis, salinity is usually of little concern because the rainfall is sufficient to leach out any accumulated salts. However, all water from surface streams and underground sources contains dissolved salts. The salt applied to the soil with irrigation water remains in the soil unless it is flushed out in drainage water or is removed in the harvested crop. Usually the quantity of salt removed by crops is so small that it will not make a significant contribution to salt removal nor will it enter into determinations of leaching requirements.

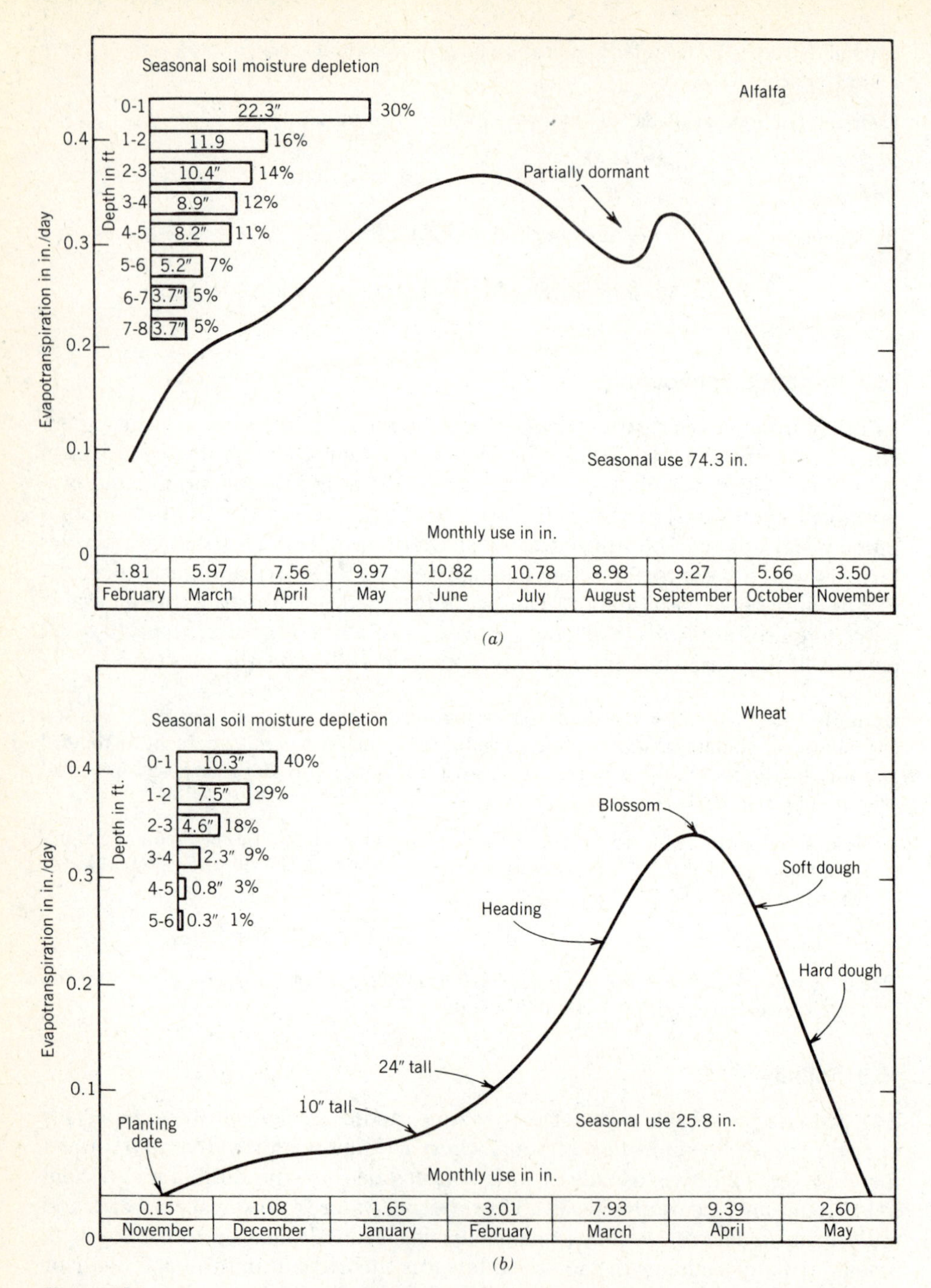

Figure 15.1. Average evapotranspiration and seasonal moisture depletion with depth for (*a*) alfalfa and (*b*) wheat at Mesa and Tempe, Arizona. (*Redrawn and adapted from Erie, French, and Harris, 1968.*)

The principal effect of salinity is to reduce the availability of water to the plant. In cases of extremely high salinity, there may be curling and yellowing of the leaves, firing in the margins of the leaves, or actual death of the plant. Long before effects like these are observed, the general nutrition and growth physiology of the plant will have been altered.

As the proportion of exchangeable sodium increases, soils tend to become dispersed, less permeable to water, and of poorer tilth. High-sodium soils usually are plastic and sticky when wet, and are prone to form clods and crusts on drying. These conditions result in reduced plant growth, poor germination and because of inadequate water penetration, poor root aeration and soil crusting.

With the necessity of using additional water beyond the needs of the plant to provide sufficient leaching, it is imperative under irrigation that there be adequate drainage of water passing through the root zone. Natural drainage through the underlying soil may be adequate. In cases where subsurface drainage is inadequate, open or pipe drains must be provided.

Water will rise 2 to 5 ft or more in the soil above the water table by capillarity. The height to which water will rise above a free-water surface depends on soil texture, structure, and other factors. Water reaching the surface evaporates, leaving a salt deposit typical of saline soils.

Tolerance to salinity varies considerably from crop to crop. Tables indicating relative tolerance to salinity for a number of crops as well as a more detailed discussion of the problem are presented in Chapter 11 and by Ayres and Westcot (1976).

15.9 Leaching

Leaching is the only way by which the salts added to the soil by irrigation water can be removed satisfactorily. Sufficient water must be applied to dissolve the excess salts and carry them away by subsurface drainage. The water-quality aspects of leaching are discussed in Chapter 11.

The traditional concept of leaching involves the ponding of water to achieve more or less uniform salt removal from the entire root zone. However, Ayers and Westcot (1976) show that salt accumulation can take place for short periods of time in the lower root zone without adverse effects. As can be noted in Fig. 15.1, most of the water transpired by the plant is taken from the upper portion of the root zone. This area will be leached to a considerable degree by the normal applications of irrigation water and by rainfall that may come at any time of the year. The same amount of water when applied with more frequent irrigations is more effective in removing salts from this critical upper portion of the root zone than from the lower root zone. Thus, high-frequency sprinkler, trickle, or surface irrigation should be effective for salinity control.

The *leaching requirement* is the fraction of water entering the soil, which must be passed through the root zone to control soil salinity at a specified level. Under conditions of adequate drainage and with no leaching by rainfall or salt removal by the crop, the leaching requirement is given by

$$LR = \frac{D_d}{D_i} = \frac{EC_i}{EC_d} \tag{15.3}$$

where LR = leaching requirement
D = depth of water applied or drained through
EC = electrical conductivity of water applied or drained through
d = subscript indicating drainage water
i = subscript indicating irrigation water

The electrical conductivity of the drainage water is that at the bottom of the root zone which the crop in question will tolerate.

Since salt is often more damaging to seedlings and younger plants than to mature plants, annual applications of leaching water are often effectively made during the preplant irrigation to reduce salt in the seed zone as well as to reduce salinity through the soil profile. In some situations excess surface water is available in the spring and may be used for leaching; in other situations leaching water may be a part of every irrigation application.

15.10 Irrigation Efficiencies

Efficiency is an output divided by an input and is usually expressed as a percentage. An efficiency figure is only meaningful when the output and input are clearly defined. Three basic irrigation efficiency concepts are the following.

(1) Water-conveyance efficiency:

$$E_c = 100W_d/W_i \tag{15.4}$$

where W_d = water delivered by a distribution system
W_i = water introduced into the distribution system

The water-conveyance efficiency definition can obviously be applied along any reach of a distribution system. For example, a water-conveyance efficiency could be calculated from a pump discharge to a given field or from a major diversion work to a farm turnout.

(2) Water-application efficiency:

$$E_a = 100W_s/W_d \tag{15.5}$$

where W_s = water stored in the soil root zone by irrigation
W_d = water delivered to the area being irrigated

This efficiency may be calculated for an individual furrow or border, for an entire field, or for an entire farm or project. When it is applied to areas larger than a field, it overlaps the definition of conveyance efficiency.

(3) Water-use efficiency:

$$E_u = 100W_u/W_d \tag{15.6}$$

where W_u = water beneficially used
W_d = water delivered to the area being irrigated

The concept of beneficial use differs from that of water stored in the root zone in that leaching water would be considered beneficially used though it moved through the soil moisture reservoir. Sometimes water-use efficiency is based on dry plant weight produced by a unit volume of water.

Another useful measurement of the effectiveness of irrigation is the uniformity coefficient,

$$C_u = 1 - y/d \tag{15.7}$$

where y = average of the absolute values of the deviations in depth of water stored from the average depth of water stored
d = average depth of water stored

This coefficient indicates the degree to which water has penetrated to a uniform depth throughout a field. Note that when the deviation from the average depth is zero, the uniformity coefficient is 1.0.

Example 15.4. If 1500 cfs are pumped into a farm distribution system and 1350 cfs are delivered to a turnout 2 miles from the well, what is the conveyance efficiency of the portion of the farm distribution system used in conveying this water?

Solution. Substituting in Eq. 15.4,

$$E_c = 100\frac{W_d}{W_i} = 100\frac{1350}{1500} = 90 \text{ percent}$$

Example 15.5. Delivery of 360 cfs to an 80-acre field is continued for 4 hr. Soil probing after the irrigation indicates that 1.10 ft of water has been stored in the root zone. Compute the application efficiency.

Solution. Apply Eq. 15.5,

$$W_d = \frac{(360)(3600 \text{ sec/hr})(4 \text{ hr})}{43{,}560} = 119 \text{ ac-ft}$$

$$W_s = (1.1)(80) = 88 \text{ ac-ft}$$

$$E_a = 100\frac{W_s}{W_d} = 100\frac{88}{119} = 74 \text{ percent}$$

Example 15.6. The depths of penetration of a probe along the length of a border strip taken at 100-ft stations were 6.5, 6.3, 6.1, 5.7, and 5.4 ft. Compute the uniformity coefficient.

Solution. Apply Eq. 15.7,

$$\text{Average depth, } d = (6.5 + 6.3 + 6.1 + 5.7 + 5.4)/5$$
$$= 30.0/5 = 6.0 \text{ ft} \quad (1.83 \text{ m})$$
$$\text{Average deviation, } y = (0.5 + 0.3 + 0.1 + 0.3 + 0.6)/5$$
$$= 1.8/5 = 0.36 \text{ ft} \quad (0.11 \text{ m})$$

$$C_u = 1 - (0.36/6.0) = 0.94$$

IRRIGATION METHODS

The methods of applying irrigation water may be classified as subsurface, surface, sprinkler, and trickle irrigation.

15.11 Subsurface Irrigation

In unique situations, water may be applied below the surface of the soil. There are two types of subsurface irrigation (also called subirrigation). The most common system develops or maintains a water table allowing the water to move up through the root zone by capillary action. This is essentially the same practice as controlled drainage, which is discussed in Chapter 14. Controlled drainage becomes subirrigation if water must be supplied to maintain the desired water table level. Water may be introduced into the soil profile through open ditches, mole drains, or pipe drains. Water table maintenance is suitable where the soil in the plant root zone is quite permeable and there is either a continuous impermeable layer or a natural water table below the root zone. Since subirrigation allows no opportunity for leaching and establishes an upward movement of water, salt accumulation is a hazard; thus the salt content of the water should be low. A special method of subirrigation introduces water into the soil through small perforated pipes. The system has been successfully applied to the irrigation of turf.

15.12 Surface Irrigation

By far the most common method of applying irrigation water, especially in arid regions, is by flooding the surface. Surface methods include wild flooding where the flow of water is essentially uncontrolled and surface application where flow is controlled by furrows, corrugations, border dikes, contour dikes, or basins. Except in the case of wild flooding, the land should be carefully prepared before irrigation water is applied. In order to conserve water, the rate of water application should be carefully controlled and the land properly graded. Surface irrigation is discussed in Chapter 16.

15.13 Sprinkler Irrigation

In recent years increasing use has been made of irrigation pipe systems for distributing water to sprinkler heads. Lightweight portable pipes with slip joint connections are common. However, in view of the high labor cost in moving these systems, such applications are becoming more limited to high value crops. Mechanical-move systems are now widely accepted. These may be either intermittent or continuous mechanical move. Solid set and permanent systems are suitable for intensively cultivated areas growing a high income crop, such as flowers, fruits, or vegetables. Sprinkler irrigation systems provide a reasonably uniform application of water. On coarse-textured soils, water application efficiency may be twice as high as with surface irrigation. Sprinkler irrigation is discussed in Chapter 17.

15.14 Trickle Irrigation

Increasing use is being made of trickle (drip) systems that apply water at very low rates, often to individual plants. Such rates are accomplished through the use of specially designed emitters or porous tubes. A typical emitter might apply water at from 0.5 to 2.5 gal per hr. Other techniques for applying water at low rates may also be called trickle irrigation. These systems provide an opportunity for efficient use of water because of minimum evaporation losses and because irrigation is limited to the root zone. Due to the high cost, their use is generally limited to high value crops. Since the distribution pipes are usually at or near the surface, the operation of field equipment is difficult. Both sprinkler and trickle systems are well adapted to the application of agricultural chemicals, such as fertilizers and pesticides with the irrigation water. Trickle irrigation is discussed in Chapter 18.

15.15 Comparison of Irrigation Methods

Table 15.5 compares different types of irrigation systems in relation to various site and situations factors. Efficient surface irrigation requires the grading of the land surface to control the flow of water. The extent of grading required depends upon the topography. In some soil and topographic situations, the presence of unproductive subsoils may make grading for surface irrigation unfeasible. The utilization of level basins where large streams of water are available generally provide high irrigation efficiencies.

Sprinkler irrigation is particularly adaptable to hilly land where grading for surface irrigation is not feasible. It is appropriate for most circumstances where the infiltration rate exceeds the rate of water application and where wind is not a continuing problem. With sprinkler irrigation the rate of water application can be easily controlled. Sprinkler irrigation systems usually have a relatively high cost of installation. With mechanical-move systems, labor can be substantially reduced. In some cases disease problems have resulted from moistened foliage. Evaporation losses with sprinkler irrigation are not excessively high even in arid regions. A well-designed sprinkler irrigation system can provide a high efficiency of water application.

A well-designed trickle irrigation system probably provides the highest efficiency of water application. It is especially well suited to tree fruit and high value crops, but not to annual row crops. Water must be clean and uncontaminated, usually achieved by a filtration system. Trickle irrigation lends itself well to automation and has a low labor requirement. Since trickle systems usually operate at low pressure, energy requirements are generally lower than with sprinkler systems. Some low-pressure sprinklers operate at pressures comparable to trickle systems.

Table 15.5. Comparison of Irrigation Systems in Relation to Site and Situation Factors

	Improved Surface Systems		*Sprinkler Systems*
Site and Situation Factors	*Redesigned Surface Systems*	*Level Basins*	*Intermittent Mechanical Move*
Infiltration rate	Moderate to low	Moderate	All
Topography	Moderate slopes	Small slopes	Level to rolling
Crops	All	All	Generally shorter crops
Water supply	Large streams	Very large streams	Small streams nearly continuous
Water quality	All but very high salts	All	Salty water may harm plants
Efficiency	Average 60–70%	Average 80%	Average 70–80%
Labor requirement	High, training required	Low, some training	Moderate, some training
Capital requirement	Low to moderate	Moderate	Moderate
Energy requirement	Low	Low	Moderate to high
Management skill	Moderate	Moderate	Moderate
Machinery operations	Medium to long fields	Short fields	Medium field length, small interference
Duration of use	Short to long	Long	Short to medium
Weather	All	All	Poor in windy conditions
Chemical application	Fair	Good	Good

Source: Fangmeier (1977).

15.16 Automation

With the current high cost of labor, much attention is given to minimizing labor requirements. Mechanical-move and solid-set sprinkler systems lend themselves well to automatic controls. The application of automatic controls to surface irrigation is also feasible in some cases. Electrically triggered, mechanically operated control gates and valves can provide a considerable reduction in labor for most irrigation systems. In some instances, water pressure may be reduced with continuous mechanical-move systems by aligning crop rows parallel with the wheel movement and by using drop pipes attached to the moving pipe to place the water into each furrow. By decreasing pressure, total energy requirements are reduced with a resultant saving in operating costs.

(Table 15.5. Continued)

Sprinkler Systems (cont.)		*Trickle Systems*
Continuous Mechanical Move	*Solid Set and Permanent*	*Emitters and Porous Tubes*
Medium to high	All	All
Level to rolling	Level to rolling	All
All but trees and vineyards	All	High value required
Small streams nearly continuous	Small streams	Small streams, continuous and clean
Salty water may harm plants	Salty water may harm plants	All—can potentially use high salt waters
Average 80%	Average 70–80%	Average 80–90%
Low, some training	Low to seasonal high, little training	Low to high, some training
Moderate	High	High
Moderate to high	Moderate	Low to moderate
Moderate to high	Moderate	High
Some interference circular fields	Some interference	May have considerable interference
Short to medium	Long term	Long term, but durability unknown
Better in windy conditions than other sprinklers	Windy conditions reduce performance; good for cooling	All
Good	Good	Very good

15.17 Temperature Control

Irrigation systems are often used as a means of frost protection. Irrigation water, especially if supplied from wells is often considerably warmer than the soil and the air near the surface under frost conditions. The heat of fusion released by water freezing on plant parts keeps the temperature from falling below 32° F as long as freezing continues.

Conversely, irrigation may be used for cooling, particularly when germination occurs under high temperatures. Sprinkler irrigation has also been used to delay premature blossoming of fruit trees when warm weather occurs before the frost danger has passed. The cooling effect of evaporation effectively lowers the temperature of the plant parts.

REFERENCES

Ayers, R. S., and D. W. Westcot (1976) "Water Quality for Agriculture," Irrigation and Drainage Paper No. 29. Food and Agriculture Organization of the United Nations, Rome.

Doorenbos, J., and W. O. Pruitt (1977) "Crop Water Requirements" (Revised), Irrigation and Drainage Paper No. 24. Food and Agriculture Organization of the United Nations, Rome.

Erie, L. J., O. F. French, and K. Harris (1968) "Consumptive Use of Water by Crops in Arizona," *Arizona Agr. Exp. Sta. Tech. Bull. 169.*

Fangmeier, D. D. (1977) "Alternative Irrigation Systems," Agricultural Engineering and Soil Science Series (Lithographed), University of Arizona, Tucson.

Hansen, V. E., O. W. Israelsen, and G. E. Stringham (1980) *Irrigation Principles and Practices*, 4th ed., John Wiley & Sons, New York.

Jensen, M. E. (1974) "Consumptive Use of Water and Irrigation Water Requirements," Drainage and Irrigation Division, Am. Soc. Civil Eng., New York.

Schwab, G. O., R. K. Frevert, T. W. Edminster, and K. K. Barnes (1981) *Soil and Water Conservation Engineering*, 3rd ed., John Wiley & Sons, New York.

U.S. Soil Conservation Service (1970) "Irrigation Water Requirements," Tech. Release No. 21 (Revised), Washington, D.C.

U.S. Bureau Reclamation (1983) *Comparison of Equations Used for Estimating Agricultural Crop Evapotranspiration with Field Research*, Denver, Colorado.

PROBLEMS

15.1 Determine the evapotranspiration and irrigation requirement for grain sorghum in eastern Colorado for the month of June. The average monthly temperature is 70° F and the percentage of daytime hours is 10 percent. Water application efficiency is 65 percent and the average rainfall is 1.0 in.

15.2 Determine the evapotranspiration and the water needed for a common crop at your present location for the entire growing season. Consult local Weather Service records and other references.

15.3 Compute the water-holding capacity of a soil in inches if the field capacity is 30 percent and the wilting point is 15 percent of the dry-soil density. This density is 80 pcf (lb/cu ft) and the soil depth to be irrigated is 2 ft.

15.4 How much water is needed at each irrigation if the moisture-holding capacity of the soil is 1.5 in./ft, and irrigation is started when 40 percent of it is depleted? The crop uses 0.25 in. per day of moisture and has a root depth of 3 ft. If there is no rain, how often will irrigation be required?

15.5 Determine the leaching requirement for wheat if it is to be irrigated with water having an electrical conductivity of 0.42 dS/m. The conductivity tolerance of wheat is 7 dS/m. (See Chapter 11 for units of electrical conductivity.)

15.6 The discharge to the farm from an irrigation canal is 3 cfs and the measured flow to a field is 2.85 cfs. What is the conveyance efficiency?

15.7 Determine the water-application efficiency if a stream of 1 cfs is delivered to a 2-acre field for 12 hr. The average depth of water added to the root zone was 4 in.

15.8 The depths of water penetration at 100-ft stations along a furrow are 3.2, 3.0, 3.6, 3.8, and 3.4 ft. Compute the uniformity coefficient.

CHAPTER 16

SURFACE IRRIGATION

In spite of the many advantages of applying water through sprinkler and trickle systems, surface irrigation is the predominant method in the United States. In the western states, where surface irrigation is especially predominant, the major portion of the water comes from surface runoff, which is usually stored in reservoirs. Since this water must be conveyed for considerable distances over rough terrain, conveyance canals and control structures are key parts of most irrigation systems in arid regions. The design of canals and control structures fall in the category of engineering design.

DISTRIBUTION OF WATER ON THE FARM

The farm water supply is normally delivered either from surface storage by conveyance ditches or from irrigation wells. Sometimes surface storage and underground supplies are combined in order to provide an adequate water supply at the farm.

16.1 Surface Ditches

A system of open ditches often distributes the water from the source on the farm to the field as shown in Fig. 16.1. These ditch systems should be carefully designed so as to provide adequate head (elevation) and capacity to supply water at all areas to be irrigated. The amount of land that can be irrigated is often limited by the head of water available as well as by the design and location of the ditch system. Where irrigation is used as an occasional supplement to rainfall, these ditches may be temporary. In order to minimize water losses there is an increasing tendency to line ditches with impermeable materials. This practice is particularly applicable in the more arid regions where irrigation water supplies are limited and crop needs are largely dependent on irrigation water.

16.2 Underground Pipe

Since surface distribution systems provide continuing problems of maintenance, constitute an obstruction to farming operations, and provide a water surface

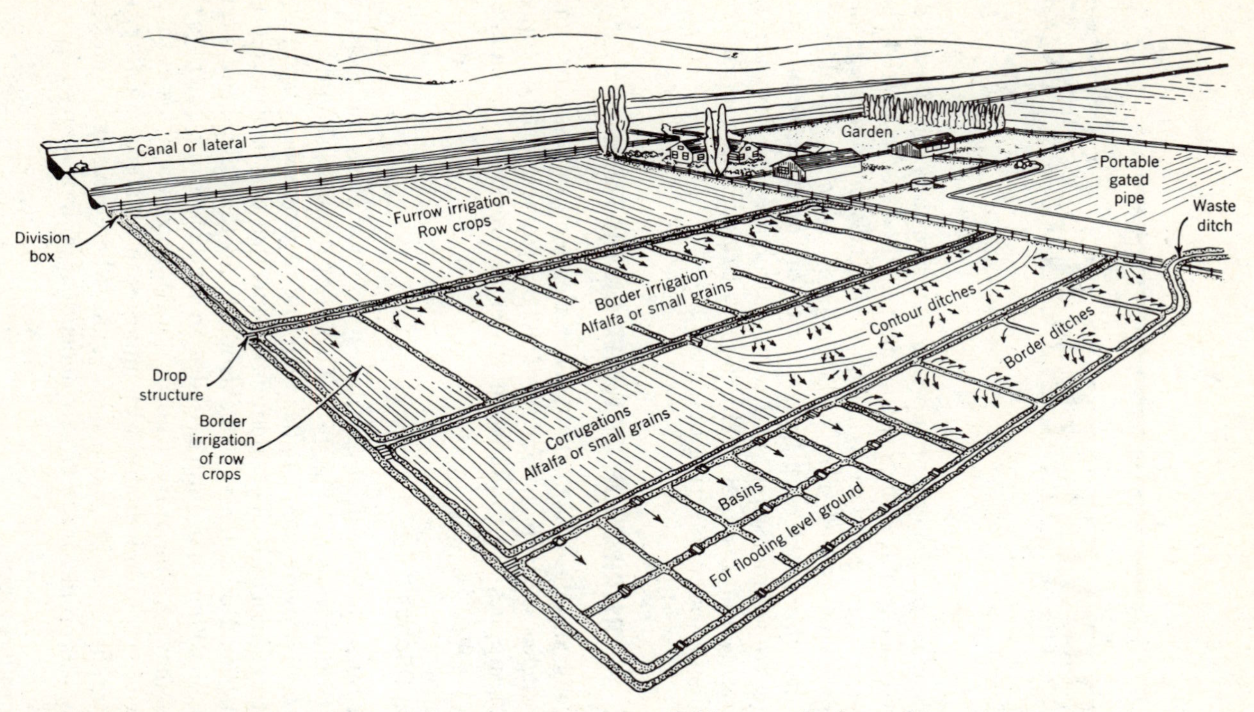

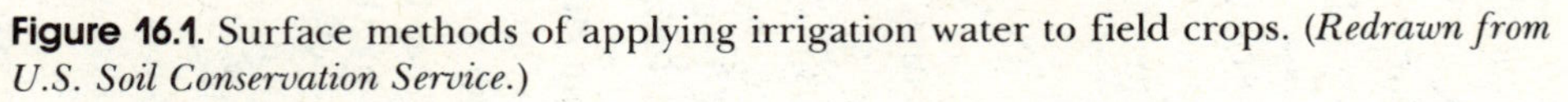
Figure 16.1. Surface methods of applying irrigation water to field crops. (*Redrawn from U.S. Soil Conservation Service.*)

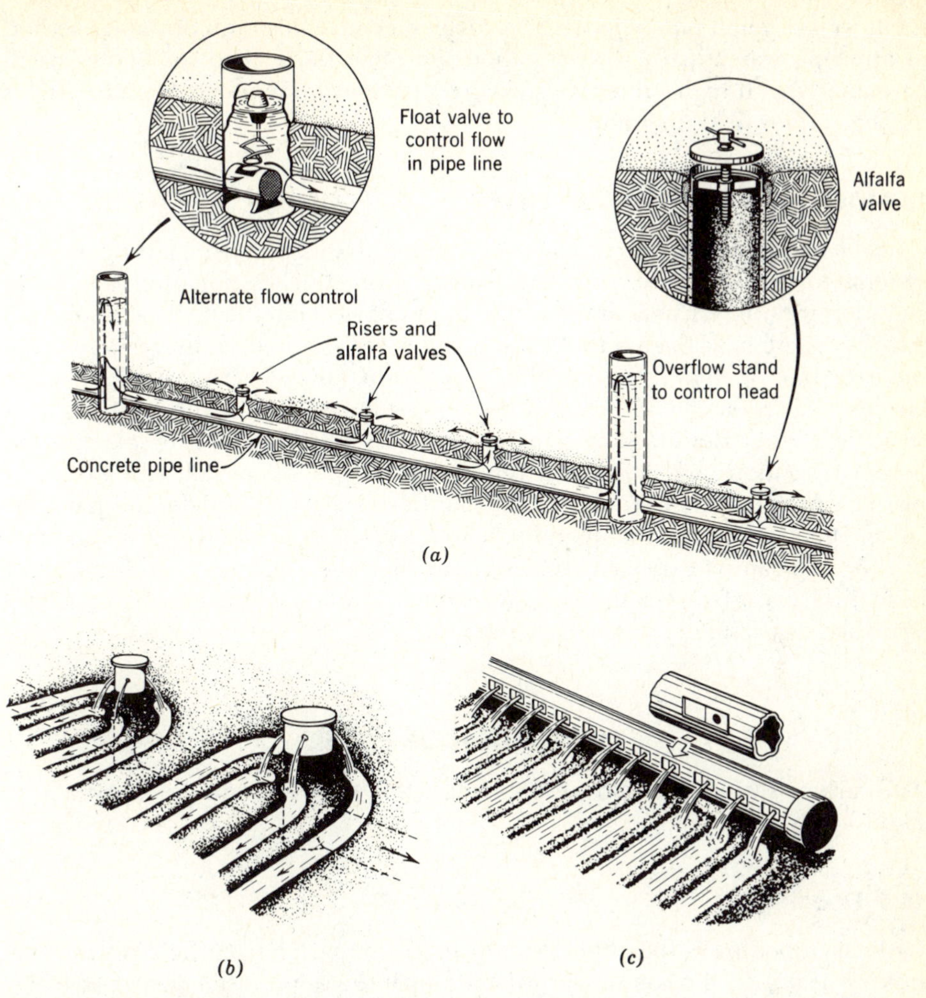

Figure 16.2. Methods of distribution of water from (*a*) low-pressure underground pipe, (*b*) multiple-outlet risers, and (*c*) portable gated pipe. (*Adapted from U.S. Soil Conservation Service and U.S. Bureau Reclamation.*)

subject to evaporation losses, underground pipe distribution systems shown in Fig. 16.2*a* are becoming increasingly popular. In these systems, water flows from the distribution pipe upward through riser pipes and irrigation valves to appropriate basins, borders, or furrows. In Fig. 16.2*b* a multiple-outlet riser controls the distribution of water to several furrows. Alfalfa valves regulate the flow to a header ditch from which water can be distributed to the field.

16.3 Portable Pipe

Surface irrigation with portable pipe or large diameter plastic tubing may be advantageous particularly in circumstances where water is applied infrequently.

Light-weight gated pipe, Fig. 16.2*c*, provides a convenient and portable method of applying water in furrow irrigation. Portable flumes are sometimes used, particularly with high value crops, where they may be removed from the fields during certain field operations.

16.4 Devices to Control Water Flow

Control structures, like those illustrated in Fig. 16.3 are essential in open ditch systems to (a) divide the flow into two or more ditches, (b) lower the water elevation without erosion, and (c) raise the water level in the ditch so that it will have adequate head for removal. Various devices are used to divert water from the irrigation ditch and to control its flow to the appropriate basin, furrow, or border. Valves may be installed in the side or bottom of the ditch during construction. Other devices shown in Fig. 16.4 are (a) spiles, (b) gate takeouts for border irrigation, and (c) siphon tubes. These siphons, usually aluminum or plastic, carry the water from the ditch to the surface of the field and have the advantage of metering the quantity of water applied. Table 16.1 gives the rate of flow that can be expected from siphons of various diameters with different head differences between the inlet and outlet. Additional information on control structures is presented in Chapter 10.

APPLICATION OF WATER

The various surface methods of applying water to field crops are illustrated in Fig. 16.1.

16.5 Flooding

Ordinary flooding is the application of irrigation water from field ditches that may be nearly on the contour or up and down the slope. After the water leaves

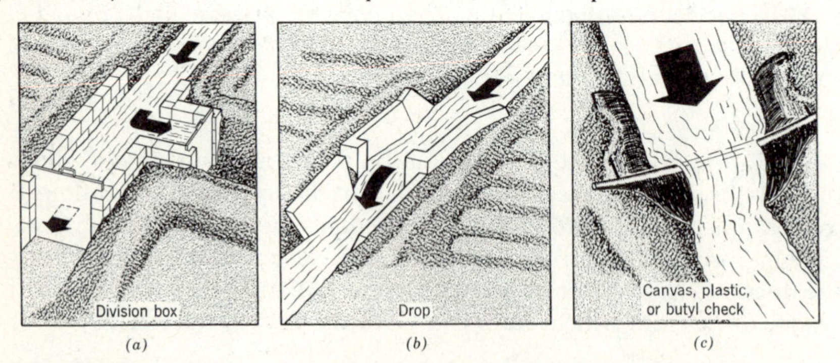

Figure 16.3. Devices to control water flow in irrigation ditches. (*a*) Division box. (*b*) Drop. (*c*) Canvas, plastic, or butyl check dam. (*Adapted from U.S. Soil Conservation Service and U.S. Bureau Reclamation.*)

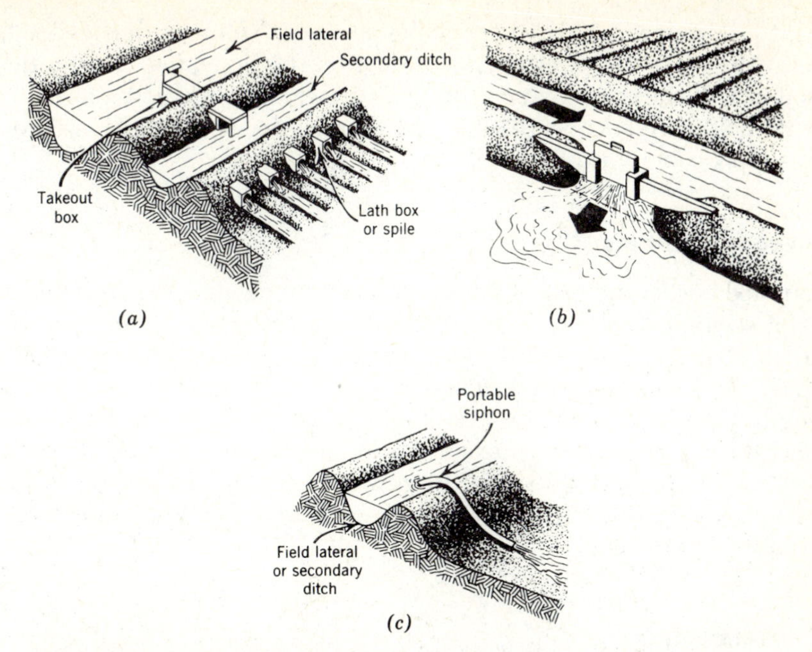

Figure 16.4. Devices for distribution of water from irrigation ditches into fields. (*a*) Spile or lathe box. (*b*) Border takeout. (*c*) Siphon. (*Adapted from U.S. Soil Conservation Service and U.S. Bureau Reclamation.*)

the ditches, no attempt is made to control the flow by means of levees or by other methods of restricting water movement. For this reason ordinary flooding is frequently referred to as "wild flooding." Although the initial cost for land preparation is low, labor requirements are usually high and the efficiency of water application is generally low. Ordinary flooding is most suitable for close-

Table 16.1. Discharge for Aluminum Siphon Tubes[a]

Outside Diameter (in)	*Length (ft)*	*Discharge in gpm*		
		Head in inches		
		2	*6*	*12*
1	5	4	7	10
$1\frac{1}{4}$	5	7	12	17
$1\frac{1}{2}$	5	11	18	26
2	5	20	35	50
3	5	50	87	125
4	8	85	145	205

[a]From U.S. Soil Conservation Service.

growing crops, particularly where slopes are steep. Contour ditches are usually spaced from 50 to 150 ft apart, depending on the slope, texture and depth of the soil, size of the stream, and crop to be grown. This method may be used on rolling land where borders, basins, and furrows are not feasible and adequate water supply is not a problem.

16.6 Graded Borders

The graded border method of flooding consists of dividing the field into a series of strips separated by low ridges. Normally, the direction of the strip is in the direction of greatest slope, but sometimes the borders are placed nearly on the contour. The strips usually vary from 30 to 60 ft in width and are 300 to 1200 ft in length. Ridges between borders should be sufficiently high to prevent overtopping during irrigation. To prevent water from concentrating on either side of the border, the land should be level perpendicular to the flow. Where row crops are grown in the border strip, furrows confine the flow and eliminate this difficulty.

16.7 Level Borders

The layout of level borders is similar to that described for graded borders, except that the surface is leveled within the area to be irrigated. These areas may be long and narrow or they may be nearly square (often called basins). Modern laser leveling techniques make possible the preparation of smooth level surfaces required for this method of irrigation. Where relatively large rates of flow are available, the field can be quickly covered resulting in high field irrigation efficiencies. Level borders also lend themselves well to preirrigation, utilizing stream flow diverted during periods of high runoff. In orchard irrigation small leveled basins may include as few as one to four trees.

16.8 Furrow

Whereas with flooding methods the water covers the entire surface, irrigating by furrows submerges only from one fifth to one half the surface, resulting in less evaporation, less puddling of the soil, and permitting cultivation sooner after irrigation. Furrows vary from large to small and are up and down the slope or on the contour. Small, shallow furrows, called corrugations, are particularly suitable for relatively irregular topography and close-growing crops, such as meadow and small grains. Furrows 3 to 8 in. deep are especially suited to row crops, since the furrow can be constructed with normal tillage. Contour furrow irrigation may be practiced on slopes up to 12 percent, depending on the crop, erodibility of the soil, and the size of the irrigation stream.

The length of furrows depends on the infiltration capacity, the slope, and size of the stream. Excessively long furrows result in deep percolation losses and erosion in the upper ends of furrows. Suggested lengths of furrows for various slopes and soil textures are given in Table 16.2.

DESIGN AND EVALUATION

A recognition and understanding of the variables involved in the hydraulics of surface irrigation is essential to effective design. Hansen (1960) has listed, with reference to Fig. 16.5, the pertinent variables as (1) size of stream, (2) rate of advance, (3) length of run and time involved, (4) depth of flow, (5) intake rate, (6) slope of land surface, (7) surface roughness, (8) erosion hazard, (9) shape of flow channel, (10) depth of water to be applied, and (11) fluid characteristics.

Since the hydraulics of surface irrigation is complex and since some of the variables involved have not been evaluated and their relationships have not been determined, empirical procedures are often employed in the design of surface irrigation systems. In designing a system it is recommended that local practices be explored and locally available data be considered. For example, extension services, experiment stations, and the Soil Conservation Service have prepared "Irrigation Guides" suggesting design procedures for many states. References, such as Hansen, Israelsen, and Stringham (1980) and Schwab et al. (1981) should be consulted. Selection of design values, such as border width, depends upon judgment and experience in managing water under specific soil, slope, and crop conditions.

16.9 Surface Irrigation Design

Probably the most commonly used reference in the United States for the design of graded and level border irrigation systems is Chapter 4 of the National Engineering Handbook on Irrigation published by the U.S. Soil Conservation Service (1974). This publication presents the derivation of equations relating the infiltration, roughness, irrigation efficiency, length of run, time of application and stream size. This information is summarized by Schwab et al. (1981).

Table 16.2. Lengths of Run for Furrows and Corrugations[a]

	Lengths of Furrows or Corrugations (ft)			
Slope (%)	*Loamy Sand and Coarse Sandy Loams*	*Sandy Loams*	*Silt Loams*	*Clay Loams*
0–2	250–400	300–660	660–1320	880–1320
2–5	200–300	200–300	300–660	400–880
5–8	150–200	150–250	200–300	250–400
8–15	100–150	100–200	100–200	200–300

[a]From U.S. Bureau Reclamation.

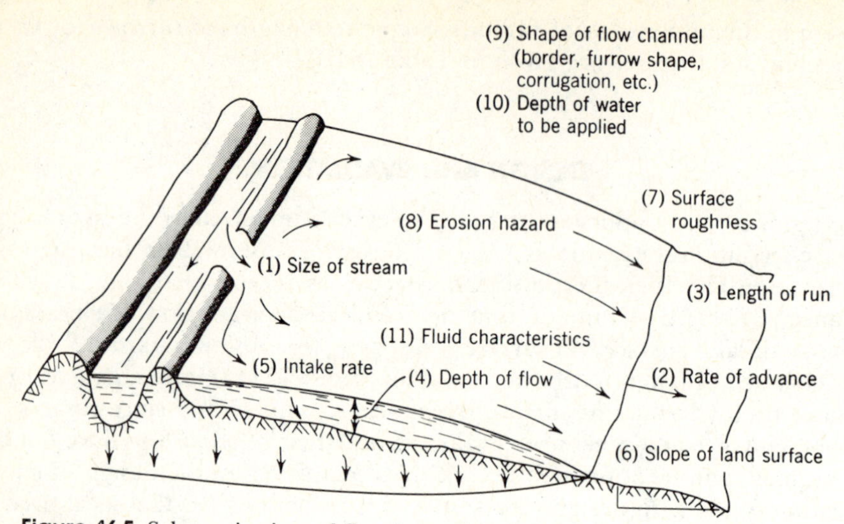

Figure 16.5. Schematic view of flow in surface irrigation indicating the variables involved. *Source*: Hansen (1960).

16.10 Evaluation of Existing Systems

Although the design of new irrigation systems constitutes an important engineering activity, it should be noted that even greater opportunities are often available in the evaluation and improvement of existing systems. Many of these older systems were designed before much was known about intake rates and water-holding capacities of soils, and when water supplies were not as limited as they are now becoming. Fields or portions of fields that do not receive enough water have limited production potential. Excessive irrigation not only wastes water, but leaches water-soluble nutrients and may cause drainage problems.

The evaluation of existing systems can be approached in a number of ways. A simple method of determining underirrigation is by use of the soil auger or tube sampler. Observations of the *opportunity time* for infiltration in various parts of the field may be helpful. More sophisticated methods involve the application of equipment such as portable flumes, meters, and water intake rings in carefully conducted diagnostic procedures, like those described by Merriam (1968).

REFERENCES

Hansen, V. E. (1960) "Mathematical Relationships Expressing the Hydraulics of Surface Irrigation," *Proc. ARS-SCS Workshop Hydraulics Surface Irrig.*, ARS 41–43.

Hansen, V. E., O. W. Israelsen, and G. E. Stringham (1980) *Irrigation Principles and Practices* (4th ed.), John Wiley & Sons, New York.

Jensen, M. E. (ed.) (1980) "Design and Operation of Farm Irrigation Systems," Publ. 1–80, *Am. Soc. Agr. Eng.*, St. Joseph, Mi.

Merriam, J. L. (1968) *Irrigation System Evaluation and Improvement*, Blake Printing and Publ. Co., San Luis Obispo, Calif.

Schwab, G. O., R. K. Frevert, T. W. Edminster, and K. K. Barnes (1981) *Soil and Water Conservation Engineering*, 3rd ed., John Wiley & Sons, New York.

Turner, J. H., and C. L. Anderson (1971) *Planning for an Irrigation System*, Am. Assoc. for Voc. Instr. Materials, Athens, Ga.

U.S. Soil Conservation Service (1974) "Border Irrigation," in *National Engineering Handbook*, Sect. 15, Chap. 4, Washington, D.C.

PROBLEMS

16.1 How many 2-in. diameter siphon tubes are required to supply 1 cfs to a border strip if the head-causing flow is 12 in. (1 cu ft = 7.5 gal)?

16.2 If a flow of 18 gpm is to be supplied to a furrow by two 1-in. diameter aluminum siphon tubes, how far below the water level in the supply ditch should the tube outlets be placed?

CHAPTER 17

SPRINKLER IRRIGATION

Sprinkler irrigation is a versatile means of applying water to any crop, soil, and topographic condition. It is particularly popular in humid regions because surface ditches and prior land preparation are not necessary and because pipes are easily transported and provide no obstruction to farm operations when irrigation is not needed. Sprinkling is suitable for sandy soils or any other soil and topographic condition where surface irrigation may be inefficient or expensive, or where erosion may be particularly hazardous. Low rates and amounts of water may be applied, such as are required for seed germination, frost protection, and cooling of crops in hot weather. Fertilizers and soil amendments may be dissolved in the water and applied through the irrigation system. The major disadvantages of sprinkling systems are the high investment cost and the high operating cost compared to surface methods.

According to the Irrigation Journal (1983) survey, 35 percent of all irrigated land in the United States is irrigated by sprinklers, of which nearly 50 percent is irrigated by center-pivot systems. The center-pivot method shown in Fig. 17.1 has increased rapidly since the 1960s. The leading states are Nebraska, Kansas, Texas, Colorado, and Georgia in decreasing order. Since about 1950, sprinkler irrigation has increased more rapidly than other methods. Its use is expected to continue to increase in the future, especially for the land application of agricultural, industrial, and municipal wastewater.

17.1 Sprinkler Systems

In general, systems are described according to the method of moving the lateral lines, on which are attached various types of sprinklers. These systems are identified and compared in Table 17.1 and Fig. 17.2. During normal operation, laterals may be hand-moved or mechanically moved. The sprinkler system may cover only a small part of a field at a time or be a solid-set system in which sprinklers are placed over the entire field. With the solid-set system, all or part

Figure 17.1. (*a*) Center pivot sprinkler system with an extension arm. (*b*) Crop pattern of several adjacent systems. (*Courtesy Valmont Industries, Inc., Valley, NB*).

of the sprinklers may be operated at the same time. Most sprinklers rotate a full circle, but some may be set to operate for any portion of a circle (part circle). Perforated pipes are suitable for distributing water to small acreages of high-value crops where a rectangular pattern is desired.

As shown in Table 17.1, hand-move laterals have the lowest investment cost but the highest labor requirement. With giant sprinklers, the spacing can be increased, thereby reducing labor, but higher pressures are required, which increase the pumping cost. These sprinklers are generally pulled or transported from one location to another. Hand-move laterals with standard sprinklers are most suitable for low-growing crops, and are impractical in tall corn because of adverse conditions for moving the pipe.

With the end-pull system shown in Fig. 17.1*a*, the lateral is moved to the next position by the pulling of one end with a tractor to the opposite side of the main. In moving the lateral, it is pulled at a slight diagonal so as to move the entire length down the field, one half the lateral spacing with each direction pulled. This system is suitable for low-growing crops and in places where adequate moving space is available. It is also practical for tree crops. Labor is greatly reduced (Table 17.1) compared with hand-move systems.

The rotating boom-type system (Fig. 17.2*b*) operates with one trailer unit per lateral at spacings up to 350 ft. The trailer and boom unit is moved to the next position along the lateral with a tractor or winch. The lateral line is added or picked up as the trailer is moved progressively through the field away from or toward the main line. The boom-type unit will cover about the same area as a

Table 17.1. Comparison of Sprinkler Irrigation Equipment

Type of System	*Relative Investment Cost per Acre*[a]	*Relative Labor Cost*	*Practical Hours of Operation per Day*
Hand-move laterals (standard sprinklers)	0.4	5.0	16
Hand-move laterals (giant sprinklers)	0.5	4.0	12–16
End-pull laterals (tractor tow)	0.5	1.4	16
Boom-type sprinklers (trailer-mounted)	0.6	3.7	12–16
Side-roll laterals (powered-wheel move)	0.7	1.7	18–20
Self-propelled (center pivot)	1.0	1.0	24
Solid set	3.0–5.0	1.0	24

[a]Based on a 160-acre field, 1000 gpm from pump, and 80% application efficiency. (Based on data from Berge and Groskopp, 1964.)

giant sprinkler, but the pressure required is not as high. For tall crops, space must be provided for moving the trailer along the lateral and at the ends.

The side-roll lateral system (Fig. 17.2*c*) utilizes the irrigation pipe as the axle of large diameter wheels that are spaced about 30 ft apart. The lateral is moved to the next position by a small gasoline engine mounted at the midpoint of the line or by hand with a lever and ratchet. The side-roll lateral is limited to crops that will not interfere with the movement of the pipe. Unless a flexible pipe is attached to the main, the lateral must be disconnected for each move. Labor requirements are about the same as for the end-pull system, but much less than for the hand-move systems (Table 17.1). Self-aligning sprinklers, which stay in a vertical position regardless of the position of the wheels, eliminate the necessity of having the sprinkler on top as required with the fixed-position sprinklers (Fig. 17.2*c*).

The self-propelled center-pivot system (Fig. 17.2*d*) consists of a radial pipeline supported at a height of 6 to 8 ft at intervals of about 100 ft. The radial line rotates slowly around a central pivot by either fluid pressure (water, oil, or air) or electric motors. The towers are supported by wheels or skids and are kept in alignment with supporting wires. The nozzles increase in size from the pivot to the end of the line, at which is placed a large sprinkler in order to obtain the maximum diameter of coverage. The nozzles are selected to provide a uniform depth of application, varying from $\frac{1}{2}$ to 4 in. per revolution. The depth of application is determined by the speed of rotation. The system is best suited to sandy soils, but it will operate in heavier soils if the depth of application is greatly reduced. A common-size system designed for a quarter section (160 acres) is 1285 ft in length. Such a system will irrigate about 135 of the 160 acres; the remaining 25 acres in the corners are not covered, unless the system is equipped with an extension arm. The center-pivot system shown in Fig. 17.1 has such an extension pipe and a steerable drive unit that allows the overall system radius to increase or decrease to correspond to the field boundaries. The steerable drive unit automatically tracks above a buried signal wire allowing the system to reach irregular areas of the field not usually irrigated. Normally, the system, once set up, remains permanently in place, but some can be towed to another field by rotating the wheels parallel to the pipe. The major advantage of the self-propelled lateral is the saving in labor, but the disadvantages are high investment cost (Table 17.1) and incomplete irrigation of a square field. The circular pattern is less objectionable where water is limited and land is plentiful.

Solid-set systems may be operated by changing the flow from one lateral to the next or by sequencing sprinklers (one operating on each line) along the lateral lines. For frost protection, all sprinklers are operated at the same time. Their operation can be made automatic with timing devices and solenoid valve controls. As with the self-propelled system, the only labor required is for setup and maintenance. Because of extremely high investment costs (Table 17.1), solid-set systems are practical only for high-value crops.

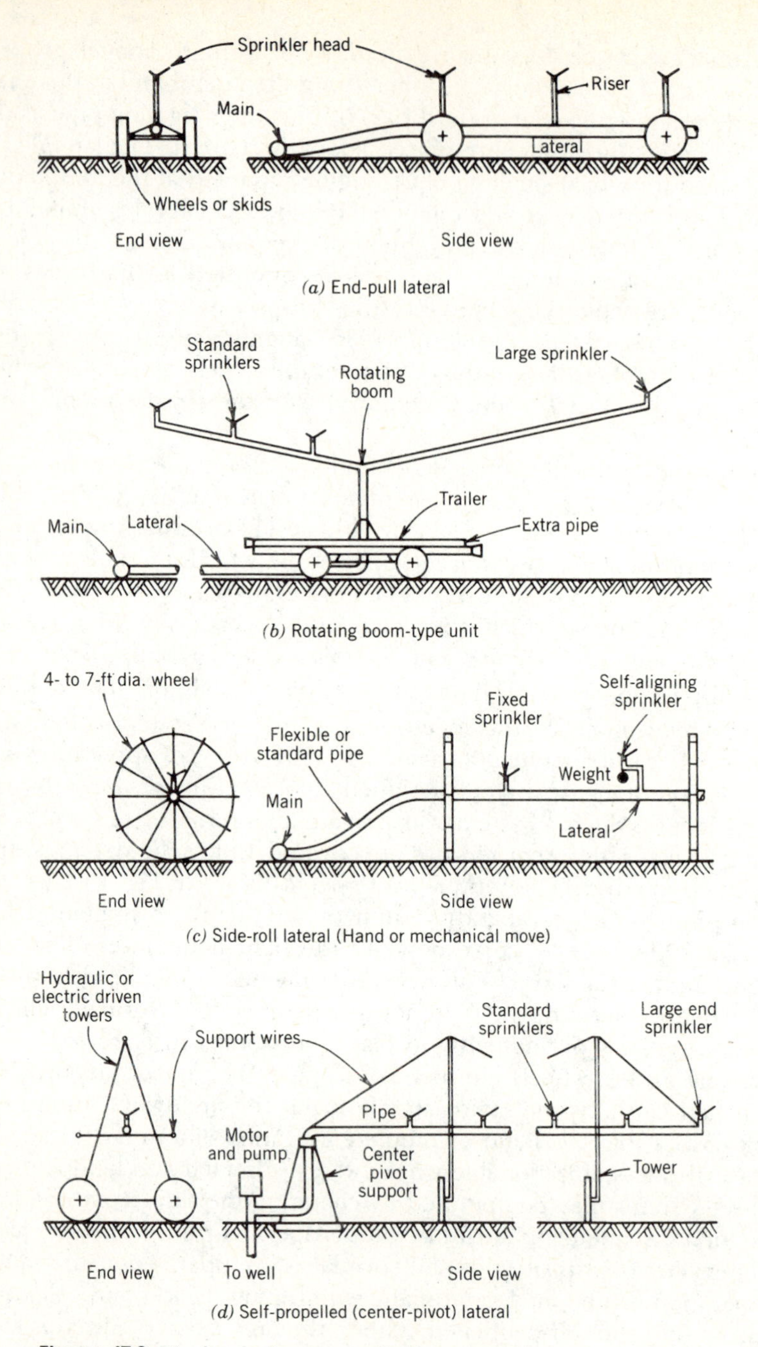

Figure 17.2. Mechanical-move sprinkler systems.

17.2 Hand-Move, Side-Roll, Rotating-Boom, and End-Pull Sprinkler Systems

The major components of a hand-move portable sprinkler system are shown in Fig. 17.3. Most components are basically the same as required for the end-pull, rotating-boom, and side-roll systems shown in Fig. 17.2. With the rotating-boom or giant sprinkler systems, the lateral spacing and the distance between sprinkler setups along the lateral are much greater than that for standard sprinklers. Latches, seals, and other details of these components vary considerably, depending on the manufacturer.

Pumps. A pump is required to lift the water from the source, push it through the distribution system, and spray it over the area. Types of pumps used for irrigation will vary with the rate of flow, the discharge pressure, and the vertical distance to the source of water. The pump should have adequate capacity for present and future needs. This capacity generally varies from 100 to 800 gpm for most small farm irrigation installations.

The characteristics of the centrifugal pump make it suitable for most installations. For wells over 20 ft in depth it is customary to install vertical turbine pumps. A pump of this type having two stages or two turbine bowls is illustrated in Fig. 17.4. The horizontal centrifugal pump shown in Fig. 17.5 is suitable if the water surface, after drawdown, is within 15 to 20 ft of the pump impeller. Horizontal centrifugal pumps are commonly used for pumping water from streams, ponds, pits, or lakes. In these cases an intake screen should be provided to avoid clogging of the sprinkler nozzles. The power is usually supplied either by an electric motor or an internal combustion engine. The choice depends upon the availability of electrical power and the relative cost of electricity and of fuel for internal combustion engines. Further details on pumps may be found in Schwab et al. (1981).

Main Lines and Laterals. The second component of the system, the main line, may be either movable or permanent. Movable mains generally have a lower first cost and can be more easily adapted to a variety of conditions, whereas permanent mains offer a saving in labor and reduced obstruction to field operations. Water is taken from the main either through a valve placed at each junction with a lateral or, in some cases, through either an ell or a tee section that has been supplied in place of one of the standard couplings in the main (Fig. 17.6).

The laterals usually consist of 20- or 30-ft lengths of aluminum tubing with attached or removable couplers. Several types of quick-connecting couplers are shown in Fig. 17.6*a*. The riser and sprinkler head are usually placed on the coupler, but some manufacturers place them in the center of a 20- or 30-ft length to facilitate moving the pipe.

Sprinkler Heads and Nozzles. The operating pressure for most sprinklers usually ranges from 30 to 100 lb per sq in. Sprinklers often have two nozzles, one to apply water at a considerable distance from the sprinkler (range nozzle), and the other to cover the area near the sprinkler (spreader nozzle). Of the

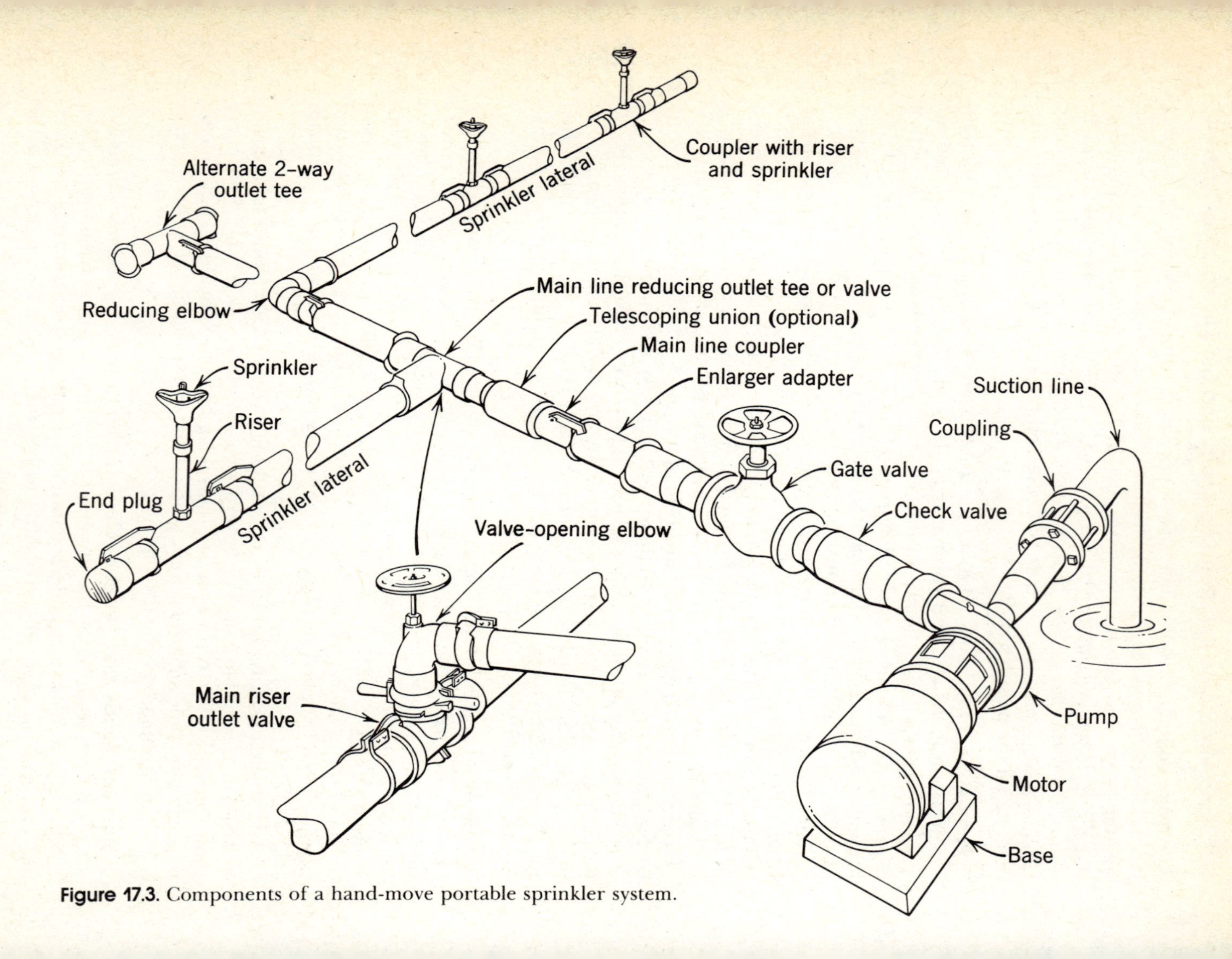

Figure 17.3. Components of a hand-move portable sprinkler system.

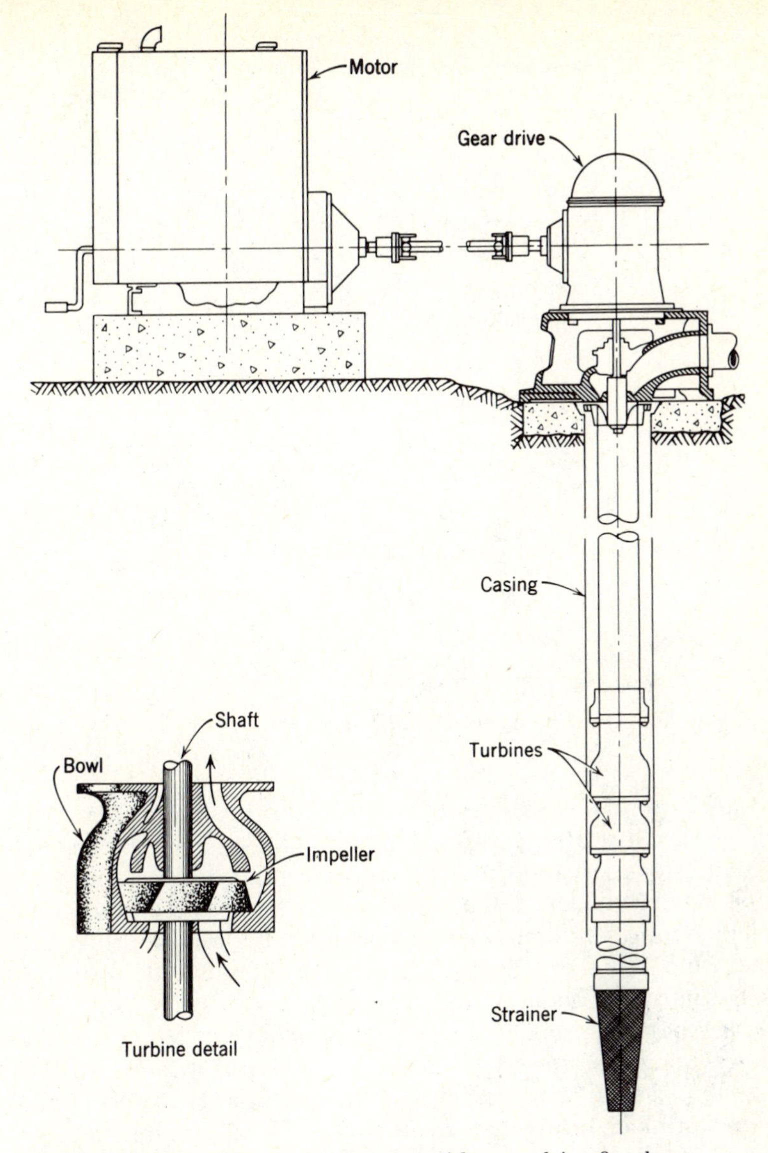

Figure 17.4. Multistage turbine pump with gear drive for deep wells.

numerous devices to rotate the sprinkler, the most usual type, shown in Fig. 17.6*b*, taps the sprinkler head with a small hammer activated by the force of the water striking against a small vane connected to it. Sprinklers designed to cover a considerable area have a slow rate of rotation, about one revolution per minute.

Since one sprinkler does not apply water uniformly over an area, the overlapping of sprinkler patterns is relied upon to provide more nearly uniform

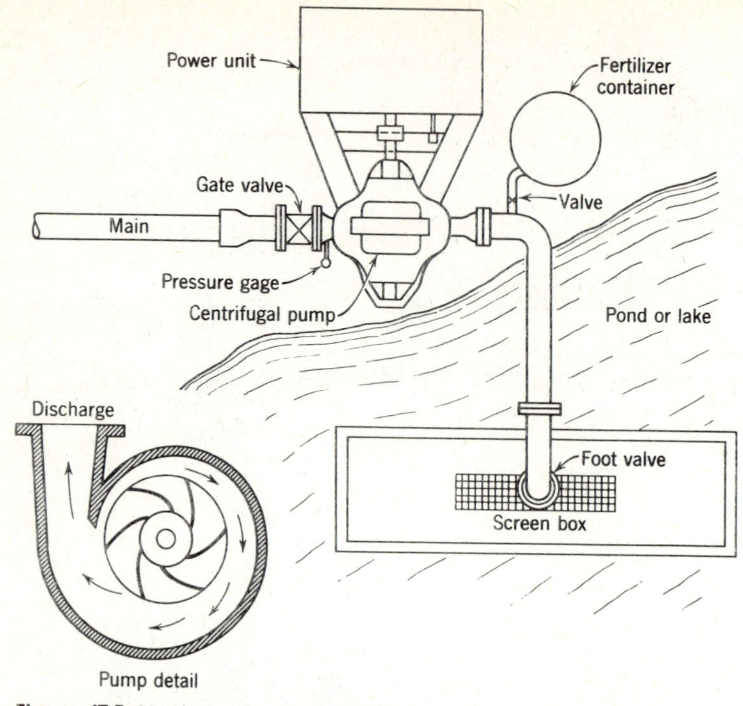

Figure 17.5. Horizontal centrifugal pump.

coverage. Figure 17.7 illustrates the way in which these overlapping patterns combine to give a relatively uniform distribution on a line between sprinklers. The extent of overlap and the depth of water that can be expected at various points between sprinklers is illustrated in Fig. 17.8. These examples are based on ideal operating conditions and may be difficult to reproduce in the field. For example, wind will skew the pattern so as to give less uniform distribution, especially winds over about 5 mph.

Several types of sprinkler heads are available for special purposes. Some provide a low angle jet for use in orchards. Some operate at especially low heads of say 20 lb per sq in., and others operate only in a part circle. There are also giant-type sprinkler heads that throw water several hundred feet. In general, these operate at higher pressures than the smaller units and result in a greater pressure loss at the sprinkler head and greater pumping costs.

System Layouts of Mains and Laterals. The number of possible arrangements for the mains, laterals, and sprinklers is practically unlimited. The most suitable layout can be determined only after a careful study of the conditions to be encountered. The choice will depend to a large extent upon the types and capacities of the sprinklers and the operating pressure. For small systems the laterals are moved 60 ft at each setting and the sprinklers are spaced every 40 ft on each lateral.

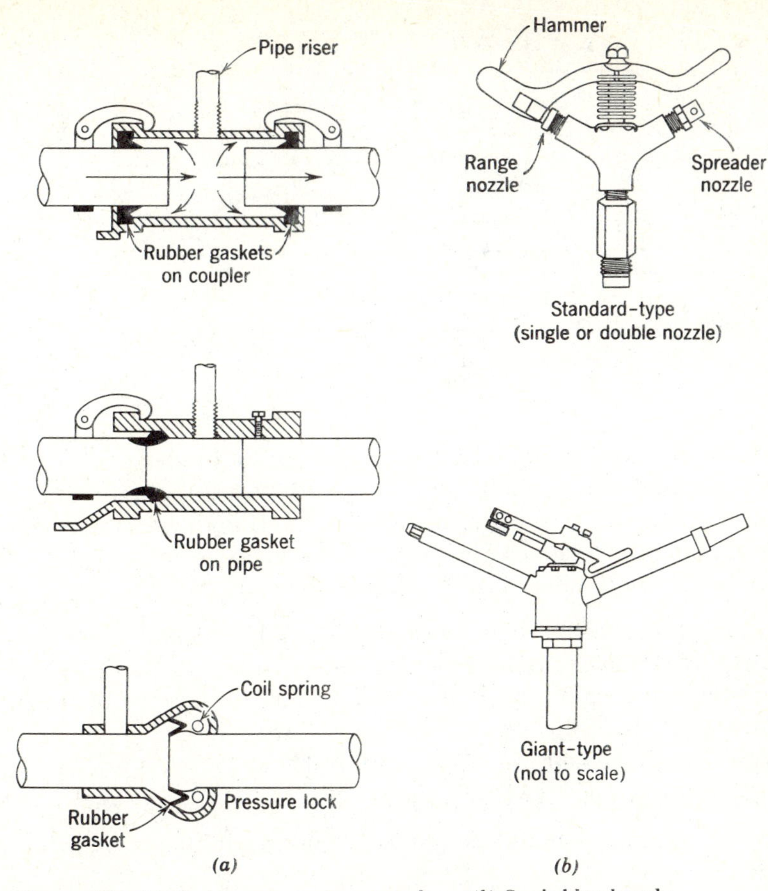

Figure 17.6. (*a*) Quick-connecting couplers. (*b*) Sprinkler heads.

A typical layout in which water is pumped from a stream or canal parallel to the edge of the field is shown in Fig. 17.9*a*. This plan permits a shorter main by resetting the pump at several positions along the length of the field. Even so, several settings of the lateral are possible for each position of the pump. This arrangement requires considerably less pipe than if one pump setting were used for the entire field, but it does have a somewhat higher labor requirement.

Figures 17.9*b* and 17.9*c* illustrate layouts that are suitable when pumping from a well or pit. The system in Fig. 17.9*b* has two laterals placed first at opposite ends of the field and then moved in opposite directions. In this way it is possible to reduce the diameter of the more distant portion of the main from the pump, because at no time will this section of the main have to serve both laterals. This arrangement is well adapted to day and night operation with an application rate such that the required amount of water can be added in perhaps 6 to 8 hr.

The system shown in Fig. 17.9*c* is designed for either a one- or two-lateral operation. While line *A* is being used, the operator turns off and moves line *B*.

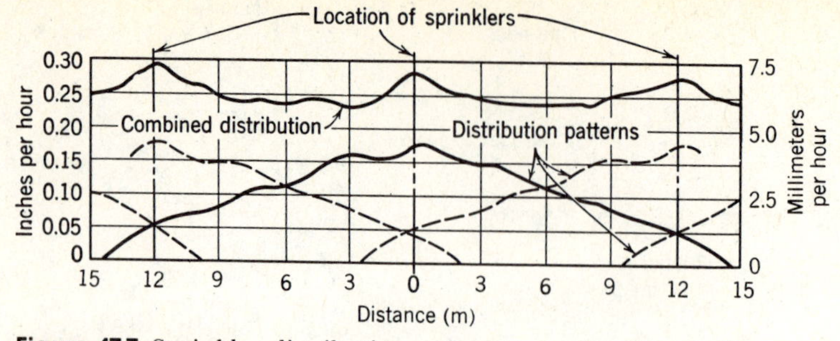

Figure 17.7. Sprinkler distribution patterns, overlapping to give relatively uniform combined distribution.

When the time needed to apply the required amount of water has elapsed, line *B* is turned on, and *A* is shut off and moved. In this way the capacity of the pump need be adequate to supply only one lateral. If both laterals are operated at the same time, pump capacity must be increased accordingly.

Some general rules that may be helpful for the layout of systems are as follows:

1. Sprinkler spacings both along the lateral and between laterals should be as wide as possible to reduce moving costs. Greater spacings require higher pressures and higher application rates, which are limited by the infiltration rate of the soil.
2. Mains should be laid up- and downhill.
3. Laterals should be laid across the slope or nearly on the contour.
4. Lateral pipe sizes should be limited to not more than two different diameters.

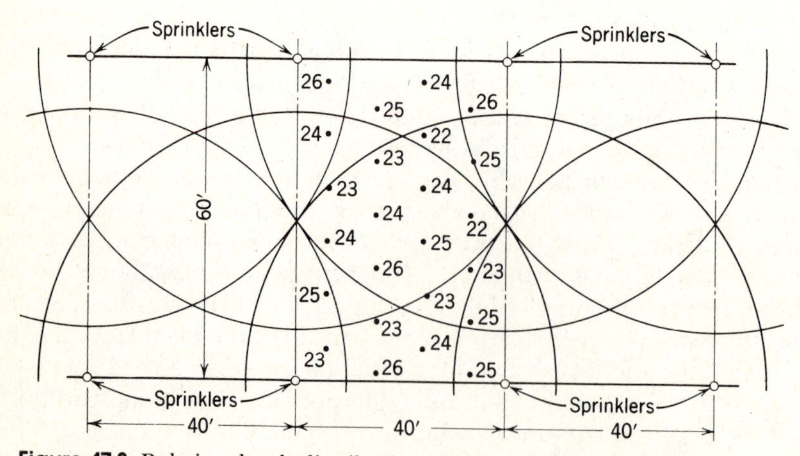

Figure 17.8. Relative depth distribution of water between sprinklers with overlapping patterns.

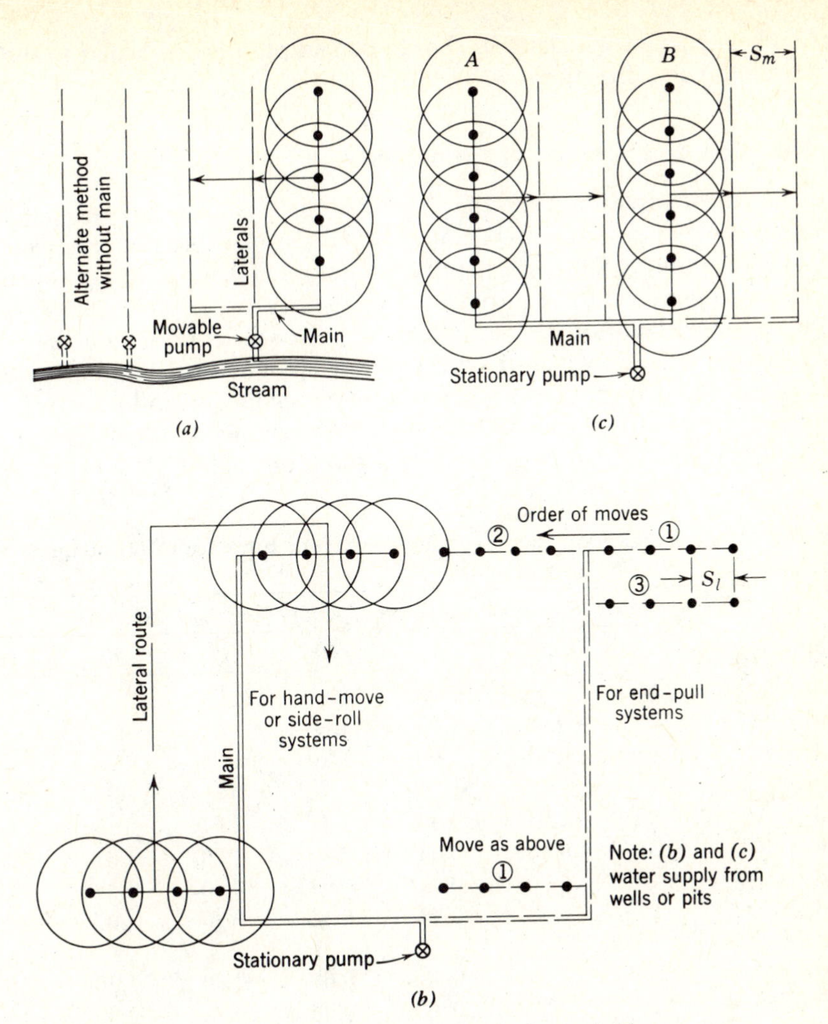

Figure 17.9. Field layouts of main and laterals for sprinkler systems.

5. For a more balanced layout, lateral operation should conform to that shown in Fig. 17.9*b*.
6. Layout should facilitate and minimize lateral movement during the season.
7. Whenever possible, water supply should be located near the center of the irrigated area.
8. Difference in the number of sprinklers operating for the various setups should be held to a minimum.
9. Layout should be modified to apply different rates and amounts of water where soils are greatly different in the irrigated area.

10. Whenever possible, laterals should be laid out perpendicular to the prevailing wind direction.

17.3 Selection and Spacing of Sprinklers

When the rate of application and the spacing of the sprinklers have been decided upon, the required capacity can be taken directly from Table 17.2. For example, with a spacing of 40 × 60 ft and an application rate of 0.40 in. per hr a sprinkler capacity of 10 gpm will be required.

The actual selection of the sprinkler will be based largely upon the design information furnished by manufacturers of the equipment. Table 17.3 gives the discharge of sprinkler nozzles for various nozzle diameters and pressures. The capacity of the sprinkler can be determined by adding the discharge of each nozzle taken from the manufacturer's data given in Table 17.3. Each manufac-

Table 17.2. Water Application Rate in Inches per Hour for Various Sprinkler Spacings and Capacities[a]

	Capacity in Gallons per Minute from Each Sprinkler						
Spacing (ft)	*4*	*6*	*8*	*10*	*15*	*20*	*30*
20 × 20	0.96	1.44	1.92				
20 × 30	0.64	0.96	1.28	1.60			
20 × 40	0.48	0.72	0.96	1.20	1.81		
30 × 30	0.43	0.64	0.86	1.07	1.61	2.14	
30 × 40	0.32	0.48	0.64	0.80	1.20	1.61	2.40
30 × 50	0.26	0.39	0.51	0.64	0.96	1.28	1.92
30 × 60	0.21	0.32	0.43	0.53	0.80	1.07	1.61
40 × 40	0.24	0.36	0.48	0.60	0.90	1.20	1.80
40 × 50	0.19	0.29	0.38	0.48	0.72	0.96	1.44
40 × 60	0.16	0.24	0.32	0.40	0.60	0.80	1.20
40 × 80	0.12	0.18	0.24	0.30	0.45	0.60	0.90
50 × 70	0.11	0.17	0.22	0.28	0.41	0.55	0.82
60 × 80		0.12	0.16	0.20	0.30	0.40	0.60
70 × 90			0.12	0.15	0.23	0.38	0.46
80 × 100			0.10	0.12	0.18	0.24	0.36

[a]Values in the table based on the following equation:

$$\text{capacity in gpm} = \frac{S_l \times S_m \text{ (ft)} \times \text{application rate (iph)}}{96.3}$$

where S_l = spacing along lateral in feet
S_m = spacing between laterals along main in feet

turer will recommend a combination of nozzle sizes and pressures to give the best breakup of the stream and distribution pattern for a uniform application.

The spacing of sprinklers along the lateral and between laterals must be selected so that water can be applied at a reasonably uniform rate. The spacing is determined largely from the diameter of coverage for each sprinkler and the expected wind velocity. Such spacings are usually in multiples of 10 ft with pipe lengths of 20 or 30 ft for hand-move systems. The sprinkler spacing along the lateral should be from 0.4 to 0.5 of the wetted diameter of sprinkler coverage for winds less than 8 mph. Spacing between laterals varies from 0.5 to 0.65 of the wetted diameter. These spacings are generally greater than those along the lateral, in order to reduce the number of moves for the laterals. For winds above about 8 mph, the maximum spacing in either direction should be 0.3 of the wetted diameter. Extremely poor distribution is likely for winds above 8 mph, and operation under these conditions should be avoided. The diameter of coverage of most sprinklers for medium pressures of 30 to 60 psi and small nozzles will vary from 60 to 140 ft. Wetted diameters should be obtained from the manufacturer's specifications for each type and size of sprinkler, and the desired operating pressure as shown in Table 17.3.

17.4 Size of Laterals and Mains

As water is forced through the irrigation system, head is lost due to the friction of the water against the sides of the conduit. The amount of head lost depends largely upon the rate of flow, the diameter of the pipe, the roughness of the inner surface of the pipe, and the number and abruptness of turns.

Table 17.3. Manufacturer's Sprinkler Characteristics

Nozzle Pressure (psi)	*Nozzle Diameter in Inches*									
	$\frac{5}{32}'' \times \frac{1}{8}''$		$\frac{5}{32}'' \times \frac{5}{32}''$		$\frac{3}{16}'' \times \frac{5}{32}''$		$\frac{7}{32}'' \times \frac{5}{32}''$		$\frac{1}{4}'' \times \frac{5}{32}''$	
	Dia.[a]	*GPM*[b]	*Dia*[a].	*GPM*[b]	*Dia.*[a]	*GPM*[b]	*Dia.*[a]	*GPM*[b]	*Dia.*[a]	*GPM*[b]
20	77	4.6	77	5.6	78	6.6	81	8.2	82	9.7
25	80	5.3	80	6.2	81	7.6	85	9.2	86	11.0
30	83	5.8	83	6.8	86	8.3	88	10.2	91	12.1
35	85	6.3	85	7.4	89	9.1	92	11.1	96	13.2
40	87	6.8	87	8.0	93	9.7	96	11.9	101	14.2
45	89	7.2	89	8.5	96	10.3	101	12.7	105	15.0
50	92	7.5	92	9.0	99	10.8	105	13.4	111	15.8
55	94	7.9	94	9.4	101	11.4	109	13.9	115	16.6
60	97	8.3	97	9.8	103	11.8	112	14.5	119	17.4

Pressures to left and below dashed line recommended for best breakup of stream.

[a]Diameter of coverage in feet.

[b]The discharge for other nozzle sizes not shown may be computed from the formula, $q = 29.85\ Cd^2\ P^{1/2}$, where q is the discharge in gallons per minute, C is the discharge coefficient, d is the nozzle diameter in inches, and P is the pressure in pounds per square inch (psi) using $C = 0.87$.

The pipe diameter chosen should be one that will provide the required rate of flow with a reasonable head loss. In the case of the lateral, the sections at the distant end of the line will have less water to carry and may therefore be smaller. Some authorities, however, advise against the tapering of pipe diameters in laterals as it then becomes necessary to keep the various pipe sizes in the same relative position. They also point out that the system may be less adaptable to other fields and situations.

The total pressure variation in the lateral should not be more than 30 percent and preferably less than 20 percent of the average pressure in the lateral. If the lateral runs up- or downhill, allowance for this difference in elevation should be made in determining the variation in head. If the water runs uphill, a greater head will be required, whereas if it runs downhill, there will be a tendency to balance the loss of head due to friction.

The friction loss in laterals may be determined by following the procedure given in Fig. 17.10. Various pipe sizes should be tried until the limits of pressure variation have been complied with. The total pressure loss in the lateral should be converted into feet of head and recorded for use in the selection of the pump. This conversion can be made from the scales or the equation in Fig. 17.11.

The diameter of the main should be adequate to provide the required pressure to the laterals in each of their positions. The rate of flow required for each lateral may be determined by multiplying the number of sprinklers by their capacity. The position of the laterals that gives the highest friction loss should be used for design purposes. The nomograph in Fig. 17.12 will give the friction loss in feet of head per 100 ft of main. Allowable friction loss in the main will vary with the cost of power and the price differential between different diameters of pipe and couplers. A reasonable maximum friction loss in the main is about 4 ft per 100 ft of pipe. The most economical size can best be determined by balancing the increase in pumping cost against the amortized cost difference in the pipe sizes. Pumping against friction presents a cost continuing for as long a time as the system will be operated.

17.5 Pumping Head and Power Requirements

The total head and the rate of pumping must be known before a pump and a power unit can be selected. The pump capacity is the sum of the discharge of all sprinklers operating at the same time, plus an allowance of about 5 percent for loss of efficiency of the pump due to wear. The total head is the sum of the friction loss and height to which water must be raised. It can be determined from the equation,

$$H = H_n + H_s + H_m + H_j + NPSH \tag{17.1}$$

where H = total head in feet

H_n = pressure head at the junction of main and lateral in feet

H_s = elevation difference between the water level at the source and the pump in feet

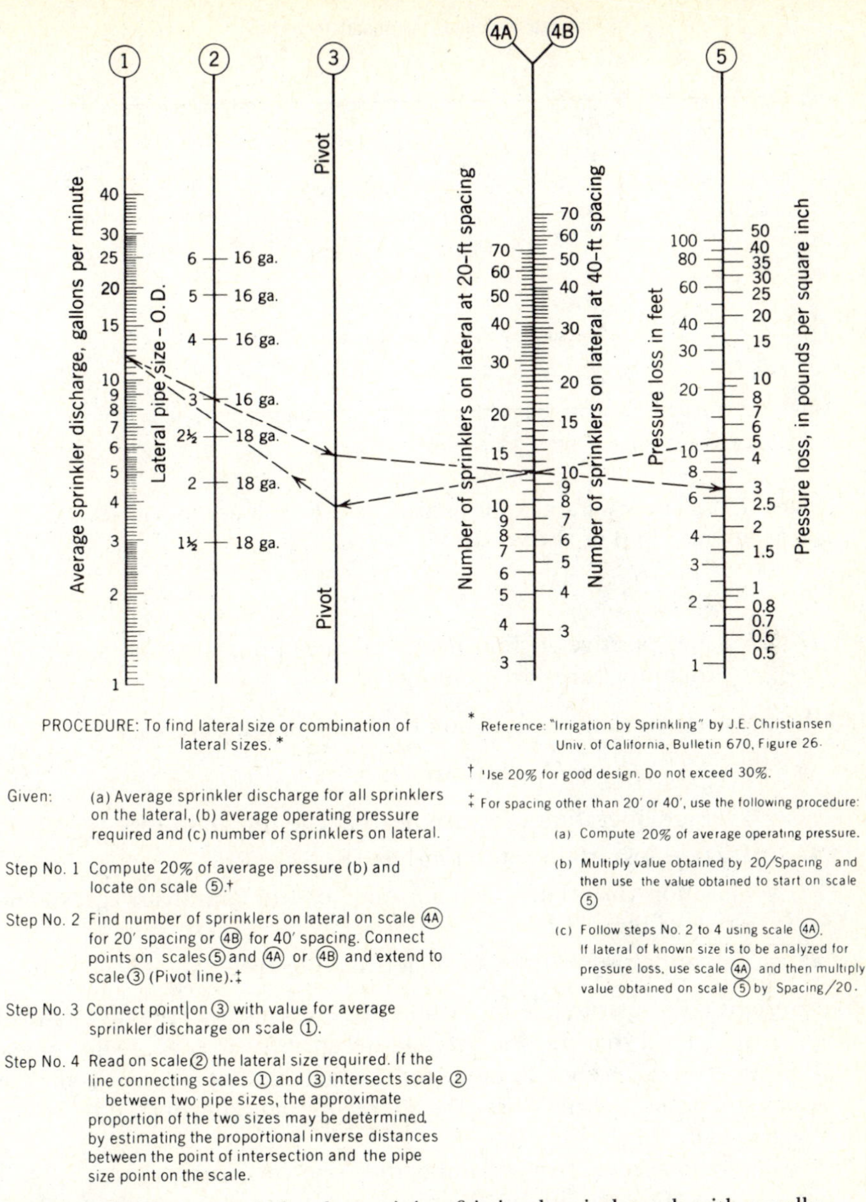

Figure 17.10. Nomograph for determining friction loss in laterals with equally spaced sprinklers. (*Redrawn from U.S. Soil Conservation Service, 1949.*)

H_m = friction loss in the main and suction line of the pump for the farthest position of the lateral in feet (Fig. 17.12)

H_j = elevation difference between the pump and the junction of the lateral and the main in feet

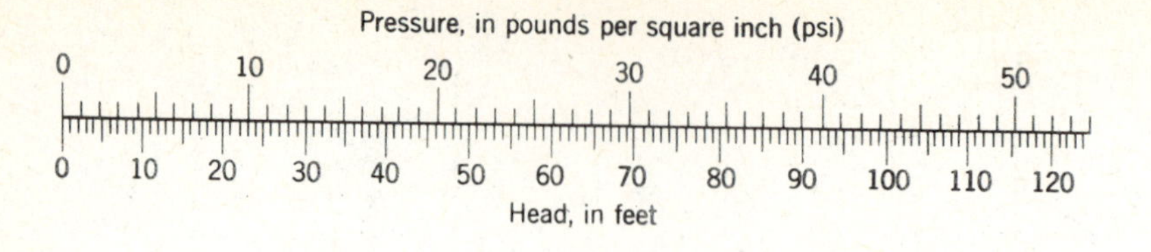

H (in feet) = 2.31 P (in psi)

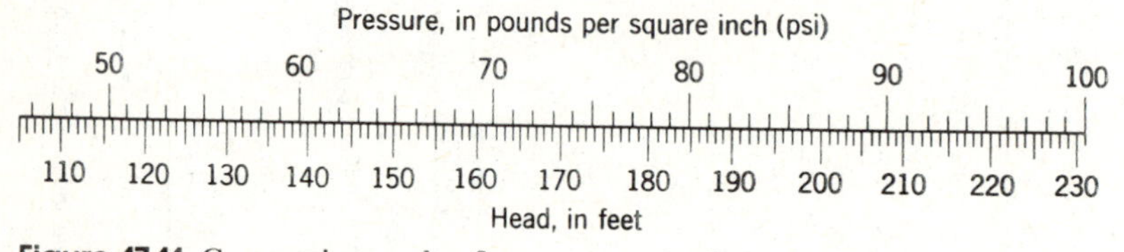

Figure 17.11. Conversion scales for pressure in feet of water to pounds per square inch (psi) or vice versa.

$NPSH$ = net positive suction head for the pump to be obtained from manufacturer's data in feet

and the pressure head at the junction of the main and lateral,

$$H_n = H_a + 0.75H_f + 0.6H_e + H_r \tag{17.2}$$

where H_a = average operating pressure head at the nozzle in feet
H_f = friction loss in the lateral in feet (Fig. 17.10)
H_e = elevation difference between ends of the lateral for the steepest slope in feet,
H_r = height of riser on the sprinkler head in feet

The constant 0.75 for the lateral friction loss corrects the friction loss to the point along the lateral that has the average design pressure, H_a. The constant 0.6 for the elevation difference accounts for the fact that not all water is pumped to the upper end of the lateral. Where the lateral operates downhill, the elevation head is negative.

Pumps should be selected from manufacturer's performance curves to provide the required head and capacity for the range of expected operating conditions at or near maximum pump efficiency. Where wells are the source of water, the drawdown-discharge capacity should be determined from long-time pumping tests. Wells sometimes limit the size and type of pump. The manufacturer can best select the size and type of pump to fit the desired operating conditions.

The size of the power unit required depends on the pump capacity, the total pumping head, and the efficiency of the pump and of the power unit. The power requirements for the pump can be determined from the nomograph in

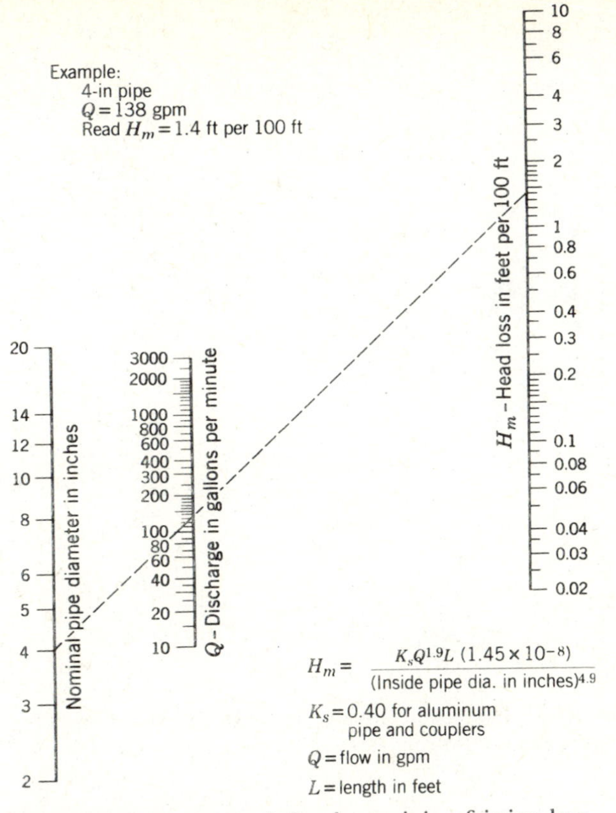

Figure 17.12. Nomograph for determining friction loss in main lines (no sprinklers). (*Redrawn from U.S. Soil Conservation Service, 1949.*)

Fig. 17.13. Pump efficiency must be obtained from manufacturer's performance curves. Since internal combustion engines should not operate continuously at full load, the size of the power unit must be greater than the horsepower requirements determined from Fig. 17.13. Multiply this horsepower by the following factors to obtain the size of the power unit:

Electric motor	1.0
Diesel engine	1.25
Gas engine, water-cooled	1.45
Gas engine, air-cooled	1.6

Electric motors operate best at full-load capacity. They have many advantages over internal combustion engines, such as ease of starting, low initial cost, low upkeep, and suitability for mounting on horizontal or on vertical shafts. Direct-drive motors are preferred because gears and belts are eliminated. Deep-well pumps are now available with watertight vertical motors. The motor is sub-

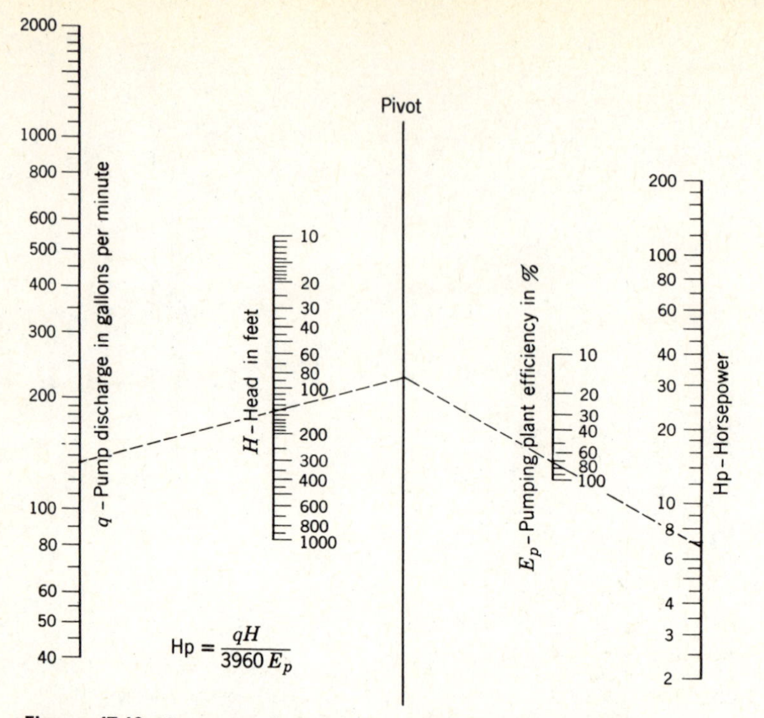

Figure 17.13. Nomograph for determining the horsepower requirements for pumping water.

merged in the well near the impellers, thus eliminating the long shaft that would be otherwise required.

17.6 Sprinkler System Design

The three basic requirements to be established before designing a sprinkler irrigation system are the maximum water-application rate, the irrigation period, and the depth of application. The maximum rate at which water can be applied will depend upon the ability of the soil to absorb it. In no case should water be applied faster than it will move into the soil under the cropping conditions encountered. If this rule is followed, runoff and erosion from irrigation will not occur. Maximum water application rates for average soil, slope, and cultural conditions are given in Table 15.3. These values should be used only where reliable local information is not available. Many states have published irrigation guides that provide such information for specific soil types.

The depth of application and the irrigation period are closely related. The irrigation period is the time required to cover the field with one application of water. The depth of application will depend on the amount of water that can be stored in the root zone of the soil and the efficiency of application. For sprinklers the water-application efficiency is usually assumed to be about 70 percent, that is, only 0.7 of each inch pumped is available to the plants, the remainder being lost by evaporation and other distribution losses.

Under humid conditions rains may bring the entire field up to a given moisture level. As plants use this moisture, the moisture level for the entire field decreases. Irrigation must be started soon enough to enable the field to be covered before plants in the last portion to be irrigated suffer from moisture deficiency.

A common criterion for beginning irrigation is when the moisture level of the field reaches 55 percent of the available moisture capacity. Thus, the effective depth of application is equal to 45 percent (100 − 55) of the available moisture. The irrigation period is set so that the entire field will be covered before the finishing end of the field reaches the wilting point. For the first irrigation cycle, the normal depth of water applied should be increased by the daily water use rate [or the evapotranspiration (ET) rate] for each day of delay of the sprinkler line setting. After the first irrigation cycle, the normal depth is applied at all settings to bring the soil moisture back to the field capacity. Using Example 17.1 as an illustration, the irrigation period is 12 days and the ET rate is 0.2 in. per day. For the first day or setting, 2.4 in. of water is applied. This depth is increased by 0.2 in. for each day delay in irrigation. By the twelfth day ($2.4 + 0.2 \times 12$) a 4.8-in. depth would be required to bring the soil to the field capacity. This procedure would reduce the soil moisture to a very low level for settings near the end of the first irrigation cycle. Another procedure would be to start the first irrigation earlier before the soil moisture is depleted to the 55 percent level (45 percent lost) or to provide additional irrigation capacity and apply the required depth in a shorter period of time. For the second irrigation cycle, assuming no appreciable rain falls, the normal depth of 2.4 in. is applied at all settings to bring the soil back to the field capacity.

Typical moisture-holding capacities for soils of different texture are given in Table 15.3. Root depths and peak moisture-use rate for some crops are given in Table 15.4.

Example 17.1. A sprinkler system is to be designed to irrigate 14 acres of corn on a deep silt loam soil with slopes less than 1 percent. Determine the maximum rate of application, the irrigation period, the depth of water pumped per application, the effective depth of application, and the required acreage to be irrigated each day.

Solution. From Table 15.3 the maximum water-application rate for bare silt loam soil is (1.0/2) 0.5 in. per hr. The available moisture-holding capacity of the soil from Table 15.3 is taken as 1.8 in. per ft and the depth of the root zone from Table 15.4 is estimated as 3.0 ft.

$$\text{Total available moisture capacity in the root depth} = 1.8 \times 3.0 = 5.4 \text{ in.}$$

$$\text{Effective depth of application} = 45\% \times 5.4 = 2.4 \text{ in.}$$

$$\text{Depth of water to be pumped} = (2.4/70\%)100 = 3.4 \text{ in.}$$

From Table 15.4 the peak rate of use for corn in a cool climate is 0.2 in. per day.

$$\text{Irrigation period} = 2.4/0.2 = 12 \text{ days}$$
$$\text{Acreage irrigated per day} = 14/12 = 1.2 \text{ acres}$$

The irrigation system must have sufficient capacity to deliver 3.4 in. of water to about 1.2 acres per day. These values thus serve as a guide in the selection of irrigation equipment.

Assistance in layout and design can be obtained from qualified irrigation-equipment dealers. The Sprinkler Irrigation Association, American Society of Agricultural Engineers, and several commercial companies have prepared layout and design sheets. The following example illustrates the design of a simple system.

Example 17.2. Design a completely portable hand-move sprinkler irrigation system for the 14-acre field described in Example 17.1. The field is 442 ft wide and 1380 ft long, with a natural stream along the boundary of the 1380-ft side. Stream flow is adequate during the driest season. An allowable variation of pressure in the lateral is 20 percent of the operating pressure.

Solution. Economic studies of several system combinations indicate that medium pressure (H_a = 40 psi) with 40 ft spacing of sprinklers along the lateral and 60 ft between the laterals is about optimum for three irrigations per year. The layout is to be similar to that shown in Fig. 17.9*a*.

A topographic survey showed that H_e = 7.0 ft (3.0 psi) and H_s = 13.0 ft (5.6 psi). The riser height for corn, H_r = 8.0 ft.

Sprinkler and Lateral Capacity. By interpolation from Table 17.2 the required sprinkler capacity for 0.5 iph application rate and 40 × 60 ft spacing is 12.5 gpm. Select a two-nozzle sprinkler head from Table 17.3 (Manufacturer's data). For the $\frac{7}{32}$- and $\frac{5}{32}$-in. sprinkler at 40 psi, read q = 11.9 gpm. For sizes not shown, use the equation in the footnote of Table 17.3. (When using the equation, compute the discharge for each size and add the two to get the total for a two-nozzle sprinkler.) From Table 17.3 read 96 ft as the diameter of coverage for a single sprinkler. Assuming a wind of less than 8 mph, the required diameter of coverage along the lateral is 80 ft (40/0.5) and along the main is 92 ft (60/0.65) for the 40 × 60 ft spacing. Since 80 and 92 are both less than 96 ft, the selected sprinkler is satisfactory for coverage. See Section 17.3 for further explanation of sprinkler spacing.

Select lateral length of 400 ft, which will allow 21 ft between the ends of the pipe and the field boundary. Install a sprinkler at both ends of the lateral.

$$\text{Number of sprinklers} = 400/40 + 1 \text{ (at end)} = 11$$
$$\text{Lateral capacity} = 11 \times 11.9 = 131 \text{ gpm}$$
$$\text{Number of moves of lateral} = 1380/60 = 23$$
$$\text{Area covered per lateral} = (11 \times 40 \times 60)/43{,}560 = 0.6 \text{ acre}$$

With two moves per day 1.2 acres can be covered, which meets the requirements in Example 17.1.

From the equation in Table 17.2, the actual application rate for the nozzle sizes selected is (11.9 × 96.3)/(40 × 60) = 0.48 iph. This rate is less than the requirement of 0.5 iph given in Example 17.1 and is satisfactory. Pumping time per lateral setting is 3.4/0.48 = 7.1 hr.

Diameter of Lateral. Allowable pressure variation along the lateral is

$$(20\% \times 40 \text{ psi}) = 8.0 \text{ psi } (18.5 \text{ ft})$$

Allowable friction loss in the lateral is

$$(8.0 - H_e) = (8.0 - 3.0) = 5.0 \text{ psi}$$

Since the first sprinkler is on the main, only 10 sprinklers are considered in computing the actual friction loss from Fig. 17.10. By following the dashed lines in Fig. 17.10, the required pipe size for 5.0 psi friction loss is seen to be about $2\frac{3}{4}$ in. Select 3-in. pipe as the next closest commercial size. By moving the straight edge on the chart to pass through 3 in., read the actual friction loss of 3.0 psi (6.9 ft).

Diameter of Main. Pump capacity for one lateral operation is the discharge of all the sprinklers, 131 gpm, plus 5 percent over capacity to allow for a loss of efficiency with use. Design $q = 131 \times 105\% = 138$ gpm.

Assuming 160 ft for the length of the main and a 4-in. diameter pipe, the friction loss is 1.4 ft per 100-ft length from Fig. 17.12, which gives the total head loss in the main

$$H_m = (1.4/100) \times 160 = 2.2 \text{ ft}$$

Pump Size. Substituting in Eq. 17.2, the pressure at the junction of the main and lateral is

$$H_n = (40 \times 2.31) + (0.75 \times 6.9) + (0.6 \times 7.0) + 8.0$$
$$= 109.8 \text{ ft}$$

Total pumping head from Eq. 17.1 assuming $NPSH = 10$ ft from the manufacturer's curve is

$$H = 109.8 + 13.0 + 2.2 + 10.0 = 135.0 \text{ ft}$$

Adding 5 percent loss of head with age, select a centrifugal pump from the manufacturer's curves to deliver 138 gpm at a head of 142 ft (105 percent $\times$ 135.0 ft).

Power Unit. Obtain pump efficiency from manufacturer's curves (assume 72 percent) and read from Fig. 17.13, the power requirements of 6.8 hp.

Select one of the following size power units, which will deliver 6.8 hp for continuous operation.

Gas engine, water-cooled	$1.45 \times 6.8 = 9.9$ hp
Gas engine, air-cooled	$1.6 \times 6.8 = 10.9$ hp

17.7 Center Pivot Systems

These systems are normally designed by the manufacturer and sold as a complete unit. The design procedure is beyond the scope of this book but is described by Schwab et al. (1981). It is more involved than that for set-operation or nonmoving lateral systems. Uniformity of application is obtained by a constant spacing of sprinklers along the lateral with variable discharge or with a combination of spacings and discharges. Uniformity coefficients for moving lateral systems are as high as 80 to 90 percent even in winds up to 20 mph. For periodic set systems, the uniformity coefficient is normally 70 to 85 percent, but it can be reduced greatly under windy conditions. The rate of lateral rotation of a center pivot system determines the depth of application, but it does not affect the sprinkler discharge. Some energy-saving systems have been developed to operate at pivot pressures as low as 20 to 40 psi. At these low pressures, the rate of application is higher because the wetted area is smaller.

Center pivot systems are usually designed for continuous operation to supply the crop at the peak rate of water use (ET) in arid regions. Where rainfall and stored soil moisture can provide some of the water, the design rate may be less than the peak.

Systems should be maintained in a high state of repair. The ability to secure parts and to make repairs quickly is an important consideration. A crop may be lost or severely damaged if the system is down during a high water-use period.

17.8 Other Moving Sprinkler Systems

The traveling gun sprinkler and traveling lateral systems are other types of moving systems. These may be moved by cable or self-powered units. Water is supplied through a flexible hose or through a suction line and inlet that remains in an irrigation canal as the unit moves along the canal. These systems can provide a complete coverage of rectangular fields, which is not possible with center-pivot systems. The reduction of labor is the principal advantage of all mechanical-move systems. The lateral-move systems require more labor than the center-pivot system because pipe must be moved at the end of the irrigation and, if water must be supplied through a flexible pipe, it must be moved or reconnected as irrigation progresses across the field. When lateral-move systems are operated at recommended pressures and the spacing of sprinklers is within 50 to 70 percent of their wetted diameter, uniformity coefficients of 77 to 83 percent can be expected.

The traveling gun that is moved with a power winch and cable or is self-propelled, usually consists of a high capacity sprinkler ranging from 100 to 1000 gpm and operating at pressures varying from 60 to 110 psi. A traveling-boom system is similar to the traveling gun, except the rotating boom usually has a gun on one arm and several smaller sprinklers on the other arm. These systems are limited to soils with a high infiltration rate, or to fields with good vegetative cover to protect the soil from surface sealing and erosion.

The traveling-lateral system has moving units ("A" frames) almost identical to those of the center-pivot lateral. Sprinkler spacing and discharge are uniform as they are with hand-move systems. Friction loss and sizing of sprinklers are similar to that described in Example 17.2. Traveling laterals may be as long as

1320 ft ($\frac{1}{4}$ mile), which would be suitable for an 80-acre field. Uniformity coefficients of 90 to 95 percent can be obtained, which are slightly higher than those for center-pivot systems because only one sprinkler size need be selected and the spacing is uniform. Operating at a reduced pressure and a reduced rate of application and still retaining good distribution has greater possibilities with this system than with the center-pivot system.

17.9 Sprinkler Systems for Environmental Control

Sprinkling has been successful for protecting small plants from wind damage, soil from blowing, and plants from frosts or freezing. It has also been successful in reducing high air and soil temperatures. Since the entire area usually needs protection at the same time, solid-set systems are required. The rate of application should generally be low or just enough to achieve the desired control. Small pipes and low volume sprinklers may be desired to reduce costs, but normal sprinkler systems may be modified for dual use. Because water is applied without regard to irrigation requirements, natural drainage should be adequate or a good drainage system should first be provided.

In organic or sandy soils where onions, carrots, lettuce, and other small seed crops are grown, the soil dries out quickly and the seed may be blown away or covered too deeply for germination. When these plants are small they are also easily damaged by wind-blown soil particles. Protection for such conditions can be provided with a sprinkler system that will apply low rates up to 0.1 iph. Operation at night when winds are usually at a minimum will provide more uniform coverage.

Low growing plants can be protected from freezing injury, which is likely either in the early spring or late fall. Sprinkling has been most successful against radiation frosts. Water must be applied continuously at 0.1 iph or greater depending on the wind and temperature, until the plant is free of ice. Sprinkling should be started before the temperature reaches 32° F at the plant level. Strawberries have been protected from temperatures as low as 21° F. Tomatoes, peppers, cranberries, apples, cherries, and citrus have been successfully protected. Tall plants, such as trees, may suffer limb breakage when ice accumulates, but in some areas low-level undertree sprinkling has provided some control. Rates of application may be reduced by increasing the normal sprinkler spacing. A slightly higher pressure may be desirable to increase the diameter of coverage and to give better breakup of the water droplets.

Sprinkling during the day to reduce moisture stress has been successful with many plants, for instance lettuce, potatoes, green beans, small fruits, turf, tomatoes, cucumbers, and muskmelons. This practice is sometimes called "misting" or "air conditioning" irrigation. Maximum stress in the plant usually occurs at high temperatures, at low humidity, with rapid air movement, on bright cloudless days, and/or with rapidly growing crops on dry soils. Under these conditions crops at a critical stage of growth, such as during emergence, flowering, or fruit enlargement may benefit greatly from the low application of water during the midday. At 81° F water loss was reduced 80 percent with an increase of humidity from 50 to 90 percent. Measured temperature reductions in the plant canopy of about 20° F were attained by misting in an atmosphere of 38 percent relative humidity. Green bean yields were increased 52 percent by midday misting during the bloom and pod-development period. Potatoes and corn respond to sprinkling, especially when temperatures exceed 86° F. The tasseling period is a critical time for corn. Small quantities of water applied frequently to strawberries increased the quality and the yield by as much as 55 percent. For low-growing crops, the same sprinkler system can be adapted for frost protection. Misting in a greenhouse or under a lath house to reduce transpiration of nursery plants for propagation increases plant growth and root development.

During periods of high incoming radiation, the soil temperatures may be 68° F greater than the ambient air temperature. Seedlings emerging through soil temperatures as high as 122° F frequently die as a result of high transpiration. Small applications of water at this critical period often assure emergence and good stands. Another benefit from sprinkling at this stage is to enhance the effectiveness of herbicides that are applied to control weeds. For further details, see Pair et al. (1983) and other current references.

17.10. Sprinkler Systems for Fertilizer, Chemical, or Waste Applications

Fertilizers, soil amendments, and pesticides may be injected into the sprinkler line as a convenient means of applying these materials to the soil or crop. This method primarily reduces labor costs and in some instances may improve the effectiveness and timeliness of the application. Liquid manure and sewage wastes are applied with sprinklers for disposing of unwanted material. A large amount of specialized commercial equipment is on the market. Cannery wastes are usually sprinkled on wooded areas or on land in permanent grass. Good subsurface drainage is required.

Liquid and dry fertilizers have been successfully applied with sprinklers. Dry material must first be dissolved in a supply tank. The liquid may be injected on the suction side of the pump, forced under pressure into the discharge line, or injected into the discharge line by a differential pressure device, such as a venturi section. The material is applied for a short time during the irrigation set. Sprinkling should be continued for at least 30 minutes after it has been applied to rinse the supply tank, pipes, and the crop. For center-pivot or other moving systems, the material must be applied continuously or until the field has been completely covered.

For waste disposal systems, the solids should be well mixed and small enough so as not to plug the nozzles. When pumping these solids, an effective non-plugging screen on the suction side of the pump is necessary. Liquid must be stored in a lagoon or other holding pond. Equipment should be resistant to corrosion from chemicals that may be present in the water. The system should be designed to apply water during subfreezing weather, or sufficient storage should be provided during the nonoperating period. Such systems are usually solid set and are operated for long periods. The application rate should be lower than the infiltration rate so that the water does not run off the surface and cause stream pollution.

REFERENCES

Berge, I. O., and M. D. Groskopp (1964) "Irrigation Equipment in Wisconsin," *Univ. Wisc. Special Cir. 90.*

Christiansen, J. E. (1948) "Irrigation by Sprinkling," *Calif. Agr. Exp. Sta. Bull. 670* (Reprinted).

Hagan, R. M., H. R. Haise, and T. W. Edminster (eds.) (1967) "Irrigation of Agricultural Lands," Monograph No. 11, Am. Soc. Agron., Madison, Wis.

Hansen, V. E., O. W. Israelsen, and G. E. Stringham (1980) *Irrigation Principles and Practices,* 4th ed., John Wiley & Sons, New York.

Irrigation Journal (1983) *1983 Irrigation Survey,* Encino, Calif.

Jensen, M. E. (Ed.) (1980) "Design and Operation of Farm Irrigation Systems," Monograph 3, Am. Soc. Agr. Eng., St. Joseph, Mi.

Pair, C. H. et al. (Eds.) (1983) *Irrigation,* 5th ed., The Irrigation Association, Silver Spring, Md.

Schwab, G. O., R. K. Frevert, T. W. Edminster, and K. K. Barnes (1981) *Soil and Water Conservation Engineering,* 3rd ed., John Wiley & Sons, New York.

U.S. Soil Conservation Service (1949) *Regional Engineering Handbook,* Chapter VI, "Conservation Irrigation," Pacific Region VII, Portland, Ore.

PROBLEMS

17.1 Determine the acreage to be irrigated per day and the depth of water to be pumped to irrigate 30 acres of corn in a hot climate. The soil has a maximum infiltration rate of 0.4 in. per day and a moisture-holding capacity of 4.2 in. in the root zone. Assume a water-application efficiency of 70 percent. Irrigation is to begin when 40 percent of the moisture has been depleted.

17.2 Determine the sprinkler capacity in gpm for a 60 × 80 ft spacing if the water application rate is 0.3 iph.

17.3 What is the application rate in iph for a 25 gpm sprinkler where the spacing is 60 × 80 ft?

17.4 Determine the discharge in gpm for a sprinkler operating at 45 psi and having one $\frac{3}{16}$- and one $\frac{7}{64}$-in. diameter nozzle.

17.5 Determine the required capacity of a lateral with 10 sprinklers that apply water at 0.4 iph where the spacing is 40 × 60 ft.

17.6 Determine the required capacity of a boom-type sprinkler unit operating at 60 psi with nozzles $\frac{3}{32}$, $\frac{1}{8}$, $\frac{1}{4}$, $\frac{3}{8}$, and 1 in. in diameter (one of each size).

17.7 For winds below 8 mph, determine the maximum sprinkler spacing for nozzles whose wetted diameter of coverage is 100 ft at 40 psi.

17.8 Determine the nearest nominal size for a lateral that has 14 sprinklers 40 ft apart operating with an average pressure of 50 psi and a discharge of 10 gpm. The maximum variation of pressure along the lateral is 20 percent of the average pressure. The difference in elevation between ends of the lateral is 7 ft.

17.9 Determine the friction loss in a 3-in. lateral with 10 sprinklers at 40-ft spacing, each discharging 15 gpm.

17.10 Prove that 1 psi of pressure equals 2.31 ft of water. Show units.

17.11 Determine the friction loss in 400 ft of a 5-in. diameter main line that carries a flow of 300 gpm.

17.12 Determine the nominal size for a main line carrying 200 gpm if the friction loss should not exceed 4 ft per 100 ft.

17.13 Determine the total head in ft to be developed by a pump to operate sprinklers at 50 psi. The elevation difference between the ends of the lateral is 10 ft and the suction lift at the pump is 12 ft. The riser height is 6 ft. The total friction loss in the main and laterals is 26 ft.

17.14 Determine the horsepower required to pump 200 gpm at a total head of 100 ft if the pump efficiency is 70 percent. What power rating would be required for a water-cooled gasoline engine for this pump?

17.15 Design a sprinkler irrigation system for a square 40-acre field to irrigate the entire field within a 12-day period. Not more than 16 hr per day are available for moving pipe and sprinkling. Two inches of water are required at each application to be applied at a rate not to exceed 0.35 iph. A 75-ft well located in the center of the field will provide the following discharge-drawdown relationship: 200 gpm—40 ft; 250 gpm—50 ft; 300 gpm—65 ft. Design for an average pressure of 40 psi at the sprinkler nozzle. The highest point in the field is 4 ft above the well site, and 3-ft risers are needed on the sprinklers. Assuming a pump efficiency of 60 percent and assuming that the engine will furnish 70 percent of its rated output for continuous operation, determine the rated output for a water-cooled internal combustion engine.

CHAPTER 18

TRICKLE IRRIGATION

Trickle irrigation is a method of applying water directly to plants through a number of low flow-rate outlets generally placed at short intervals along small tubing. At these outlets, specially designed orifices may apply water to individual plants or to a row of plants. Trickle irrigation, sometimes referred to as *drip irrigation*, is similar to watering a plant with a slowly leaking bucket. Unlike sprinkler or surface irrigation only the soil near the plant is watered rather than the entire area.

According to Karmeli and Keller (1975), trickle irrigation research began in Germany about 1860. In the 1940s it was introduced in England primarily for watering and fertilizing plants in greenhouses. With the increased availability of plastic pipe and the development of emitters in Israel in the 1950s, it has since become an important method of irrigation in Australia, Europe, Israel, Japan, Mexico, South Africa, and the United States (California, Hawaii, and Florida). According to the Irrigation Journal (1983) survey, California alone had 260,000 acres and the total in the United States was more than 568,000 acres.

Trickle irrigation has been accepted mostly in the more arid regions for watering high value crops, such as fruit and nut trees, grapes and other vine crops, sugar cane, pineapples, strawberries, flowers, and vegetables. Although successfully used on cotton, sorghum, and sweet corn, trickle irrigation is not as well adapted to field crops.

18.1 Advantages and Disadvantages of Trickle Irrigation

With trickle irrigation only the root zone of the plant is supplied with water. With proper system management deep percolation losses are minimal. Soil evaporation is lower because only a portion of the surface area is wet. Like solid-set sprinkler systems, labor requirements are less and the systems can be readily automated. Reduced percolation and evaporation losses result in a greater economy of water use. Weeds are more easily controlled, especially for the soil area that is not irrigated. Bacteria, fungi, and other pests and diseases that depend on a moist environment are reduced as the above-ground plant parts normally are completely dry. Because soil is kept at a high moisture level and the water does not contact the plant, the use of more saline water may be possible with less stress and damage to the plant, such as leaf burn. Field edge losses and spray evaporation, such as occurs with sprinklers, are reduced with trickle systems. Low rates of water application at lower pressures are possible so as to eliminate runoff. With some crops, yields and quality are increased probably because of the maintenance of a high temporal soil moisture level adequate to meet transpiration demands. Crop yield experiments have shown wide differences varying from little or no difference to a 50 percent increase compared with other methods of irrigation. Crop quality may also be improved.

The major disadvantages of trickle irrigation are high cost and the clogging of system components, especially emitters by particulate, biological, and chemical matter. Emitters are not well suited to certain crops and special problems may be caused by salinity. Salt tends to accumulate along the fringes of the wetted surface strip (Fig. 18.1). Rainfall could move the salt to the plant roots to cause injury. Since trickle systems normally wet only part of the potential soil-root volume, plant roots may be restricted to the soil volume near each emitter as shown in Fig. 18.1. The dry soil area between emitter lateral lines may result in dust formation from tillage operations and subsequent wind erosion. Compared to surface irrigation systems, more highly skilled labor is required to operate and maintain the filtration equipment and other specialized components.

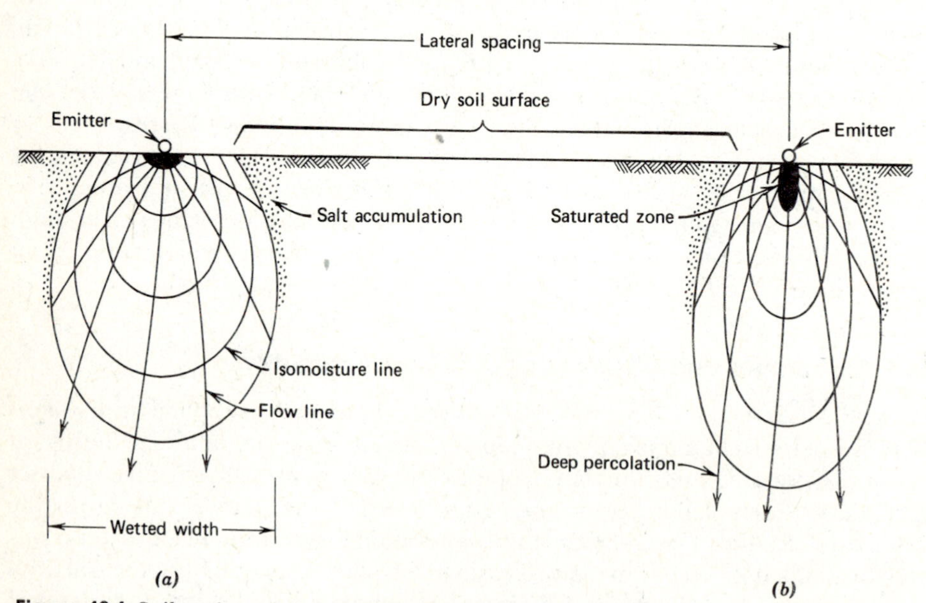

Figure 18.1. Soil moisture distribution pattern with trickle irrigation for (*a*) medium and heavy-textured soils, and (*b*) sandy soils. (*Adapted from Karmeli and Keller, 1975 and United Nations, 1963.*)

18.2 Components of Trickle Systems

Trickle system layouts are similar to sprinkler systems (Chapter 17). As with sprinkler systems many arrangements are possible. The one shown in Fig. 18.2 shows the split-line operation for a 3.75-acre orchard. The well is located at the south side of the field, centrally located for all areas to be irrigated. Tree rows are next to the 17 trickle laterals at a spacing of 20 ft. The entire field is to be irrigated at the same time as a solid set system. For a larger field with longer tree rows, several submains and manifolds could be provided as in Fig. 18.3.

As shown in Fig. 18.3, the primary components for a trickle system are an efficient filter, a main and submain, a manifold, and a lateral line to which the emitters are attached. The manifold is a line to which the trickle laterals are connected. Pressure regulators, pressure gages, a water meter, flushing valves, time clocks, and automatic control devices are other desirable components. The manifold, submain, and main may be laid on the surface or buried underground. The manifold is usually flexible pipe if laid on the surface and rigid pipe if buried. The main lines may be any type of pipe, such as polyethylene (PE), polyvinylchloride (PVC), galvanized steel, or aluminum. The lateral lines that have emitters are usually flexible PVC or PE tubing. They generally range from $\frac{3}{8}$ to $1\frac{1}{4}$in. in diameter and have emitters spaced at short intervals appropriate for the crop to be grown.

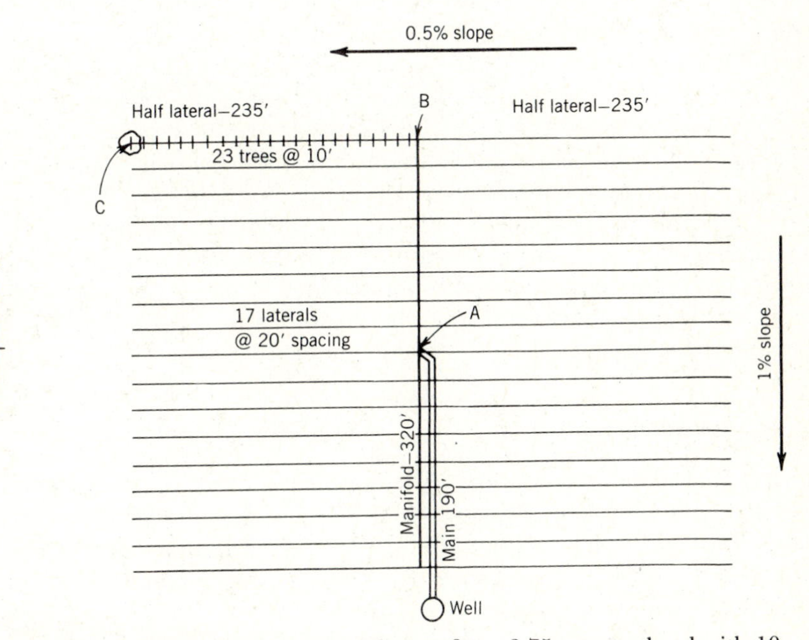

Figure 18.2. Trickle irrigation system layout for a 3.75-acre orchard with 10 × 20 ft tree spacing.

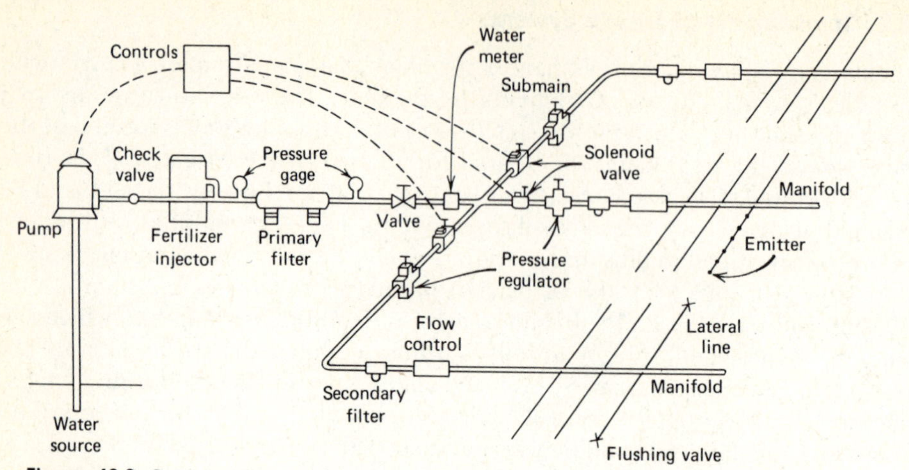

Figure 18.3. Components and nomenclature for a trickle irrigation system.

A filter is one of the most important components of the trickle system because of emitter clogging. Most water should be cleaner than drinking water. Trickle irrigation systems generally require screen, gravel, graded sand, or diatomaceous earth filters. Recommendations of the emitter manufacturer should be followed in selecting the filtration system. In the absence of such recommendations the net opening diameter of the filter must be smaller than $\frac{1}{4}$ to $\frac{1}{10}$ of the emitter opening diameter. For clean ground water an 80 to 200 mesh screen filter may be adequate. This filter will remove soil, sand, and debris, but it should not be used with high algae water. For high silt and high algae water, a sand filter backed up with a screen filter may be required. A sand separator ahead of the filter may be necessary if the water contains considerable sand. In-line strainers with replaceable screens and clean-out plugs may be adequate with small amounts of sand. Secondary filters may be installed at the inlet to each manifold. These are recommended as a safety precaution should accidents during cleaning or filter damage allow particles or unfiltered water to pass into the system. Filters must be cleaned and serviced regularly. Pressure loss through the filter should be monitored as an indication for maintenance.

18.3 Layout of Trickle Systems

Lateral lines may be located along the row of trees with several emitters required for each tree as shown in Fig. 18.4. Many laterals have multiple emitters, such as the "spaghetti" tubing or "pigtail" lines shown in Fig. 18.4*c*. One or two laterals per row (Fig. 18.4*a* or 18.4*b*) may be provided, depending on the size of the trees. With small trees a single line is adequate.

Many types of designs of emitters are commercially available, some of which are shown in Fig. 18.5. The emitter controls the flow from the lateral. The

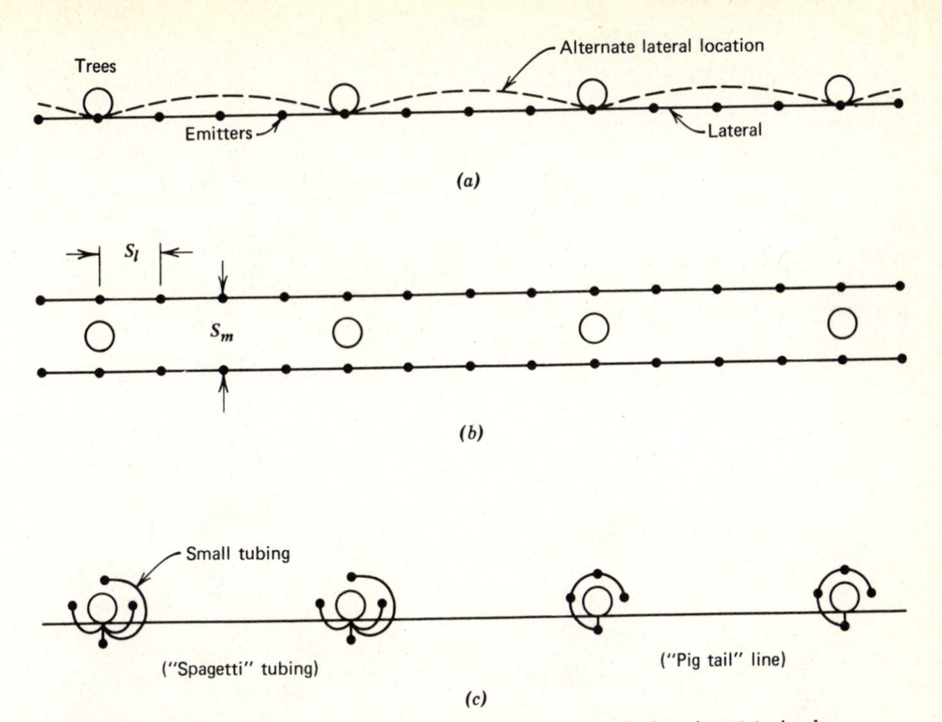

Figure 18.4. Lateral and emitter locations for an orchard showing (*a*) single lateral for each row of trees, (*b*) two laterals for each row of trees, and (*c*) multiple-exit emitters.

pressure is greatly decreased by the emitter; this loss is accomplished by small openings, long passageways, vortex chambers, manual adjustment, or other mechanical devices. Some may be pressure regulated by changing the length or cross section of passageways or the size of orifice. These emitters (Fig. 18.5*c*) give nearly a constant discharge over a wide range of pressures. Some are self-cleaning and flush automatically. Porous pipes or tubing may have many small openings as shown in Figs. 18.5*e* to 18.5*g*. The actual hole size is much smaller than the one indicated in the drawing. Some holes are barely visible to the naked eye. The double-tube lateral shown in Fig. 18.5*g* has more openings in the outer channel than in the main flow channel. Such tubes have thin walls and are low in cost. In Hawaii they are often discarded after the sugar cane is harvested and they are replaced with new lines. Most emitters are placed on the soil surface, but they may be buried at shallow depths for protection.

18.4 Emitter Discharge

The discharge of any emitter may be expressed by the power-curve equation (Karmeli and Keller, 1975) in which

$$q = Kh^x \tag{18.1}$$

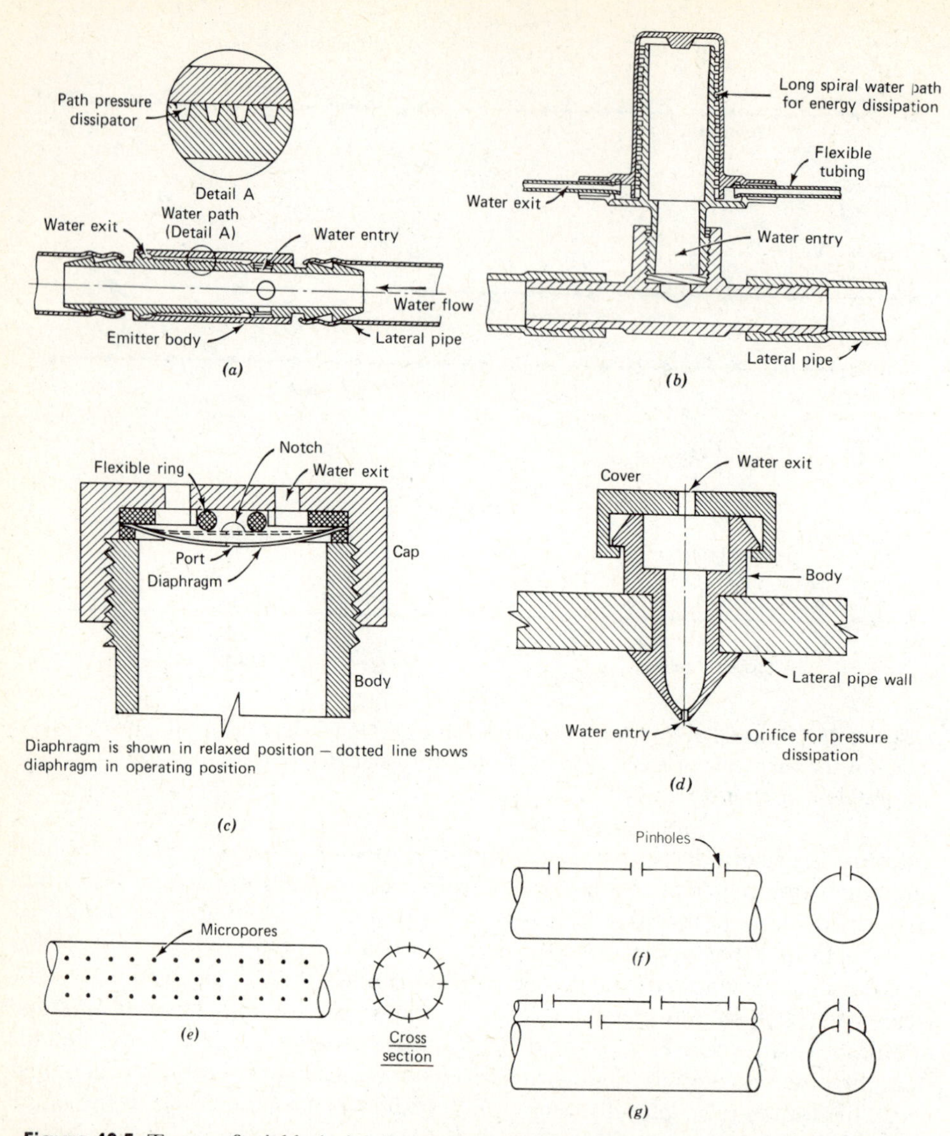

Figure 18.5. Types of trickle irrigation emitters and emitter laterals showing (*a*) in-line long-path single-exit emitter, (*b*) in-line long-path multiple-exit emitter, (*c*) flushing-type emitter, (*d*) orifice-type emitter, (*e*) porous tubing lateral, (*f*) single-tube emitter lateral, and (*g*) double-tube or double-wall emitter lateral. (*Figs. 18.5*a–*18.5*d *redrawn from Karmeli and Keller, 1975.*)

where q = emitter discharge, *gph*
K = constant for each emitter
h = pressure head in feet
x = emitter discharge exponent.

The discharge from several types of emitters for pressure heads from 10 to 80 ft are shown in Table 18.1. The two Rain Bird models are commercial emitters. Emitter discharge usually varies from about 0.3 to 8 *gph*, and the pressures range from about 5 to 90 ft (2 to 40 psi). The average diameters of openings for emitters range from 0.0001 to 0.01 in.

Emitters made from thermoplastic material may vary in discharge depending on the temperature. Thus, the discharge should be corrected for temperature.

18.5 Water Distribution from Emitters

An ideal trickle system should provide a uniform discharge from each emitter. Application efficiency depends on the variation of emitter discharge, pressure variation along the lateral, and seepage below the root zone or other losses, such as soil evaporation. Emitter discharge variability is greater than that for sprinkler nozzles because of smaller openings (lower flow) and lower design pressures. This variability may be due to the design of the emitter, materials, and care in manufacture. The uniformity for emitters is approximately (Nakayama et al., 1979)

$$EU = 1 - \frac{0.8C_v}{n^{0.5}} \tag{18.2}$$

Table 18.1. Trickle Emitter Discharge in gph

Head (ft)	*Double Tube 12″ × 72″*[a]	*Long-Path (Turbulent)*			*Rain Bird*		
		Small	*Medium*	*Large*	*EM-L10*	*EM-L20*	*EM-M05*
	$K = 0.60$[b]	0.076	0.138	0.341	0.17	0.34	*Pressure*
	$x = 0.5$[b]	0.63	0.63	0.63	0.50	0.50	*Compensating*
10	0.24	0.34	0.59	1.45	0.54	1.08	0.35
15	0.29	0.44	0.76	1.88	0.66	1.32	0.40
20	0.34	0.52	0.91	2.25	0.76	1.52	0.47
25	0.38	0.60	1.05	2.59	0.85	1.70	0.50
30	0.41	0.68	1.18	2.90	0.93	1.86	0.52
35	0.45	0.75	1.30	3.20	1.00	2.00	0.53
40	0.48	0.81	1.41	3.48	1.08	2.16	0.54
50	0.54	0.93	1.63	4.01	1.20	2.40	0.54
60	0.59	1.05	1.82	4.49	1.32	2.64	0.53
70	0.63	1.16	2.01	4.95	1.42	2.84	0.52
80	0.68	1.26	2.18	5.39	1.52	3.04	0.51

[a] Discharge for 12-in. length. Hole spacing 12 in. on outer wall and 72 in. on inner wall.

[b] Values in $q = Kh^x$, where q = discharge in gph and h = head in feet.

Source: Adapted from Karmeli and Keller, 1975, Walker, 1979, Davis, 1976, and Rain Bird Corporation.

where EU = emission uniformity
C_v = manufacturer's coefficient of variation
n = number of emitters per plant

The application efficiency for trickle irrigation can thus be defined as

$$E_{ea} = EU \times E_a \times 100 \tag{18.3}$$

where E_{ea} = trickle irrigation efficiency
E_a = application efficiency as defined for sprinkler or surface irrigation (water stored in the root zone divided by the water delivered)

A reasonably good design value for E_a is 90 percent.

Example 18.1. Determine the trickle irrigation efficiency of a system using emitters that have a manufacturer's coefficient of variation of 0.125 and an application efficiency of 90 percent.

Solution. Assuming one emitter per plant and substituting in Eq. 18.2,

$$EU = 1 - \left(\frac{0.8 \times 0.125}{1^{0.5}}\right) = 0.90$$

Substituting in Eq. 18.3,

$$E_{ea} = 0.90 \times 0.90 \times 100 = 81 \text{ percent}$$

18.6 Trickle System Design

Trickle systems may be designed by using the same general rules and procedures outlined in Chapter 17 for sprinkler systems. The primary differences are that the spacing of emitters is much less than the spacing required for sprinkler nozzles and that the water must be filtered and treated to prevent clogging of the small emitter openings. Another major difference with trickle irrigation, especially for widely spaced tree crops, is that not all of the area will be irrigated. Karmeli and Keller (1975) suggest that a minimum of 33 percent of the potential root volume should be irrigated. For closely spaced plants a much higher percentage may be necessary to assure sufficient water to the plants. In design, the water-use rate or the area irrigated may be decreased to account for this reduced area. Karmeli and Keller (1975) suggest the following water-use rate for trickle irrigation design,

$$ET_t = ET \times \frac{P}{85} \tag{18.4}$$

where ET_t = average evapotranspiration rate for crops under trickle irrigation, inches per day (ipd)
P = percent of the total area shaded by the crop
ET = conventional ET rate for the crop, ipd

For example, if a mature orchard shades 70 percent of the area and the conventional ET is 0.25 ipd, the trickle irrigation design rate is 0.21 ipd ($0.25 \times 70/85$). The shaded area is that which contributes to ET. For P greater than 85 percent, $ET_t = ET$.

The diameter of the lateral or of the manifold should be selected so that the difference in discharge between emitters operating simultaneously will not exceed 10 percent. This allowable variation is the same as for the sprinkler irrigation laterals discussed in Chapter 17. To stay within this 10 percent variation in flow, the head difference between emitters should not exceed 10 to 15 percent of the average operating head for long-path emitters, or 20 percent for in-line flow emitters. The maximum difference in pressure is the head loss between the control point at the inlet and the pressure at the emitter farthest from the inlet. The inlet is usually at the manifold where the pressure is regulated. In Fig. 18.2 the maximum difference in head loss to the farthest emitter is that for one half the lateral length plus one half the manifold length.

For small systems on nearly level land, 50 percent of the allowable friction loss should be allocated to the lateral and 50 percent to the manifold. As in sprinkler laterals (Chapter 17), allowable head loss should be adjusted for elevation differences along the lateral and along the manifold.

The friction loss for mains and submains (without emitters) can be obtained from the nomograph in Chapter 17 for aluminum tubing of 2-in. diameter and larger. For tubing of less than 2-in. diameter, select the friction loss from Table 18.2, applicable for polyethylene tubing. For manifolds and laterals, which have uniformly spaced outlets, the friction loss in Table 18.2 should be multiplied by the factor F given in Table 18.3. For in-line emitters (Figs. 18.5*a* and 18.5*b*) the head loss should be increased. Such losses may be expressed as an equivalent length of lateral pipe.

The water application rate for various emitter spacings may be read from Table 18.4 or computed from the equation below the data. For spacings greater than 20×20 ft use Table 17.2. The system shown in Fig. 18.2 will illustrate the design procedure.

Example 18.2. Design a trickle solid-set irrigation system for the tree orchard shown in Fig. 18.2. The tree spacing is 10 by 20 ft, the normal ET rate is 0.25 ipd, the shaded area per plant is 50 percent of 10×20 ft, and the trickle application efficiency is 90 percent.

Solution. Substituting in Eq. 18.4,

$$ET_t = 0.25\left(\frac{50}{85}\right) = 0.15 \text{ ipd}$$

1. Required application for 90 percent application efficiency is $(0.15/0.9) = 0.167$ ipd or 0.007 iph
2. Emitter discharge per tree from the equation in Table 18.4 is

$$q = 0.623 \times 10 \times 20 \times \frac{0.167}{24} = 0.87 \text{ gph (24-hr operation)}$$

Table 18.2. Friction Loss in Feet of Head per 100 Feet of Length for Polyethylene Mains or Submains[a]

Flow, q (gph)	Nominal Diameter (in.)						
	$\frac{3}{8}$	$\frac{1}{2}$	$\frac{3}{4}$	1	$1\frac{1}{4}$	$1\frac{1}{2}$	2
	Actual Inside Diameter (in.)[b]						
	0.375	*0.622*	*0.824*	*1.049*	*1.380*	*1.610*	*2.067*
20	2.1	0.2					
30	4.4	0.4					
45	8.8	0.8	0.2				
60	14.6	1.3	0.4				
120		4.4	1.2	0.4			
180		9.0	2.4	0.8	0.2		
240		15.0	3.9	1.3	0.3		
300		22.1	5.8	1.9	0.5	0.2	
360		30.4	8.0	2.5	0.7	0.3	
480		50.3	13.2	4.2	1.1	0.6	0.2
600			19.6	6.2	1.7	0.8	0.3
900			39.7	12.6	3.4	1.7	0.5
1200			65.7	20.9	5.7	2.7	0.8
1500				30.9	8.4	4.0	1.2
1800				42.5	11.5	5.6	1.7
2100				55.6	15.1	7.3	2.2
2400					19.1	9.2	2.8
3000					28.2	13.6	4.1
$d^{4.75}$ =	0.0095	0.105	0.399	1.26	4.62	9.60	31.5

[a]Based on equation $H_f = 0.000107\, Lq^{1.75}\, d^{-4.75}$, where H_f = friction loss in feet per 100 ft length, L = length of tubing in 100 ft, q = flow in gph, and d = inside diameter in inches.

[b]Inside diameter for polyethylene (any grade) tubing as per ASTM 2104, Schedule 40.

Source: NRAES, 1981.

Table 18.3. Correction Factor for Friction Loss in Laterals or Manifolds with Multiple Outlets

Number of Outlets	*F*	*Number of Outlets*	*F*
1	1.00	7	0.44
2	0.65	8–11	0.42
3	0.55	12–19	0.40
4	0.50	20–30	0.38
5	0.47	31–70	0.37
6	0.45	70+	0.36

Source: NRAES, 1981.

Table 18.4. Water Application Rate in Inches per Hour (iph) for Various Emitter Spacings[a]

Spacing (ft)	*Capacity in Gallons per Hour (gph) per Outlet (Plant)*						
	0.5	*1*	*1.5*	*2*	*3*	*4*	*5*
1 × 3	0.27	0.53	0.80	1.07	1.60	2.14	
1 × 5	0.16	0.32	0.48	0.64	0.96	1.28	
1 × 10	0.08	0.16	0.24	0.32	0.48	0.64	0.80
2 × 3	0.13	0.27	0.40	0.54	0.80	1.07	1.34
2 × 5	0.080	0.16	0.24	0.32	0.48	0.64	0.80
2 × 10	0.040	0.080	0.12	0.16	0.24	0.32	0.40
5 × 5	0.032	0.064	0.10	0.13	0.19	0.26	0.32
5 × 10	0.016	0.032	0.048	0.064	0.096	0.13	0.16
5 × 20		0.016	0.024	0.032	0.048	0.064	0.080
10 × 10		0.016	0.024	0.032	0.048	0.064	0.080
10 × 15		0.010	0.016	0.021	0.032	0.043	0.054
10 × 20					0.024	0.032	0.040
10 × 25					0.020	0.026	0.032
10 × 30						0.021	0.027
20 × 20and greater see Table 17.2							

[a]Values in the table computed from the equation:

$$\text{emitter capacity, gph} = 0.623\, S_l S_m \times \text{application rate in iph},$$

where S_l = spacing along lateral in feet
S_m = spacing along manifold or main in feet

3. For 10 hours operation per day, assumed

$$q = 0.87 \text{ gph} \times \frac{24}{10} = 2.08 \text{ gph}$$

4. Assuming two emitters per tree, q/emitter = 2.08/2 = 1.04 gph
5. Total trees irrigated = 46/lateral × 17 = 782 trees
6. Total number of emitters = 782 × 2 = 1564
7. For split line operation, use half lateral and half manifold lengths for friction loss and tubing sizes.

Line	*Number Trees*	*Number Emitters*	*Length (ft)*	*Total Flow (gph)*
Half lateral	23	46	235	48
Half manifold	184	368	160	383
Main	782	1564	190	1627

8. From Table 18.1 select Rain Bird EM-L10 emitter to discharge 1.04 gph at 37-ft head as required.
9. The total allowable pressure variation in the lateral and manifold is assumed as 20 percent of average operating head of 37 ft or 7.4 ft. Allowing 50 percent

variation in the lateral (3.7 ft) and 50 percent for the manifold (3.7 ft), the allowable loss for friction is the total less the variation in elevation. Elevation difference along the half lateral is (235 × 0.005) 1.2 ft and along the half manifold is (160 × 0.01) 1.6 ft.

Line	*Allowable Loss (ft)*	*Elevation Difference (ft)*	*Friction Allowance (ft)*
Half lateral	3.7	1.2	2.5
Half manifold	3.7	1.6	2.1

10. Select tubing sizes from Table 18.2

Line	*Total q (gph)*	*Tubing Size (in.)*	*Table 18.2 H_f/100 ft*	*Table 18.3 F*	*Actual[a] H_f (ft)*	*Allowable H_f (ft)*
Half lateral	48	$\frac{1}{2}$	0.9	0.38	0.8	2.5
Half manifold	383	1	2.8	0.42	1.9	2.1
Main	1627	2	1.4	1.0	2.7	7.6[b]

[a]H_f/100 ft × F × length in 100 ft.
[b]Based on a maximum of 4 ft per 100 ft for reasonable energy pumping cost compared with tubing cost.

11. Pressure required at lateral inlet to manifold in Fig. 18.2. Substituting into Eq. 17.2 and assuming $H_r = 0$,

$$H_n = 37 + (0.75 \times 0.8) + (0.6 \times 1.2) = 38.3 \text{ ft}$$

12. Pressure required at mainfold inlet to main at A in Fig. 18.2. Substituting in Eq. 17.2, where H_a is H_n for the lateral,

$$H_{nm} = 38.3 + (0.75 \times 1.9) + (0.6 \times 1.6) = 40.7 \text{ ft}$$

where H_{nm} = pressure at inlet to manifold at the main.

13. The total head to be developed by the pump at the well, including a 1.9 ft (190 × 0.01) elevation drop along the main and assuming the water level in the well is 20 ft below the soil surface, the sand filter head loss is 15 ft, and the friction loss in the suction line, pump, and valves is 8 ft. Substituting in Eq. 17.1 (NPSH = 7 ft),

$$H = 40.7 + 20 + 2.7 + 1.9 + 7 + 15 + 8 = 95.3 \text{ ft (41.3 psi)}$$

The pump should be selected to develop about 44 psi and a flow rate of (1627/60) 30 gpm. (About 5 percent should be added for loss with age.)

18.7 Maintenance

The plugging of emitters caused by physical, chemical, or biological materials is the biggest maintenance problem. Main filters and screens should be cleaned periodically to prevent emitter clogging. The discharge of emitters should be uniform and sufficient to provide adequate moisture to the crop. Some emitters are supposed to be self-cleaning. Secondary filters and screens on manifolds and laterals should be routinely checked. Sulphuric and hydrochloric acids are commonly used to reduce the chemical precipitation where irrigation water has a high pH and high concentrations of calcium and magnesium carbonates (Pair et al., 1983). Accumulations of sediment, bacterial slime, and iron in the tubing should be flushed out as required. Flushing is an attempt to cure rather than prevent plugging and is not always successful. Flushing can be done manually by opening the ends of the lines, using as high a velocity as possible. Where lines are used for more than one season, they should be flushed prior to each season's use and thereafter as frequently as required. Chlorine or copper sulphate are common chemicals to kill bacteria and algae. Chlorine and copper sulphate cause less problems when the nutrient level in the water is low.

18.8 Fertilizer, Pesticides, and Other Chemical Applications Through Trickle Systems

As with sprinkler irrigation systems, soluble chemicals can be applied with the water in trickle systems. Various injection methods and equipment are discussed in Chapter 17. The application of chemicals in the water reduces labor, energy, and equipment costs compared with conventional soil surface spreading methods. Less chemicals are required because they are added at the point of use when they are most effective. Any material that causes precipitation and clogging of the emitters should be avoided. Some fertilizers are made specifically for this application. The fertilizer injection rate depends on the concentration of liquid fertilizer and the desired quantity needed. This concentration can range from 4 to 10 ppm (parts per million) (Howell et al., 1980). Nitrogen injection will usually not significantly increase clogging. With some nitrogen sources, aqua ammonia and anhydrous ammonia will increase pH, which can result in precipitation of insoluble calcium and magnesium carbonates. Phosphorous, potassium, micronutrients, and herbicides have been successfully applied through trickle irrigation systems.

REFERENCES

Davis, D. D. (1976) *Rain Bird Irrigation Systems Design Handbook*, Rain Bird Sprinkler Manufacturing Corporation, Glendora, Calif.

Hagan, R. M., H. R. Haise, and T. W. Edminster (Eds.) (1967). "Irrigation of Agricultural Lands," Monograph No. 11, Am. Soc. Agron., Madison, Wis.

Howell, T. A. et al. (1980) "Design and Operation of Trickle (Drip) Systems," Chapter 16 in Jensen, M. E. (1980) "Design and Operation of Farm Irrigation Systems," Monograph No. 3, ASAE, St. Joseph, Mi.

Howell, T. A., and E. A. Hiler (1974) "Trickle Irrigation Lateral Design," *ASAE Trans.* **17**(5):902–908.

Irrigation Journal (1983) *1983 Irrigation Survey*, Encino, Calif.

Karmeli, D., and J. Keller (1975) *Trickle Irrigation Design*, Rain Bird Sprinkler Manufacturing Corporation, Glendora, Calif.

Nakayama, F. S., D. A. Bucks, and A. T. Clemmens (1979) "Assessing Trickle Emitter Application Uniformity," *ASAE Trans.* **22**(4):816–821.

Northeast Regional Agricultural Engineering Service (NRAES) (1981) *Trickle Irrigation in the Eastern United States*, Cornell Univ., Ithaca, N.Y.

Pair, C. H. et al. (Eds.) (1983) *Irrigation*, 5th ed., The Irrigation Association, Silver Springs, Md.

Solmon, K. (1979) "Manufacturing Variation of Trickle Emitters," *ASAE Trans.* **22**(5):1034–1038, 1043.

United Nations, Food and Agriculture Organization (1973) *Trickle Irrigation*, FAO, Rome, Italy.

Wu, I. P., and D. D. Fangmeier (1974) "Hydraulic Design of Twin-Chamber Trickle Irrigation Laterals," *Arizona Agr. Exp. Sta. Tech. Bull. 216*.

Wu, I. P., and H. M. Gitlin (1974) "Design of Drip Irrigation Lines," *Hawaiian Agr. Exp. Sta. Tech. Bull. 96*.

PROBLEMS

18.1 Determine the trickle irrigation efficiency for a trickle system using emitters with a manufacturer's coefficient of variation of 15 percent. Assume two emitters per plant and an application efficiency of 90 percent.

18.2 Determine the daily *ET* rate for a young orchard where the tree canopies shade only 40 percent of the total land area. The conventional *ET* rate for a mature orchard with 85 percent cover is 0.30 ipd (7.6 mm/d).

18.3 In Example 18.2 determine the lateral and manifold diameters if medium long-path emitters from Table 18.1 were selected so the pressure could be reduced to 25 ft (7.6 m). The total allowable pressure loss is 20 percent of the average emitter pressure.

18.4 In Example 18.2 select an emitter from Table 18.1 that will provide sufficient water for 24-hr operation. Determine the average operating head for the emitters and select tubing sizes for the lateral, manifold, and main to keep the allowable pressure loss within 15 percent of the average emitter pressure. Compute the friction loss for the sizes selected.

18.5 Design a trickle irrigation system from data supplied by your instructor giving the information shown in Example 18.2.

CHAPTER 19

CONSERVATION PLANNING

In broad terms, conservation means the wise use of natural resources. When conservation is applied to soil and water management on farms, it should lead to the maximizing of income from the land over a long period of time. The reduction of erosion and the efficient use and disposal of water are the most important practical aspects.

Cultivated land is a limited resource, amounting to about one acre for each person on earth. The net loss of rural land in the United States is about three million acres each year. Forests, wildlife, and minerals are other natural resources of concern to conservationists. All natural resources have recreational values. Recreation in recent years has received more attention but primarily from governmental agencies.

Conservation planning of soil, water, and vegetation involves private land in farms, cities, and suburban developments. Large-scale planning encompasses the entire area of a watershed. City, state, and federal lands may be included in the watershed. Changing land-use patterns may severely alter runoff characteristics. Watershed planning should take into account the potential land-use changes associated with urban and industrial development. This chapter deals with some of the legal aspects and other requirements for good planning, the necessary prelude to sound action programs.

On individual farms, soils, crops, climate, water, type of farming enterprise, economic capabilities of the farmer, and other factors are important aspects of farm planning. Such planning is essential for the success of a national conservation program. In carrying out conservation practices on farms, the application of engineering practices as described in previous chapters is an integral part of a sound conservation program.

The intended land use will largely dictate the type of planning to be carried out. The major land uses are agricultural, woodland, urban, and recreational. Each of these land uses is planned with different criteria in mind. Agricultural uses are either pasture, range, or cultivated land. In arid areas one of the major considerations is the capability of the land to produce water.

19.1 Economics of Conservation

Of the several soil and water conservation problems listed in Chapter 1, soil erosion by wind and water is often considered the most critical. Failure to control erosion results in the loss of a valuable irreplaceable resource, which impacts on the environment and the well-being of society as a whole. Impairment of water quality is one of the most serious off-farm consequences of erosion. The long-term effects of erosion result in the destruction and eventual abandonment of the land and could result in the loss of civilization itself.

Failure to drain, irrigate, or conserve soil moisture usually reduces crop production, but the land resource is generally not degraded or permanently affected. Water resource development is likewise a water problem, which is closely associated with irrigation. Flooding is a naturally occurring problem, which would take place regardless of man's activities. Failure to control flooding may cause damage to the land or, in some instances, soil deposition may be a benefit. Damages to property and man-made structures can often be reduced by proper construction or by avoiding the occupation of the floodplain. As an indication of the concern of conservationists for these problems, the Soil Conservation Society of America (1972) developed a general policy on the economics of soil and water resource conservation.

Economists define conservation as the maintenance of future production per unit area from a given input of labor and capital. A farmer must decide whether to make expenditures for nonproduction goods, machinery, and buildings or for conservation practices such as drainage, irrigation, or terraces. The financial status of the farmer, general economic conditions, and extent to which he is convinced of the need for conservation will largely determine the degree to which he is able to make specific improvements and land-use adjustments. Where farm lease arrangements do not make adjustments for the length of time between the adoption of conservation practices and the realization of returns, and where off-site benefits must come from on-the-farm practices, the individual may not be able to justify expenditures under such conditions. Farm tenancy with absentee owners is generally not conducive to good soil conservation.

The long-term net farm income and intangible benefits to man and the environment are the best basis for evaluating conservation practices. The results of a study of high- and low-conservation farms are shown in Table 19.1. The data were obtained in McLean County, Illinois, but the results are typical of other studies. The lower initial income in the first year for the high-conservation farms is most probably due to the greater investment in conservation practices. The 5- and 10-year averages show that the high-conservation farms have 21 to 31 percent higher incomes than the low-conservation farms. This difference is accounted for mostly by increased yields. However, conservation costs were higher on the high-conservation farms than on the low. In addition to higher net income, the conservation farms in this study have a greater expenditure for conservation improvements, have more land in legumes and grasses, produce higher crop yields, grow better quality crops, and have higher livestock pro-

Table 19.1. Effect of Conservation Farming on Net Income[a]

Time After Conservation Practices Were Started	*Net Income in Dollars per Acre*		*Change in Net Income (percent)*
	Average of 20 Low-Conservation Farms	*Average of 20 High-Conservation Farms*	
First year	$16	$14	−12
First 5 years, average	19	25	+31
Second 5 years, average	49	60	+21
First 10 years, average	35	44	+25

[a]Based on data from Sauer (1949), which were increased by a factor of 2.5 to reflect 1980 prices.

duction. Conservation practices normally increase the net income within 1 to 4 years after establishment.

Other studies have shown that erosion control practices do not increase the net income of the farmer. In Missouri, Ervin and Washburn (1981) showed that minimum tillage always gave the highest net income. With continuous corn and corn-soybean rotations, highest annualized net income for four soils were obtained for up and downhill farming 75 percent of the time and for contouring 25 percent of the time. Tax incentives and off-site damages of erosion were not considered. Average reductions in corn yields per inch of soil depth removed were 1.3 to 2.4 bu/a. In general, strip cropping and terracing costs significantly outweighed the benefits of these practices. Studies by Boggess et al. (1979) in the loess soils of western Iowa found that farmer's cash income was lowered 20 percent from a base level to obtain a soil loss of 10 t/a and by 28 percent for a 5 t/a limit. The reduction in crop yield for this deep soil is less affected by erosion than the soils with a shallow topsoil.

As shown in Table 19.2, drainage of poorly drained soils in a humid area is a good investment. Compared to undrained land, surface drainage increased corn yields 20 bu/a, tile drainage without surface drainage 44 bu/a, and combination tile plus surface drainage 54 bu/a. Because of increased costs, the benefit/cost ratios were 2.2, 2.0, and 1.7, decreasing in the same order as the intensity of drainage increased. Interest rates represent a major portion of the drainage costs.

The economics of irrigation varies greatly depending on the reliability, seasonal distribution, and amount of precipitation. In the arid West, where rainfall is less than 15 in. per year, the benefits of irrigation are self-evident. Crop production is generally not possible without it. In humid areas where rainfall exceeds about 30 in. per year, irrigation is usually economical for vegetable and high-value crops, but questionable for field crops, except on low water-holding capacity soils. Where rainfall is from 15 to 30 in. per year, irrigation is desirable and usually economical, but benefits and cost must be carefully considered. Dry land farming is often an alternative, where water is not available or costly. Irrigated acreage in the United States has been gradually increasing especially

Table 19.2. Annual Benefit-Cost Analysis for Drainage of Toledo Silty Clay Soil in Northern Ohio

	Undrained	*Surface Drainage Only*	*Tile Only*	*Combination Tile Plus Surface Drainage*
Average corn yield* (bushels per acre)	70	90	114	124
Benefits[a]		$60	$132	$162
Costs[b]		$27	$ 67	$ 94
Benefits/costs ratio		2.2	2.0	1.7

[a]Benefits are based on corn at $3 per bushel for the increase in yields over those from undrained plots.

[b]Cost estimates do not include fertilizer and other production costs. All costs were computed using a depreciation of 5 percent per year for the tile (20-year economic life), interest on the average investment at 14 percent, and annual maintenance cost of 0.2 percent of investment cost per acre for tile and 4 percent of investment cost for surface drains. Initial investment costs per acre were assumed to be $150 for surface, $550 for tile, and $700 for combination drainage.

*Average for 13 years.

Source: Schwab et al. (1981).

since 1950. It will probably continue to do so until it is limited by the availability of water, land, and/or high energy costs.

FARM PLANNING

19.2 Basic Data for Farm Planning

Basic data required to develop a farm plan include soil characteristics, topographic data, climatic information, and present farm practices. The past history of the farm is also desirable. Farm planners who have lived in the community for several years are usually sufficiently informed concerning the climatic conditions that prevail. Most of the remaining data must be obtained in the field or by consultation with the farmer. Such problem areas as poorly drained fields can be easily pointed out by persons who have cultivated the land for several years. The collection of field data is generally known as the soil conservation survey. It should include soil series and type, degree of erosion, length and degree of slope, and present field boundaries. This survey of the farm should be made in considerable detail, and the information may be recorded on a suitable soil map. Aerial photographs make very satisfactory base maps (Chapter 4). As shown in Fig. 19.1, the three items generally included on the map are soil type, slope class, and degree of erosion.

The classification of the soil can best be done by experienced soil surveyors. Conservation survey maps for use in farm planning should be made to a sufficient scale to permit adequate delineation of soil-type boundaries, degree of erosion, and topographic characteristics.

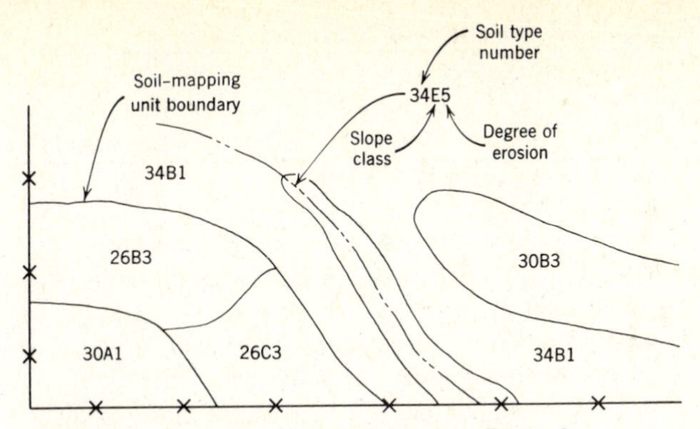

Figure 19.1. Soil-mapping units and method of recording soil type, erosion, and land slope.

The degree of erosion is designated by the present depth of topsoil as compared with the depth of virgin topsoil. On the soil map it is convenient to express the degree of erosion by symbols, such as "1" for 0 to 25 percent of the topsoil removed, etc. Erosion on cultivated land is influenced principally by the cropping history of the field, the length and degree of slope, and extent of such conservation practices as contouring and terracing.

On the soil map the slope class is represented by a capital letter indicating a range of slope, that is A, nearly level; B, gently sloping, and so forth. An area may have the same soil type and degree of erosion but would be designated separately due to the difference in slope (see soil 26 in Fig. 19.1).

19.3 Land-Capability Classification

The purpose of collecting basic data is to classify the land according to its capability for crop production. As shown in Table 19.3, each of the eight land-capability classes is determined by the general degree of limitations and hazards. The first four classes are suitable for cultivation, but the last four are not. Except for class I land, each capability class may be designated with a subclass, such as II(*e*), III(*c*), etc., as described in Table 19.3.

The decreasing order of land classification is the soil-mapping unit, the capability unit, the capability subclass, and the capability class. The land-capability unit includes all land requiring the same kind of conservation treatment and the same kind of management. The soil-mapping unit is the smallest subdivision mapped by the soil conservation survey (see Fig. 19.1). However, the only category designated in the farm plan is the land-capability class. In determining the land-capability class, the soil type, the slope, the degree of erosion, the extent of drainage, the presence of rocks, stones, and other impediments to cultivation, the present productivity of the soil, the water-holding capacity, and the amount and distribution of rainfall are considered. It is possible by drainage, the removal of trees and rocks, and other practices to change the land-capability class to a higher level, that is, from class III to class II. Several classes of land may be found on the same farm. In Fig. 19.2 all classes are shown.

Table 19.3. Land-Capability Classification[a]

Land-Capability Class	*General Description and Limitations*
Suited for Cultivation	
I	Few limitations that restrict its use. No subclasses.
II	Some limitations that reduce the choice of plants or require moderate conservation practices.
III	Severe limitations that reduce the choice of plants or require special conservation practices or both.
IV	Very severe limitations that restrict the choice of plants or require very careful management or both.
Not Suited for Cultivation (except by costly reclamation)	
V	Little or no erosion hazard, but has other limitations impractical to remove that limit its use largely to pasture, range, woodland, or wildlife food and cover.
VI	Severe limitations that make it generally unsuited to cultivation and limit its use largely to pasture, range, woodland, or wildlife and cover.
VII	Very severe limitations that make it unsuited to cultivation and that restrict its use largely to grazing, woodland, or wildlife.
VIII	Limitations that preclude its use for commercial plant production and restrict its use to recreation, wildlife, or water supply or to esthetic purposes.

Note. Except for class I land, the following subclasses are recognized where the dominant limitations for agricultural use are the result of soil or climate: (*e*) *Erosion*, based on susceptibility to erosion or past damage; (*w*) *excess water*, based on poor soil drainage, wetness, high water table, or overflow; (*s*) *soil limitations within the rooting zone*, based on shallowness, stones, low moisture-holding capacity, low fertility, salinity, or sodium; and (*c*) *climate*, based on temperature extremes or lack of moisture.

[a]From Klingebiel and Montgomery (1966).

19.4 Farm Plan

A farm plan is the organized and systematic schedule of all soil and water management practices integrated with a well-balanced farming program. In brief, the procedure for developing a farm plan consists of: (1) the collection of basic data, (2) the land-capability class determination and field layout, and (3) a soil and water management program that will support a well-balanced farming program. Much of the information necessary for farm planning is shown in Fig. 19.3. On the left is the farm before planning and in the center the land is classified according to its capability class (Roman numerals). Frequently, these land classes are indicated by color. The final map of the conservation farm plan as shown on the right in Fig. 19.3 may include field designation, new field boundaries, type of crop and acreage, and supporting conservation practices, such as waterways and terraces. Since such a map shows the overall picture of the physical

LAND CAPABILITY CLASSES			
SUITABLE FOR CULTIVATION		NO CULTIVATION – PASTURE, HAY, WOODLAND, AND WILDLIFE	
I	Requires good soil management practices only	V	No restrictions in use
II	Moderate conservation practices necessary	VI	Moderate restrictions in use
III	Intensive conservation practices necessary	VII	Severe restrictions in use
IV	Perennial vegetation – infrequent cultivation	VIII	Best suited for wildlife and recreation

Figure 19.2. Land-capability classes. (*Courtesy U.S. Soil Conservation Service.*)

changes and the principal features of the farm, it becomes an important part of the farm plan. Note that, if necessary, several closely associated classes of land may be included in one field to obtain a desirable size for efficient operation. For example, field 9 includes both class II and class III land.

The procedure for determining soil loss for given conditions, as described in Chapter 6, can serve as a guide in farm planning. By selecting suitable rotations, fertility practices, and conservation practices soil loss can be reduced to a permissible value.

The farm plan should be developed in considerable detail. It is desirable to list in tabular form schedules for the application of conservation practices in each field, the planned production for each crop, feed requirements for planned livestock, and the distribution of pasture-carrying capacity for each month in the growing season.

Standard forms to meet local conditions should be developed in each area where soil, climate, crops, type of farming enterprise, and other conditions vary. Any procedure or method of presentation that is clear and concise is satisfactory. Once the complete plan has been developed, it is a ready reference for the future

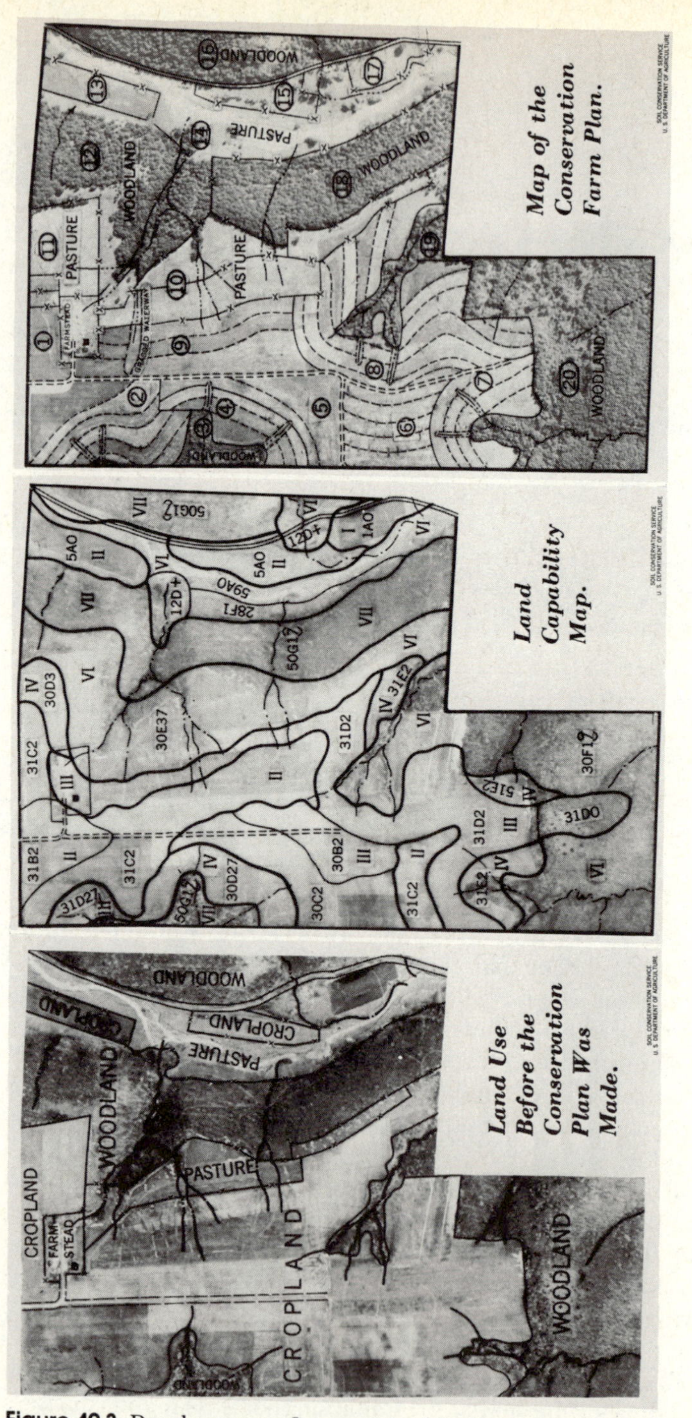

Figure 19.3. Development of a conservation farm plan. *(Courtesy U.S. Soil Conservation Service.)*

planning of all farm operations. The plan should be flexible so that as conditions change and new techniques and crops are developed they may be incorporated into the revised plan. The farmer may request technical assistance from soil- and water-conservation districts, cooperating state and federal agencies, and professional people engaged in private practice.

WATERSHED PLANNING

Because watersheds are natural hydrologic units, overall programs for flood control, erosion control, irrigation, and drainage can be accomplished more advantageously by considering watershed areas rather than geographical subdivisions, such as a farm or township. Watershed planning is essential for flood control. Examples of such projects are the Miami Conservancy District in Ohio, the Trinity Flood Control Project in Texas, and the Little Sioux Flood Control Project in Iowa. More emphasis has been given in the watershed approach in recent years, particularly by federal agencies.

Enabling legislation has been passed by most states to provide for the planning and construction of projects involving several landowners. Where such cooperative projects are undertaken, individuals benefited should contribute their equitable share of the cost. Because many of these projects have public benefits, state and federal funds are often provided for this purpose.

19.5 Drainage Enterprises

Many states have laws that provide for the organization of mutual drainage enterprises, also called drainage associations. To establish such an enterprise the landowners involved must be fully in accord with the plan of operation and with the apportionment of the cost. After the agreement has been drawn up and signed, it must be properly recorded in the drainage record of the county or other political subdivision. The local court may be asked to name the district officials, sometimes called commissioners, who are responsible for the functioning of the district, or they may be named in the agreement. The principal advantage of the mutual district is that less time is required to establish an organization and the costs are held to a minimum. Because it may be difficult for several landowners to come to an agreement, particularly on the division of the costs, such districts are difficult to organize where the number of landowners is large or where considerable area is involved. Another disadvantage is that it cannot assess taxes. However, there are a large number of these small enterprises, and much drainage has been accomplished in this manner.

An organized drainage enterprise, often called a county ditch or a drainage district, is a local unit of government established under the state laws for the purpose of constructing and maintaining satisfactory outlets for the removal of excess surface and subsurface water. It is different from a mutual enterprise in that minority landowners can be compelled to go along with the project. The organizational procedures for districts are similar in most states. The first step is a petition signed by the prescribed number of landowners; a preliminary report giving location, benefits, costs, and other details is filed with the proper officials, and then one or more hearings are held before either the district is established or the petition is dismissed. The distribution of assessments acceptable to all landowners is the most troublesome and difficult aspect of district administration. The method for making assessments should be simple, equitable, and above all easily explained and understood. Damages to property or land due to the improvement are considered costs to the district and are handled separately from benefits. Financing is normally done by issuing bonds that are paid off by levying taxes on the assessed land.

19.6 Irrigation Enterprises

Private enterprises for developing water supplies and irrigation delivery systems include partnerships, commercial companies, and various types of cooperative enterprises, which are locally designated by a variety of names. The two principal types of cooperative enterprises are the mutual association, which is an unincorporated organization, and the mutual irrigation company, which is incorporated. The mutual company is a nonprofit organization established for the purpose of delivering water to its members only. The amount of water delivered is in proportion to the stock owned. It is the dominant type of organization in many of the irrigated areas of the West. Commercial companies may furnish water on an annual rental basis, sell the water right along with an additional charge for the water, or sell the water right that carries with it a perpetual interest in the company. Those of the last type may eventually become irrigation companies. Partnerships and mutual associations are organized and function similarly to the mutual drainage enterprises, previously discussed.

The irrigation district is established under state laws and is a political subdivision of the state. The district is sometimes referred to as a quasi-public enterprise. In practically all respects it is organized, financed, and operated similarly to the drainage districts, previously discussed. It can issue bonds, levy and collect taxes, sue and be sued in court, and take the necessary action to provide irrigation water. Almost without exception, it has the authority to provide for drainage and in some states it may develop electric power. Many of the large irrigation systems in the West are developed in this manner.

The first major use of federal funds for irrigation development was authorized by the Reclamation Act of 1902. These funds were provided without interest for the construction of large irrigation projects with a repayment period of normally 40 years. After a substantial part of the construction costs are paid, the water users may assume control of the project. Either an irrigation district or a mutual irrigation company is created to conduct the administration affairs of the project with the irrigators and the government. Many of these Bureau of Reclamation projects are multiple-purpose. Examples of these major reservoir projects include Hoover, Grand Coulee, and Fort Peck, as well as many smaller projects.

19.7 Conservancy Districts

Conservancy districts are those enterprises generally organized for the purpose of soil conservation and flood control. They are often organized in the same general manner as drainage districts. In some states flood control works constructed by the U.S. Corps of Engineers or the U.S. Department of Agriculture under legislation, such as Public Law 566, are sponsored by and function under a conservancy district established for that purpose. The district may arrange for the necessary local support and cooperation among landowners and provide the necessary maintenance after the project is completed. The Miami Conservancy District in Ohio is the outstanding example of a major privately financed project organized under the conservancy laws. Conservancy districts are known by a variety of names in different states, such as watershed districts, water-management districts, and other special-purpose titles. Conservancy districts are authorized in a number of states, including Colorado, Ohio, Oklahoma, Indiana, Florida, Iowa, Minnesota, Kansas, and Texas.

19.8 Soil and Water Conservation Districts

As a condition to receiving benefits under the Soil Conservation and Domestic Allotment Act, passed by Congress in 1935, the states were required to enact suitable laws providing for the establishment of soil-conservation districts. The first district was organized in 1937, and all states have now enacted district laws.

This legislation has been patterned largely after the standard state soil conservation districts law prepared by the U.S. Soil Conservation Service (1936). Most districts are called soil and water conservation districts. After these districts are established in local communities, they may request technical assistance from agencies such as the Soil Conservation Service, State Cooperative Extension Service, and county officials for carrying out erosion control and other land-use management activities.

Soil and water conservation districts differ considerably from drainage, irrigation, and conservancy districts. Their boundaries are usually the same as a political subdivision, such as a county; operating funds are obtained from state or local appropriations; and in most states they cannot levy taxes or issue bonds. The election of officials is by all voters rather than being limited to only landowners. Although the powers of these districts have been broadened in many states, they have little authority and tend to be voluntary in nature. The development of conservation farm plans and technical assistance in adopting practices have been the principal results of the districts' program.

REFERENCES

Boggess, W., J. McGrann, M. Boehlje, and E. O. Heady (1979) "Farm-Level Impacts of Alternative Soil Loss Control Policies," *J. Soil and Water Conservation* **34**(4):177–183.

Ervin, D. E., and R. A. Washburn (1981) "Profitability of Conservation Practices in Missouri," *J. Soil and Water Conservation* **36**(2):107–111.

Hudson, N. (1981) *Soil Conservation*, 2nd ed., Cornell Univ. Press, Ithaca, N.Y.

Klingebiel, A. A., and P. H. Montgomery (1966) "Land-Capability Classification," U.S. Soil Conservation Service, *Agr. Handb. 210*.

Sauer, E. L. (1949) "Economics of Soil Conservation," *Agr. Engin.* **30**:226–228.

Schwab, G. O., B. H. Nolte, and M. L. Palmer (1981) "Drainage—What's it Worth on Corn Land?" Ohio Ext. Service, Agr. Eng. Soil and Water No. 23. (Mimeographed)

Soil Conservation Society of America (1972) "Economics of Soil and Water Resource Conservation," *J. Soil and Water Conservation* **27**(1):44–46.

U.S. Soil Conservation Service (1936) *A Standard State Soil Conservation Districts Law*, U.S. Government Printing Office, Washington, D.C.

U.S. Soil Conservation Service (1965) "What is a Farm Conservation Plan?" Pam. 629, U.S. Government Printing Office, Washington, D.C.

GLOSSARY

Definitions of words and phrases used in the text are given below, some of which are unique to this field of study.

Abney level. A small hand level for leveling or measuring slope in percent or degrees.

Alfalfa valve. A screw-type valve placed on the end of a pipe to regulate the flow of (irrigation) water.

Alidade. A combined peep-sight or telescopic-sight and straight-edged instrument for surveying with a plane table.

Appropriation doctrine. See Water rights.

Aquifer Underground water-bearing geologic formation or structure.

Available Soil Moisture Capacity. Amount of available moisture held by soil for use by plants, usually the moisture held between field capacity and the wilting point, also called water holding capacity.

Backfurrow. A small ridge formed when soil plowed out of two adjacent furrows is thrown together.

Backsight. A level rod reading on a point of known elevation, that is always added to the elevation to obtain the height of a surveying instrument, also called a plus sight.

Backslope. The land area on the downhill side or opposite the water side of a terrace or earth embankment.

Base height. The vertical distance from the sight bar or laser detector on a plow or trenching machine to the bottom of the digging mechanism (grade line).

Base line. A reference line taken as a base for measurement, also a parallel of latitude for public land surveys.

Basin irrigation. A method of flood irrigation using level plots or fields surrounded by dikes, also called level border irrigation.

Beaman arc. A special vertical scale on a surveying instrument for computing horizontal and vertical distances, usually on a transit or alidade.

Bedding angle. The acute angle of a V-groove in the bottom of a trench in which pipe drains are laid.

Bed load. Coarse sediment moving on or almost in continuous contact with the bottom of a flowing stream, similar to surface creep in wind erosion.

Bedding. A method of surface drainage consisting of narrow-width plow lands in which deadfurrows are parallel to the prevailing land slope and serve as field ditches, also called crowning or ridging.

Bench mark. A permanently established reference point, the elevation of which is assumed or known.

Bench terrace. A level strip of land built in a stair-step pattern on very sloping land having a steep or vertical embankment from one step to the next.

Boom-type sprinkler system. A mechanical system using two large rotating arms usually mounted on a movable trailer and having several sprinklers of varying sizes.

Border irrigation. A method of surface irrigation by flooding between small border dikes.

Break tape. A term to describe a procedure for measuring horizontal distances with a tape on steep sloping land.

Broadbase terrace. An earth embankment constructed across the slope, wide enough to allow both the embankment and the channel to be farmed.

Buffer strip cropping. The practice of growing different crops in consecutive strips across the slope with grass or other erosion-resisting vegetation grown between or below cultivated strips.

Center-pivot sprinkler system (selfpropelled). An irrigation system with sprinklers mounted on a long elevated pipe which rotates about a fixed point in the field.

Centrifugal pump. A mechanical device that develops water pressure by centrifugal force from an impeller.

Chute spillway. A steep stable channel for conveying water to a lower level without erosion.

Claypan. A dense and heavy soil horizon underlying the upper part of the soil. It is hard when dry and plastic or stiff when wet.

Conservancy district. A legally organized enterprise for the purpose of soil conservation and flood control.

Conservation bench terrace. An earth embankment with a level bench for the purpose of conserving soil moisture and reducing soil erosion.

Conservation practice factor. The ratio of soil loss for a given practice to that for up and down the slope farming, as used in the soil loss equation.

Consumptive use. See evapotranspiration.

Contouring. A conservation practice in which all farm operations are performed on the contour.

Contour strip cropping. A conservation practice in which all farm operations are on the contour and different crops are grown in consecutive strips to provide barriers for controlling soil erosion.

Convective storm. Rainfall caused by condensation of heated air that moves upward, being cooled both by the surrounding air and by expansion.

Creep distance. The longitudinal length along the outer surface of a pipe or conduit within an earth dam.

Critical runoff. The maximum runoff rate which is economically feasible to provide for in control channels

Cropping management factor. The ratio of soil loss for a given condition to that from cultivated continuous fallow land, as used in the soil loss equation.

Curve number. An arbitrary number varying from 0 to 100 that is related to the quantity of runoff from a watershed.

Cut-fill ratio. The ratio of cut volume to fill volume in relation to land grading and earthwork operations.

Cutslope. The uphill side slope of a broadbase terrace channel.

Deadfurrow. The double furrow left between two areas or lands due to plowing in opposite directions.

Differential leveling. A method of leveling in which the difference in elevation between two or more points is determined.

Diversion. A channel constructed across the slope for the purpose of intercepting surface runoff.

Division box. An irrigation structure for dividing the flow into two or more streams.

Drainage area. The land area from which runoff or drainage water is collected and delivered to an outlet.

Drainage coefficient. The rate of removal of water expressed as the depth to be removed in one day (24 hours).

Drain plow. A machine with a large vertical blade and point for installing subsurface drains.

Drawdown. The lowering of the water surface or water table resulting from the withdrawal of water.

Drop spillway. An overfall structure in which the water drops over a vertical wall onto an apron at a lower elevation.

Dumpy level. A surveying level in which the telescope is rigidly attached to the instrument.

Electrical conductivity. A measure of the salt content of irrigation or drainage water or solution extract of a soil.

Emergency spillway. A water conveying structure for a dam to carry runoff exceeding a given design flood, also called a flood spillway.

Emission uniformity. A measure of the uniformity of discharge of a trickle system emitter.

Emitter. A device that controls the discharge of water from a trickle irrigation system.

Engineer's scale. Usually a 12-inch boxwood rule divided into 10, 20, 30, 40, 50, and 60 parts to the inch.

Envelope filters. Granular material or geotextile sheet fabrics which surround subsurface pipe drains to prevent soil inflow.

Erosion. The wearing away of the land surface by running water, wave action, glacial scour, wind, and other physical processes.

Erosion pavement. Soil surface layer as a result of an accumulation of rock fragments at the surface caused primarily by the removal of fine material by water or wind.

Erosivity index. A measure of rainfall and runoff erosion by geographical areas, a factor in the universal soil loss equation.

Evapotranspiration. The combined loss of water from the soil surface by evaporation and from plant surfaces by transpiration.

Evapotranspiration coefficient. A measure of evapotranspiration as affected by type of crop and temperature, a factor in the Blaney-Criddle equation.

Eyepiece. The small end of a telescope where the eye is placed in order to read the level rod.

Fallow. The practice of allowing cropland to lie idle, either tilled or untilled, during the whole year or the greater part of the growing season.

Farm planning. An organized and systematic scheduling of all soil and water management practices integrated with a well-balanced farming program.

Farm reservoir. A multiple-use conservation structure for storing water for irrigation, livestock, spray water, fish production, recreation, fire protection, or any combination of uses, also called ponds, tanks, or lakes.

Field capacity. Water content of a soil after it has been saturated and allowed to drain freely, expressed as a percentage of its oven-dry weight or volume.

Field ditch. A constructed open channel within a field for either drainage or irrigation.

Field strip cropping. A conservation practice in which grass or close-growing crops are grown alternately with cultivated crops in parallel strips laid out normally to the prevailing wind or approximately on the contour.

Filter drain. A pipe drain surrounded with an envelope filter.

Flood. An overflow or inundation from runoff in a river or other body of water, which causes or threatens to cause damage.

Flood spillway. A water conveying structure for a dam to carry runoff exceeding a given design flood, also called an emergency spillway.

Flood storage depth. The vertical distance between the normal water level and the bottom of the flood spillway of a dam.

Flume. An open conduit for conveying water down a slope or across an obstruction, also an elevated channel or a device for measuring the flow of water in an open channel.

Foresight. The level rod reading on a turning point or on some other point of unknown elevation.

Freeboard. The vertical distance from the top of the embankment to the maximum expected water level in the reservoir or channel.

Frisco rod. A type of telescoping surveying rod, usually consisting of three $4\frac{1}{2}$-ft. sections.

Frontslope. The upslope or channel side of a terrace ridge.

Frontal storm. A type of weather pattern causing precipitation along the boundary of cold and warm air masses.

Furrow irrigation. A method of surface irrigation in which water is distributed and infiltrated from small ditches or furrows.

Gated pipe. A portable irrigation pipe with small outlet gates for distributing water to corrugations or furrows.

Geotextile. A fabric or synthetic material placed at the boundary between granular material or an impermeable surface and the soil to enhance water movement and retard sediment inflow.

Grade. The degree of slope expressed as a percent and equal numerically to the rise or fall in a horizontal distance of 100 feet.

Grade breaker. A special mechanical device to change the grade line established by an earth-moving machine from that made by a laser beam.

Graded terrace. A terrace having a constant or variable grade along its length.

Grade line. A sloping line along an established direction, such as along a ditch, terrace, or land surface.

Gully erosion. A type of soil removal by water which produces well-defined channels that usually cannot be obliterated by normal tillage.

Guard stake. A flag, lath, or stake placed beside a hub stake (elevation) for location and identification.

Hardpan. A hardened or cemented soil layer in the lower A or in the B horizon.

Height of instrument. The elevation of the line of sight of a surveying instrument.

Herbicide A chemical substance for killing plants, especially weeds.

Horizontal interval. The horizontal distance between consecutive terraces.

Hub stake. A short stake *driven almost flush* with the ground surface and used for an elevation reference in surveying.

Hydraulic conductivity. The rate at which water will move through soil under a unit gradient.

Hydraulic radius. The cross-sectional area of a channel or conduit divided by the wetted perimeter.

Hydrograph. A graphical or tabular representation of the flow rate of a stream with respect to time.

Hydrologic cycle. The movement of vapor or water from the atmosphere to the earth, over or through the soil, returning to the atmosphere through various phases and processes.

Hydrologic soil group. A special grouping of soil types into four major categories for the purpose of estimating runoff.

Impeller meter. A rotating mechanical device for measuring water flow rates in a pipe or open channel.

Infiltration (intake rate). The downward rate that water can enter the soil, usually expressed in inches per hour.

Inversion. An increase of temperature with altitude.

Irrigation district. A cooperative, self-governing corporation set up as a subdivision of the state and organized primarily to provide irrigation water to several farms in a local area.

Irrigation requirement. Quantity of water, exclusive of effective precipitation, that is required for crop production.

Key terrace. A staked terrace line that is selected as a reference for laying out other terraces.

Land capability. The suitability of land for cropping, grazing, woodland, wildlife, or other use without damage, usually designated as classes I to VIII.

Land grading. The operation of shaping the land surface, also called land forming, land leveling, or land shaping.

Land plane. A machine for land smoothing operations.

Land smoothing. The process of removing minor differences in elevation and depressions without changing the general contours of the land (the finishing operation after land grading).

Laser beam. A collimated beam of light from a laser source.

Laser grade control. A method for controlling the grade of an earth-moving machine from a laser plane or beam.

Laser plane. A rotating laser beam that produces a level or sloping plane of reference for grade control.

Lathe box (spile). A device placed through a ditch bank for transferring irrigation water from the ditch to the field.

Leaching fraction. The ratio of the water passing through the soil for removal of soluble substances to the irrigation water depth, also called leaching requirement.

Lenker rod. A surveying rod with an inverted movable scale from which elevations can be read directly.

Level. A surveying instrument for establishing a horizontal line.

Level rod. A surveying rod, normally graduated in tenths and hundredths of a foot, also called a Philadelphia or Frisco rod.

Level terrace. A terrace with zero grade along its channel.

Manifold. The submain of a trickle irrigation system to which laterals with emitters are attached.

Metes and bounds. A method of public land survey, in which land is described by direction and distance.

Mist irrigation. A method of sprinkler irrigation in which the water is distributed by very small droplets.

Net positive suction head. Friction losses within a centrifugal pump and the velocity head.

Nonpoint pollution. Water pollution from diverse sources, such as a field or watershed.

Orifice. An opening with a closed perimeter and of regular form through which water flows.

Orographic storm. Precipitation caused by air masses which rise to high elevations, such as over mountain ranges.

Parallel field ditch. Small surface channel, larger than a deadfurrow, in a parallel surface drainage system.

Parallel lateral ditch. A surface channel deeper than a field ditch in a parallel surface drainage system.

Pesticide. Chemical agents, such as insecticides, herbicides, fungicides, etc. for control of specific organisms.

Philadelphia rod. A type of level rod for surveying.

Phreatophyte. A nonbeneficial water-loving plant, which derives its water from subsurface sources.

Pipe drain. A buried drain made from clay, concrete, plastic, and other materials, also called a subsurface or tile drain.

Pipe riser. A vertical conduit for carrying water from the ground surface to a horizontal drain below.

Pipe spillway. A structure for carrying water through an earth embankment, also a culvert.

Point row. A crop row which forms an acute angle with another row or with the field boundary.

Prime farm land. The best suited land for producing food, feed, forage, fiber, and oilseed crops with the least damage to the soil.

Profile leveling. A method of surveying to secure the elevation of a series of points located along a line.

Pump efficiency. The ratio of water power produced by the pump based on the total head to the power input to the pump.

Raindrop erosion. Soil splash resulting from the impact of raindrops on the soil.

Rainfall factor. A relative value for comparing rainfall intensities at two geographical locations.

Rainfall intensity. Rate at which rain falls, usually expressed as inches per hour.

Radius of influence. Maximum distance from a well at which the natural water table or piezometric surface is affected by pumping.

Random field ditch. A constructed open channel in a field usually connecting low areas to an outlet drain.

Range lines. True meridan lines which are the east and west boundaries of townships in the rectangular system of public land survey.

Range pole. A sight stake for marking points to be seen from a long distance.

Ratio of error. The difference of two measurements between the same points divided by the average distance.

Rational runoff equation. A simple equation for estimating the peak runoff rate from a watershed.

Relief pipe. A vertical riser from a pipe drain to the ground surface to relieve hydrostatic pressure.

Retardance class. A relative resistance to flow in a vegetated channel depending on the length and condition of the vegetation and the hydraulics of the channel.

Return period. The time in years which a given event can be equalled or exceeded, once on the average, also called recurrence interval.

Rill. A small, but well-defined channel which can be obliterated by normal tillage.

Riparian doctrine. A water right principle which gives the landowner the right to use and control water by virtue of ownership of the stream bank.

Rotating-boom sprinkler system. A system with two rotating arms having a large nozzle and several smaller nozzles.

Roughness coefficient. A constant in the Manning velocity equation representing channel roughness.

Runoff. The portion of precipitation on a drainage area that is discharged from the area into stream channels.

Runoff coefficient. The ratio of the runoff rate to the rainfall intensity in the rational runoff equation.

Saltation. Soil movement by wind or water in which particles skip or bounce along the soil surface or stream bed.

Sheet erosion. Removal of a fairly uniform layer of soil from the land surface by runoff.

Shelterbelt. A long barrier of living trees and shrubs for protection of farmland from wind erosion.

Side-roll sprinkler system. A type of system in which sprinkler laterals are moved on wheels whose axles are the lateral pipe.

Siphon tube. A short curved pipe to remove water over the side of an irrigation ditch.

Slope. The rate of rise or fall from the horizontal, expressed in percent or degrees, also called grade or gradient.

Slope factor. A relative number for evaluating the land slope in the soil loss equation.

Slope length factor. A relative number for evaluating the length of slope in the soil loss equation.

Slope stake. A survey stake to mark the edge of an earth dam or ditch.

Sodding. The practice of establishing grass sod on a bare soil surface.

Sodium-adsorption-ratio. The proportion of sodium ions in the soil water extract in relation to the calcium and magnesium ions.

Soil and water conservation district. About a county-size area organized under state laws to conserve natural resources and to promote better land use.

Soil-erodibility factor. A numerical value by soil type for estimating soil erodibility in the soil loss equation, also called erosivity index.

Soil permeability. A soil characteristic indicating the rate at which water moves through the soil.

Soil series. A group of soils having horizons similar in differentiating characteristics in the soil profile, except for the texture of the surface soil.

Soil structure. The arrangement of individual soil particles into larger crumbs, granules, or aggregates.

Soil texture. A term related to the size of primary mineral particles in the soil, such as sand, silt, and clay.

Soil type. A subdivision of a soil series based on surface soil texture.

Solid-set sprinkler system. A system which covers the entire field without moving any pipe and usually remains there until the end of the irrigation season.

Specific yield. The fraction of a unit volume of an aquifer which drains by gravity.

Spile. A device placed through a ditch bank for transferring irrigation water from the ditch to the field.

Spillway. A channel or structure for conveying runoff through or around an earth embankment.

Splash erosion. Soil movement caused by impact of raindrops on the soil.

Sprinkler head. A device for distributing water under pressure, consisting of a body, rotating mechanism, and nozzles.

Sprinkler riser. A vertical pipe connecting the lateral pipe to the sprinkler head.

Stadia. A method of measuring distance by reading the rod interval between stadia hairs in a telescope.

Steel tape. A narrow strip of steel marked for measuring distances, also called a chain.

Stick. A command by the rear tapeman, when measuring with a steel tape, to have the head tapeman place a pin the the ground.

Strip cropping. A conservation practice in which grass or close-growing crops are grown alternately with cultivated crops in parallel strips.

Stuck. A command by the head tapemen, when measuring with a steel tape, to have the rear tapeman drop the tape and move forward.

Subirrigation. Application of water to the root zone by raising the water table through a pipe or ditch distribution system.

Subsoil. That part of the soil beneath the topsoil.

Subsurface drain. An underground collector for removing excess soil water, also called a pipe, tile, or underdrain.

Summer fallow. The practice of tilling uncropped land during the summer to control weeds and store moisture in the soil for a later crop.

Surface creep. Coarse sediment which moves in almost continuous contact with the soil surface during wind erosion.

Surface inlet. A structure for conveying surface water to a pipe or subsurface drain.

Suspended sediment. Sediment which remains in suspension in flowing water or in air for a considerable period of time.

Taping. The practice of measuring distances with a steel tape.

Terrace. A broad surface channel or embankment constructed across the land slope at suitable spacings and grades to reduce erosion and/or to increase soil moisture.

Terrace grade. Slope along the terrace channel to an outlet.

Terrace spacing. The vertical or horizontal distance between adjacent terraces, except for the top terrace.

Tile depth. The vertical distance from the soil surface to the grade line or bottom of the tile.

Tile drain. A subsurface or pipe drain made from clay or concrete.

Tilting level. A special tripod level in which the bubble tube can be viewed from the eyepiece.

Time of concentration. The time required for water to flow from the most remote point in the watershed to the outlet, once the soil has become saturated and minor depressions are filled.

Toe drain. A subsurface drain located at the downstream toe of an earth embankment.

Topographic factor. The product of the slope factor and the slope length factor in the soil loss equation.

Topographic map. A map which shows contour lines, stream channels, and other topographic features.

Topsoil. A nonspecific term for the top surface layer of the soil, normally containing the most organic matter.

Township lines. Parallels of latitude at 6-mile intervals which are the north and south boundaries of townships in the rectangular system of public land survey.

Transit. A versatile surveying instrument for measuring horizontal and vertical angles as well as for leveling.

Traverse table. A small drafting board that is mounted on a tripod, capable of being rotated about the vertical axis, and suitable for field mapping.

Trickle (drip) irrigation. Low rate and frequent application of water to soils through small holes or mechanical devices, called emitters, usually placed at short intervals along small tubing and operated at relatively low pressure.

Trickle spillway. A pipe or other conduit through an embankment to carry low flows and to maintain the normal water level, also called a mechanical spillway.

Tripod. A three-legged stand to which a survey instrument is attached.

Turbine pump. A type of pump having a combination of axial and centrifugal flow through the impeller, surrounded by stationary and usually symmetrical guide vanes.

Turning point. An identifiable, temporary point whose elevation is determined by substracting the foresight from the height of instrument in leveling.

Uniformity coefficient. A numerical value less than one for evaluating the uniformity of application of irrigation water.

Vegetated waterway. A shaped and grassed channel for carrying water at relatively high velocities, usually of broad width and shallow depth.

Vertical interval. The vertical distance between corresponding points on adjacent terraces or from the top of the hill to the channel of the first terrace.

Water-application efficiency. The ratio of irrigation water stored in the soil root zone to the water delivered to the area being irrigated.

Water-conveyance efficiency. The ratio of irrigation water delivered by a distribution system to that introduced into the system.

Water harvesting. Any practice which increases the runoff from a watershed, such as covering the surface with plastic, applying sealants, paving, etc.

Water holding capacity. Amount of soil moisture, usually field capacity less wilting point amounts, also called available soil moisture capacity.

Water rights. Legal rights to water derived from common law, court decisions, or statutory enactments.

Watershed (catchment). Land and water area that contributes runoff to a given point on a waterway or stream.

Watershed gradient. The average slope from a given point on a waterway, along the path of water flow, to the most remote point on the watershed.

Watershed planning. Formulation of a program of land and water use practices for optimum long-range benefits.

Water table. The maximum height to which water will rise in a vertical hole in the soil.

Water-use efficiency. The ratio of water beneficially used to the water delivered to the area being irrigated.

W-ditch. Two closely spaced parallel field drainage ditches in which the spoil from construction is placed between them.

Weir. A device or structure in a channel for measuring or regulating the flow of water.

Wetted perimeter. Length of the wetted contact between a liquid and its containing conduit, measured along a plane at right angles to the direction of flow.

Wild flooding. Relatively uncontrolled flooding of fields which have not been prepared for surface irrigation.

Wilting point. Soil moisture content below which plants permanently wilt, also called permanent wilting point, wilting coefficient, or wilting percentage.

Windbreak. Any type of barrier for protection from winds, especially for buildings, gardens, orchards, and feed lots.

Wind erosion. Accelerated soil movement caused primarily by the action of wind.

INDEX

ENGLISH—SI LENGTH CONVERSION CONSTANTS

Length	*in.*	*ft*	*yd*	*mile*	*cm*	*m*	*km*
1 in.	1	0.083	0.027	—	2.54	—	—
1 ft	12	1	0.333	—	30.48	0.305	—
1 yd	36	3	1	—	91.44	0.914	—
1 mile (statute)	—	5280	1760	1	—	1609	1.61
1 cm	0.394	0.033	0.011	—	1	0.1	—
1m	39.37	3.281	1.094	—	100	1	0.001
1 km	—	3281	1094	0.621	—	1000	1

ENGLISH—SI AREA CONVERSION CONSTANTS

Areas	*in.*^{2}	*ft*^{2}	*yd*^{2}	*acre*	*cm*^{2}	*m*^{2}	*ha*
1 in.^{2}	1	0.007	—	—	6.45	0.00064	—
1 ft^2	144	1	0.1111	—	—	0.0929	—
1 yd^2	1296	9	1	—	—	0.8361	—
1 acre	—	43 560	4840	1	—	4047	0.405
1 cm^2	0.155	—	—	—	1	0.0001	—
1 m^2	1550	10.76	1.20	—	10 000	1	0.0001
1 ha	—	107 650	11 961	2.47	—	10 000	1

ENGLISH—SI VOLUME CONVERSION CONSTANTS

Volume	*in.*^{3}	*ft*^{3}	*Am. gal*	*liter*	*m*^{3}	*ac-ft*	*ha-m*
1 cu in.	1	—	0.0043	0.0164	—	—	—
1 cu ft	1728	1	7.481	28.32	0.0283	—	—
1 Am. gal	231	0.134	1	3.785	0.0038	—	—
1 liter	61.02	0.0353	0.2642	1	0.001	—	—
1 cu m	61 022	35.31	264.2	1000	1	0.00081	0.0001
1 ac-ft	—	43 560	325 872	—	1233.4	1	0.1233
1 ha-m	—	353 198	—	10^7	10 000	8.108	1

1 cu yd = 0.765 cu m   1 cu m = 1.308 cu yd